STOREY'S GUIDE TO

GROWING ORGANIC ORCHARD FRUITS

Market or Home Production
Site & Crop Selection
Planting, Care & Harvesting
Business Basics

Danny L. Barney

Storey Publishing

The mission of Storey Publishing is to serve our customers by publishing practical information that encourages personal independence in harmony with the environment.

Edited by Deborah Burns and Sarah Guare
Art direction by Cynthia N. McFarland
Book design by Kent Lew
Text production by Jennifer Jepson Smith

Cover photography by © Alexander Selektor/iStockphoto.com (plum), © Marek Mnich/iStockphoto.com (apple), © RedHelga/iStockphoto.com (peach), © Yuri Shirokov/iStockphoto.com (pear), and © Zone Creative/iStockphoto.com (cherries)
Illustrations by © Elayne Sears
Tables and infographics by Kent Lew

Indexed by Christine R. Lindemer

Storey Publishing is committed to making environmentally responsible manufacturing decisions. This book was printed in the United States on paper made from sustainably harvested fiber.

Storey Publishing
210 MASS MoCA Way
North Adams, MA 01247
www.storey.com

Printed in the United States by McNaughton & Gunn, Inc.
10 9 8 7 6 5 4 3 2 1

LIBRARY OF CONGRESS CATALOGING-IN-PUBLICATION DATA

Barney, Danny L.
Storey's guide to growing organic orchard fruits for market / by Danny Barney.
p. cm.
Guide to growing organic orchard fruits for market
Includes index.
ISBN 978-1-60342-570-4 (pbk. : alk. paper)
ISBN 978-1-60342-723-4 (hardcover : alk. paper)
ISBN 978-1-60342-905-4 (e-book)
1. Fruit. 2. Organic gardening. 3. Orchards.
I. Title. II. Title: Guide to growing organic orchard fruits for market.
SB357.24.B37 2013
634—dc23

2012027707

Contents

Chapter 1 **Where We Came From and Where We Are** 5

Where We Came From • Developing Definitions and Standards • What Does "Organic" Mean for Orchardists?

Chapter 2 **Selecting a Winning Site and Crops to Match** 17

Climate • Soils • Irrigation • Topography • Site History and Neighborhood • Pests and Diseases • Access and Utilities • Labor and the Law

Chapter 3 **Designing or Redesigning Your Orchard** 42

Tree Size • Training Systems • Orchard Layout Design • Sample Orchard Design

Chapter 4 **Preparing or Upgrading Your Orchard Site** 76

Drainage • Staking the Orchard • Setting Your Soil Standards • Mineral Nutrients • Organic Amendments • Microorganisms • Other Soil Amendments • Weed and Pest Control

Chapter 5 **Selecting Pome Fruit Crops and Varieties** 118

Apple • Pear • Quince • Medlar • Mayhaws • Loquat or Japanese Medlar • Saskatoon

Chapter 6 **Selecting Stone Fruit Crops and Varieties** 177

Apricot • Cherry • Peach • Nectarine • Plum

Chapter 7 **Obtaining and Planting Your Trees** 229

Purchasing and Handling Your Trees • Preparing for Planting • Planting and Training • First Year Care and Management

Chapter 8 **Nutrient Management** 249

Essential Elements • Determining the Nutrient Status of Your Crop • Approved Materials • Applying Fertilizers

Chapter 9 Orchard Floor Management 286
History of Orchard Management • Soil Organic Matter • Controlling Weeds • Cover Crops • Managing Fruit Crop Rows

Chapter 10 Disease Management 325
Disease Management Strategies • Pome Fruit Diseases • Stone Fruit Diseases • Diseases of Both Pome and Stone Fruits • Root Diseases

Chapter 11 Pest Management 368
Survey of Pests • Pest Management Strategies • Insect and Mite Pests

Chapter 12 Pruning and Training 408
Types of Pruning Cuts and Plant Responses to Them • Where Should You Make Pruning Cuts? • Pruning Tools • Training Systems • Training Trees, Crop by Crop

Chapter 13 Fruit Thinning and Harvesting 465
Types of Thinning • Thinning, Crop by Crop • Harvesting

Chapter 14 Marketing an Orchard Business 493
Selecting Products • Selecting a Farm Design and Selling Venue • Setting Prices and Terms of Sales • Creating an Enterprise Budget • Publicizing Your Enterprise • Packaging and Selling • Developing a Business Plan

Appendix A Sample Planning, Preparation, and Planting Year Calendar 521

Appendix B Sample Establishment and Production Years Calendar 523

Resources 525

Index 527

CHAPTER 1

WHERE WE CAME FROM AND WHERE WE ARE

- Where We Came From | 6
- Developing Definitions and Standards | 10
- What Does "Organic" Mean for Orchardists? | 14

People have grown tree fruit crops throughout the world for thousands of years. Production systems have ranged from harvesting fruit from wild trees to sophisticated, highly managed orchards in which the fruit is harvested by robots. Within that range, each fruit grower can find a niche that is comfortable and meets his or her needs. This chapter provides insight into the evolution of farming methods, particularly as they relate to the organic production of orchard crops.

Where We Came From

BEFORE 1900, VIRTUALLY ALL FOOD was grown by what today we would call "organic" methods. Farms were small and often integrated crops and livestock. Labor requirements were high, and a typical farmer could feed only a small number of people. During the early 1900s, products of the industrial revolution began spilling over into agriculture. Tractors and combines reduced labor needs and made large, specialized farms possible.

The development of modern agriculture accelerated after World War I as new technology was transferred from military to civilian use. Synthetic, petroleum-based nitrogen fertilizers were discovered in the 1800s, but the production methods used for munitions explosives made large-scale manufacture of synthetic fertilizers feasible and profitable. Research on chemical weapons helped in the development of synthetic pesticides. Advances in plant breeding led to the creation of high-yielding hybrid crops that flourished in heavily fertilized monocultures.

In 1945, the Mexican government began research and educational programs directed at becoming self-sufficient in food production in the face of a rapidly growing population. These efforts, funded largely by U.S. philanthropic organizations, were highly successful. Within 11 years, Mexico was self-sufficient in wheat production; 8 years later they were a significant wheat exporter. Government and agricultural industry policies and goals emphasized developing hybrid crop varieties, synthetic pesticides, and fertilizers, and increasing mechanization. These practices rapidly spread around the world, fueling what came to be called the "green revolution."

Many people benefited from the green revolution, particularly those in less developed countries, where populations were rising rapidly and hunger was all too common. In developed countries, modern agricultural practices spurred enormous economic growth. With fewer farmers and farm workers required to feed the population, people were able to pursue education, vocational training, and business opportunities. Huge agribusinesses and food companies thrived, spurring more employment and wealth.

A few voices of warning could be heard, however, against the onrush of agricultural development. In the early 1900s, soil scientist Franklin H. King emphasized the impacts of soil chemistry on fertility and crop yields and wrote about traditional Oriental farming practices that had produced sustainable crops for millennia. From 1909 to 1924, Sir Albert Howard (sometimes called the father of organic farming) served as a British agricultural advisor in India. Although he went to India to teach modern agricultural methods, he came to appreciate and improve upon traditional Indian agricultural practices, such as composting. His 1940 book, *An Agricultural Testament*, served as an early organic agricultural text.

Worldwide Research

In the 1920s, Austrian-born philosopher Rudolf Steiner began teaching an integrated approach to farming that came to be called biodynamic agriculture. His approach emphasized the integration and harmony among soils, crops, and animals. Steiner's recommended practices included the use of composts, manures, and herbal preparations, with the goal of creating farms that were holistic, self-nourishing organisms. Today we refer to such practices as system or ecological approaches, and they are at the forefront of organic research.

Building on Sir Albert's teachings, Lady Eve Balfour began the Haughley Experiment in England in 1939 to test the claims of organic farming advocates and fill in the gaps in knowledge of how such systems worked. This remarkably well-designed ecological experiment was conducted on farm-scale plantings. The experiments included three side-by-side sections of land, each large enough to operate a full farm rotation. The study examined the food chains involved on a farm — soil, plant, animal, and back to the soil through many crop successions and generations of plants and animals. The goal, according to Lady Balfour, was to determine "interdependences between soil, plant and animal, and also any cumulative effects could manifest." Lady Balfour described her work in the book *The Living Soil*, which helped lead to the formation of the organic advocacy group The Soil Association in the United Kingdom.

Japanese microbiologist and plant pathologist Masanobu Fukuoka turned from "scientific agriculture" to "natural farming," which he considered went one step further than organic practices. Growing cover crops, mulching with native vegetation, not using tillage, and integrating crops and livestock were among his recommended practices. His books *One Straw Revolution* (1978) and *The Natural Way of Farming* (1985) are classic texts in the organic movement.

In the United States, J. I. Rodale became convinced of the truth and importance of the work done by Sir Albert and Lady Balfour and began publishing *Organic Farming and Gardening* magazine. In 1947, the Soil and Health Foundation was established and became the forerunner to the Rodale Institute. Among other things, Rodale emphasized building natural soil fertility after nitrogen fertilizers became unavailable during World War II, when ammonium supplies were diverted from agriculture to munitions.

Rodale, the Champion for Organic

J. I. Rodale was known for his missionary-like zeal in promoting organic agriculture. In 1954, Rodale wrote, "Organics is not a fad. It has been a long-established practice – much more firmly grounded than the current chemical flair. Present agricultural practices are leading us downhill."

Turning Point

Perhaps the key turning point in the organic movement came in 1962 with the publication of *Silent Spring* by Rachel Carson. Educated in marine biology and zoology at Woods Hole Marine Biological Laboratory and Johns Hopkins University, Carson served as a scientist and writer for the U.S. Fish and Wildlife Service. Following World War II, she became concerned about the overuse and misuse of pesticides and spoke out against pesticide practices that she believed were threatening all life by destroying the ecosystems to which we belong.

Carson did what all of the organic advocates could not do by reaching out to everyone, everywhere. People who had never heard of organic agriculture and who had never been on a farm suddenly became vitally concerned with food safety and agricultural practices that protected the environment. Rather just than a small number of farmers and gardeners, people around the world became environmental advocates. Here was born the widespread demand for commercially available organic produce, meats, and dairy products. Without that demand for organic farm products by consumers, organic production and research, as we know them today, would not exist.

Milestones and Growing Pains

J. I. Rodale's son, Bob, and Bob's wife, Ardath, carried on his work and founded the Rodale Institute. Bob Rodale's impassioned testimony before the U.S. Congress helped convince lawmakers to include funds for regenerative agriculture in the 1985 Farm Bill. Unwilling to adopt the term "organic," however, the U.S. Department of Agriculture coined the name "low-input sustainable agriculture" which was known by the acronym LISA.

One of the few things that mainstream scientists and organic farmers at the time agreed on was that we hated the name, which was flawed and misleading. At that time, however, many mainstream farmers and agricultural scientists still considered organic farming to be pseudoscience and were put off by some organic advocates' inclusion of mysticism, astrology, and spiritualism as key components of organic production. In a few years, "low-input sustainable agriculture" became simply "sustainable agriculture."

Farmers and researchers made many advances as they developed more sustainable practices. Initially, we looked for ways to reduce pesticide use and develop "softer" pesticides. The more we learned, however, the more we discovered that our tried and true method of reducing research to a single question and a few variables did not explain much of what was taking place on a farm. During the past 25 years, we have learned to recognize the intricate relationships between plants and their environments. In the 1980s, most fruit researchers worked alone or with a small number of like-minded colleagues on narrowly focused projects. Today we work in teams that include horticulturists, breeders, entomologists, weed specialists, soil scientists, and, very importantly, private fruit growers.

In particular, we are discovering what organic advocates have taught for a century. Plant health is absolutely dependent upon healthy, chemically balanced, and biologically active soil. In turn, healthy soils require healthy, diverse populations of plants, insects, and other invertebrates, and microorganisms.

Developing Definitions and Standards

ALTHOUGH WE WERE MAKING PROGRESS in developing more sustainable production systems, the organic movement was hindered by confusion over just what the term "organic" meant. Without uniform food production and handling standards, consumers were unsure whether the "organic" food that they were buying at premium prices was any safer than conventionally produced food. Organic growers were also hindered by not knowing what practices and materials were or were not acceptable. Grassroots efforts produced many regional definitions and standards related to organic farming, which only added to the confusion and frustration. They did, however, spur national demand for uniform standards.

The National Organic Program (NOP)

The Organic Food Production Act of 1990 required the U.S. Department of Agriculture to develop uniform standards for the production and handling of food products marketed as organic. Despite an initially tumultuous period of confusion and distrust, representatives from the organic movement, commercial farmers, government specialists, lawmakers, and scientists worked together to create the National Organic Program (NOP). The NOP became law in October 2002. According to the law:

> This national program will facilitate domestic and international marketing of fresh and processed food that is organically produced and assure consumers that such products meet consistent, uniform standards. This program establishes national standards for the production and handling of organically produced products, including a National List of substances approved for and prohibited from use in organic production and handling. This final rule establishes a national-level accreditation program to be administered by AMS [U.S. Department of Agriculture Agricultural Marketing Service] for state officials and private persons who want to be accredited as certifying agents. Under the program, certifying agents will certify production and handling operations in compliance with the requirements of this regulation and initiate compliance actions to enforce program requirements.

Three key provisions of the NOP are:

1. Food producers and processors who wish to market their goods under an organic label must be certified.

2. A program was established to certify agencies in the United States for domestic food production and abroad for producers and processors wishing to market organic food in the United States.

3. Producers were provided with lists of materials and practices that they could and could not use.

The NOP was a huge step forward for the organic movement. Consumers could be confident that food labeled organic was actually being produced using environmentally safe practices. Sales for legitimate organic producers benefited because unscrupulous individuals could no longer market food as organic when it had actually been produced using toxic chemicals and environmentally damaging practices. Furthermore, growers and food processors finally had specific production guidelines to follow.

Organic Materials Review Institute (OMRI)

More work was, and still is, needed to shape organic guidelines. The Organic Foods Production Act of 1990 required the Secretary of Agriculture to establish a National List of Allowed and Prohibited Substances for organic producers. While helpful, the original list, which was amended many times, was generic. For example, under allowed substances, the list included insecticidal soaps, but it did not specify what an insecticidal soap was or which soaps could be used in organic production. Likewise, the microbial insecticide *Bacillus thuringiensis* (Bt) would be allowed as a natural product, although certain formulations might contain prohibited inert materials.

The nonprofit Organic Materials Review Institute (OMRI) (see Resources) began in 1997 to address some of these shortcomings. According to their website, the OMRI "provides organic certifiers, growers, manufacturers, and suppliers an independent review of products intended for use in certified organic production, handling, and processing. . . . When companies apply, OMRI reviews their products against the National Organic Standards. Acceptable products are OMRI Listed and appear on the OMRI Products List. OMRI also provides subscribers and certifiers guidance on the acceptability of various material inputs in general under the National Organic Program." Other enterprises similar to OMRI are now carrying on similar activities.

Recently, for example, the OMRI website listed 14 approved insecticidal soaps and 20 *Bacillus thuringiensis* products by brand and product name. Such a list makes selecting a product much easier for food producers and certifiers.

For More Information

Complete information on the National Organic Program, including a list of certifiers and the list of allowed and prohibited synthetic materials, is available on the USDA Agricultural Marketing Service website (see Resources).

The Canadian Perspective

A similar organic certification program exists in Canada, based on the Canadian Standard for Organic Agriculture that was first approved in 1999. The "Organic Production Systems General Principles and Management Standards for Canada" and the "Permitted Substances List" were first published in 2006 and are available online (see Resources).

In 2009, Canada implemented the Organic Products Regulation to regulate the certification of organic products. Also in 2009, the Canadian and U.S. governments agreed that their national organic programs would be equivalent, with a few exceptions. For more information on Canadian organic certification, contact your provincial Ministry of Agriculture.

In this book, we will refer to the National Organic Program or NOP. For Canadian readers, interpret this as the Canadian organic program and refer to the appropriate regulations and approved substances and practices lists. In general, the two programs are quite similar.

Certification

Before you can market fruit as organic, you must first be certified by the agency responsible for your location. Depending on where you live, certification programs are administered by state departments of agriculture, provincial offices, grower organizations, and/or private companies.

Certification is required only when you market fruit under the organic label and when your annual sales total a certain amount. Home fruit growers are free to use whatever cultural practices they wish. If you want to produce fruit organically, however, commercial organic standards provide an extremely valuable resource and can serve as a starting point for developing or improving your orchard.

What Does "Organic" Mean for Orchardists?

UNLIKE CURRENTLY POPULAR TERMS such as "ecofriendly," "low-spray," and "grown naturally," "organic" has a specific, legal definition when applied to commercially grown fruits and other produce. The national lists of allowed and permitted substances are only a first step, however. Likewise, the OMRI's and other testing organizations' lists of approved soil amendments, pesticides, and the like are valuable, but more is needed.

Growing fruit organically goes far beyond simply using certain fertilizers or pesticides. As conventional orchardists began transitioning to organic production in the 1990s, a standard practice was to substitute "softer" organic pesticides for conventional materials. Little else in the orchards changed, and as you might predict, the results were not often great. The substitution approach alone failed to recognize that farms were miniature, yet highly complex, ecosystems. Focusing only on controlling codling moth, for example, can kill off predatory insects and mites that normally keep mites or aphids under control. Suddenly, you have many serious pests instead of one as natural control systems in the orchard are disrupted.

Another example of creating problems is overusing sulfur as a fungicide to control fruit diseases. While sulfur can effectively control some diseases, it is toxic to insects and mites. Again, overuse or use at the wrong time can disrupt natural controls and create serious pest problems.

One of the greatest challenges in organic fruit production is weed management. With very few herbicides available to organic orchardists, extensive cultivation became the rule for many growers. In one fruit growers' guide from the 1980s, the author recommended rototilling at least 12 times each growing season. In this case also, the results were predictable. Soil erosion and compaction increased. Soil organic matter declined, as did populations of beneficial micro- and macroorganisms. Tree health and fruit yields also declined.

Looking at the Entire Farm Ecosystem

To successfully grow tree fruits using organic methods, you must have an entirely different viewpoint and approach than do commercial growers who rely on industrial fertilizers and pesticides. Instead of focusing on one practice or problem at a time, you must consider the entire farm ecosystem and how everything in that system interacts.

For example, in place of tillage I might choose to plant a permanent crop of grass in the alleys between the trees and apply bark mulch around the trees in the rows. In this case, annual weeds become less troublesome, but perennial quack grass and thistles flourish under the mulch. With the grass competing for nutrients, the trees grow poorly and produce reduced yields of small, poor-quality fruit. Being weakened, the trees become more susceptible to pests. Some pests that pupate in the soil under the trees, such as oblique-banded leaf roller, become more troublesome as the pupae are protected by the mulch from predators, heat, and drought. At the same time, concentrations of organic matter in the alleys and under the trees slowly begin to increase and irrigation needs decrease as the mulch helps retain soil moisture.

Grass alley crops and organic mulches are not, in themselves, bad practices and have their roles in organic fruit production. The trick is to think holistically. You must understand that everything you do in your orchard affects every living organism in the trees, on the ground, and in the soil. The way in which those organisms interact with each other has a profound influence on the health and productivity of your trees.

Pome and Stone Fruits Defined

Pome fruits include apple, pear, quince, medlar, loquat, and other fruits that contain multiple seeds in the core. Pome fruits can usually be stored longer than stone fruits – up to 2 years for apples in controlled atmosphere (CA) storage. Stone fruits have a single, large seed contained within a hard pit at the center of the fruit. They are often referred to as "soft fruits" and have short storage lives. Apricot, cherry, nectarine, peach, and plum are stone fruits.

Even understanding and acting on this concept is not enough to be a successful organic orchardist. We must avoid practicing "organic by neglect" where we literally mine the soil and leave it depleted. Organic programs today emphasize that soil building is a critical component to sustainable organic production.

To be blunt, you will find growing most fruits organically is much more complicated than relying on conventional fertilizers and pesticides and requires far more planning and work. Comparisons across the country also show that growing fruit organically, at least commercially, costs more than it does for conventional orchards.

Furthermore, you cannot reduce organic fruit growing methods to a few recipes or recommendations. Each orchard site is unique in terms of soils, climate, pests, diseases, and beneficial organisms. As organic orchards mature, both they and the environments they are part of change. Soil fertility, pH, and soil organisms shift in response to orchard floor management practices. Beneficial and pest organism populations undergo changes. As an orchardist, you must continually evaluate what is going on and what you need to do to maintain a healthy orchard ecosystem.

So Why Grow Fruits Organically?

There are many reasons for having an organic fruit orchard. For some, marketing organic tree fruit is highly profitable. Demand by commercial organic growers for improved production practices and materials has driven much of the research and led to many of the advances in fruit production during the past decade. According to market statistics, there are still great commercial prospects for organic orchards, from local sales to international exports.

For some, producing food organically is a matter of being a good steward and caring for the environment. Others consider that organically grown fruit tastes better or is healthier for you than conventionally grown produce.

Whatever your reason, now is a terrific time to be an organic orchardist. So let's get started.

SELECTING A WINNING SITE AND CROPS TO MATCH

- Climate | 18
- Soils | 27
- Irrigation | 34
- Topography | 37
- Site History and Neighborhood | 38
- Pests and Diseases | 39
- Access and Utilities | 40
- Labor and the Law | 41

The ideal orchard site is one where the climate, soils, irrigation, topography, and surrounding land uses are well suited to producing the crops and varieties that you want to grow. For commercial fruit growers, access to the site, support services, and legal restrictions are also important.

Many orchard problems come from growing crops and varieties in locations where they are poorly or marginally adapted. On a small scale, this is not usually a serious issue. Experimenting is part of the fun of a hobby orchard, and the worst that can happen is that you may have to replant a few trees. Even for the smallest orchard, however, growing crops and varieties that are healthy, productive, and relatively easy to produce is more fun than struggling for mediocre results.

The situation becomes more serious as orchard size and investment costs increase. Purchasing land and establishing a large orchard are long-term decisions that are difficult and expensive to change. Loss of trees and crops and the extra care required to produce varieties poorly matched to a site can make your enterprise unprofitable. Fruit growing should be fun and, for some, profitable. That starts by picking a winning site.

This chapter is designed to help you pick out a great orchard site and to match crops to your growing conditions. For those who already own their orchard sites and have trees planted, this chapter is still important because it will help you evaluate your site with the goal of improving it. At this time, you should also consider how you plan to market your crops because this will impact which crops you choose and how you set up your orchard. See chapter 14 for more information about marketing.

Climate

CLIMATE SETS THE LIMITS within which all other factors operate. Summer and winter temperatures, length of the growing season, the amount of heat available during the growing season, precipitation, humidity, and, depending on where you live, wind, all play key roles in growing fruit. The important thing to recognize is that "climate" is not the same as "weather." Climate refers to temperature and precipitation patterns measured over decades. Weather, on the other hand, is what you see out the window at any given moment.

Fortunately, the information you need to evaluate your climate is available online for Canadian and U.S. locations. The National Oceanic and Atmosphere Administration (NOAA) Regional Climate Centers provide localized climate records for the entire United States (see Resources). The National Climate Data and Information Archive provides similar information for Canada (see Resources). Unfortunately, the Canadian database does not provide information on average frost dates, growing season lengths, and heat units. Some regional Environment Canada websites help fill in this information. Agricultural specialists at provincial Ministry of Agriculture offices can be good sources of information on climate as it relates to crop production in a given area. In the United States, state and county Cooperative Extension offices often have information on climate and fruit crop selection and production.

Think long-term in choosing a site and crops. Orchard crops are very long-lived. Depending on the crop, some trees and bushes can easily remain productive for 75 years or more. In a commercial orchard, 15 to 20 years is a good productive life expectancy, due more to shifting consumer preferences than the productivity of the trees.

Winter Temperatures

Winter temperatures are the first factors to evaluate. Northern growers need to consider the risk of freezing injury to tree trunks, branches, and buds from late fall through early spring. Southern growers are concerned with the accumulation of chilling units, which temperate zone fruit trees need to meet their dormancy requirements and develop new shoots, leaves, and fruit each spring.

Temperate zone plants require a dormant or rest period during the winter. Shorter days and cool night temperatures trigger the trees' entry into dormancy. Physical and chemical changes take place in the buds, bark, and wood that allow them to survive winter temperatures.

Once trees are dormant, they must spend a certain amount of time at temperatures between about 32 and 55°F (0 and 13°C), during which time chilling units accumulate. Chilling units accumulate most rapidly near 45°F (8°C). Temperatures below 32°F (0°C) do not contribute to chilling, and temperatures of 70°F (21°C) or more for as little as 4 hours can actually reverse chilling that has already occurred. Without the necessary chilling, the trees will be unable to bloom, leaf out, or grow normally.

Different varieties, even within the same crop, can have very different chilling requirements. High-chilling apples, for example, require more than about 800 hours of chilling; medium-chilling apples need 400 to 800 hours; and low-chilling apples require 100 to 400 hours. Of good news to warm-climate growers, fruit breeders have made excellent progress in developing "low-chilling" varieties, extending the range for traditionally northern fruits further south. Be very careful to match your crops and varieties to your site. We'll cover varieties in detail in chapters 5 and 6.

Several different models have been developed to calculate chilling units for a given location based on weather records. A discussion of those methods, however, goes far beyond what we can or need to do here. The U.S. Department of Agriculture (USDA) Plant Hardiness Zone Map (see box on page 23), combined with extreme minimum winter temperature data for your site and the recommended zones for different varieties, will give you a good start on selecting crops and varieties. Fruit growers, gardener groups, and Cooperative Extension educators and specialists can help you fine-tune your selections.

A good strategy is to select plants rated one or two zones hardier than your location. For example, Mansfield, Pennsylvania, is designated USDA Zone 5b, with average minimum temperatures between –10 and –15°F (–23 to –26°C). An orchardist in Mansfield might want to select varieties rated to Zone 4 (–20 to –30°F [–29 to –34°C]) to reduce the risk of freeze damage during an unusually severe winter. In the case of the Pennsylvania grower, 'Honeycrisp' apple (Zones 3 to 6) may prove more reliable in the long term than 'Jonafree' (Zones 5 to 8). (Note that while the map is broken down into two ranges per zone [5a, 5b, etc.], plants are typically rated to whole zones [4, 5, 6, etc.].)

A word of caution! I have found that growers typically overrate how warm their sites are and how cold-hardy their plants are. If you are conservative and objective in evaluating your site and selecting your crops, you will have a more successful and enjoyable time growing fruit.

Spring and Summer Temperatures

Spring temperatures are critically important to orchardists. Of primary concern are killing freezes prior to bloom and frosts during bloom. During deep dormancy, tree fruit buds can tolerate temperatures ranging from about −15°F to −40°F (−26 to −40°C). During spring, the plants quickly lose their cold hardiness and open blossoms are killed at about 28°F (−2°C).

Sites with warm springs but frequent frosts make tree fruit production difficult. This is especially true for stone fruits and Asian pears, which bloom earlier than European pears and apples. Trees located on north-facing slopes bloom later than those in sunnier locations — an advantage where spring frost damage is a problem. If your orchard is located on a frosty site, late-blooming varieties provide a distinct advantage. You may also need to provide frost protection during and shortly after bloom.

Locations near the ocean or along the eastern shores of large lakes are often excellent for orchards. The water acts as a thermal shock absorber, cooling the orchard and slowing bud development in the spring while reducing the risk of spring frost. Because weather typically moves from west to east in the Northern Hemisphere, western shores of lakes have less influence on orchard temperatures than eastern shores. In northwestern Montana, for example, sweet cherries are grown commercially in a narrow strip along the eastern shore of Flathead Lake. Just a few miles away, sweet cherry production becomes very difficult due to spring frosts. Likewise, the eastern shores of the Great Lakes host varied and abundant orchards, made possible by the moderating influence the water has on nearby temperatures. In general, the larger the body of water, the greater its influence on temperatures.

Summer temperatures are important factors for cool-climate orchardists in terms of the heat units available to produce and ripen fruit crops. While most stone fruit varieties ripen wherever they can be grown in North America, some pome fruit varieties require long, warm summers to mature. In short-season areas, early-maturing fruits are generally more reliable and productive.

USDA Plant Hardiness Zone Map

The map (see below and available online) divides the United States into 13 hardiness zones based on average minimum winter temperatures. While other hardiness zone systems have been published, the USDA map is the most comprehensive and universal for North America.

This updated map, released in January 2012, expands from 11 to 13 hardiness zones and updates the zones based on recent weather data. Unfortunately, the 2012 edition does not include Canada, as earlier versions of the map did. For Canadian growers, the 1990 version of the map is still available for downloading on the USDA site under the tab "Map and Data Downloads." (see Resources). The zones may have shifted slightly but will still give you a starting point for selecting crops and varieties. In general, you may find that your location would now be rated about one-half zone warmer than previously.

The new map is based on 30 years of temperature data, rather than 13 years used for the original map. Also, it turns out that the original 13-year period was unusually cold. Hardiness zone classifications in the new map tend to be slightly warmer than under the original map for many locations.

Fruit trees are typically rated for a range of hardiness zones. 'Harglow' apricot, for example, is recommended for hardiness Zones 5 to 8 and is well suited for orchards from the mid-southern United States into Canada. At the other extreme, tropical and subtropical loquat trees are rated to Zones 8 to 12 and are grown commercially and in home orchards in the United States in the warmer parts of Florida, coastal Georgia, Hawaii, and southern California.

When using hardiness zones to select crops, remember that the system is based on average minimum temperatures. For a long-lived plant, such as an apple or cherry tree, the average temperature has little meaning. The real question is, what is the likelihood of a killing freeze during the expected life of your orchard? By using both the hardiness zone map and extreme minimum winter data available from the climate databases mentioned above, you can make an informed choice (see box on page 25).

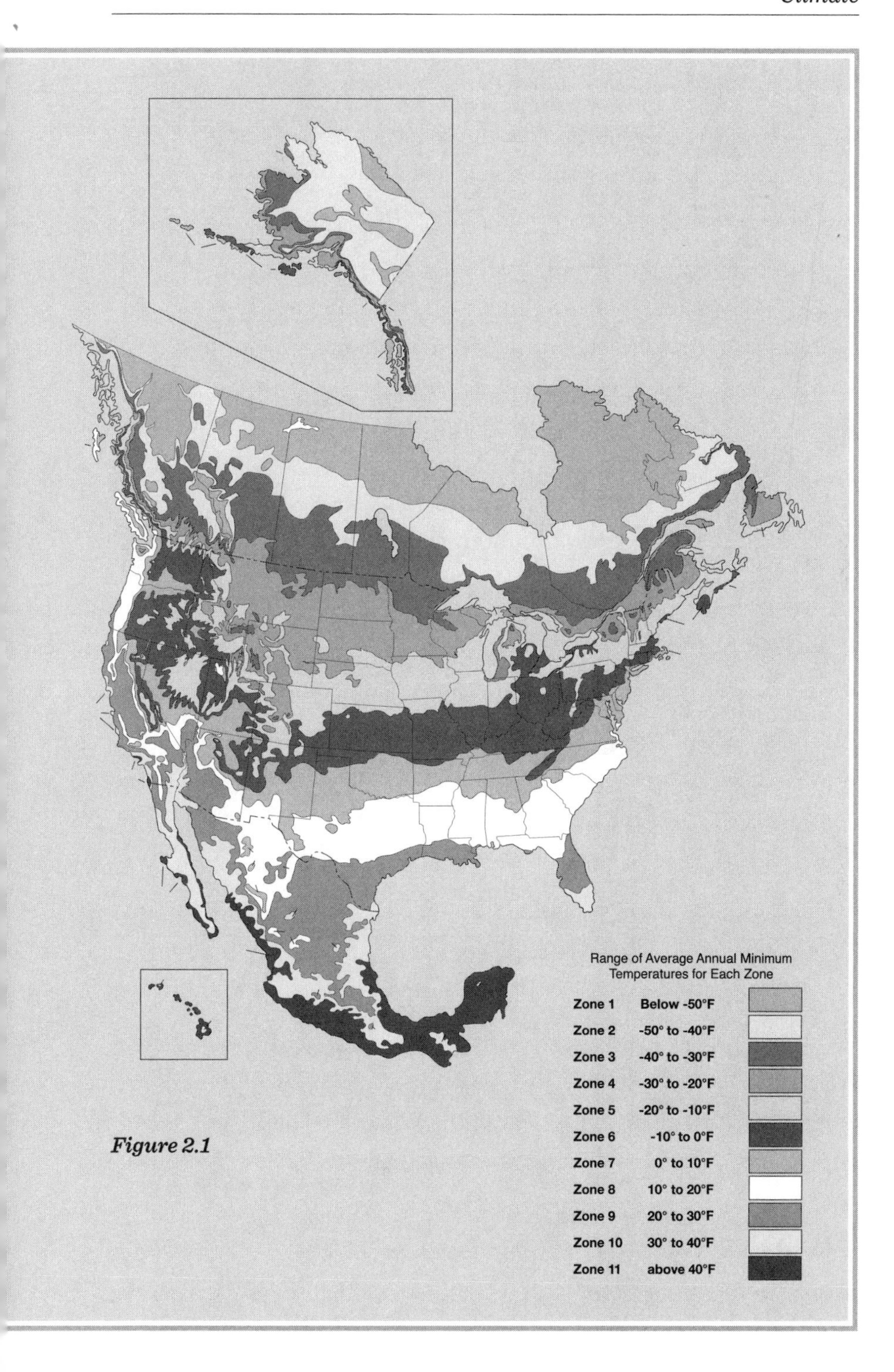
Range of Average Annual Minimum
Temperatures for Each Zone
Zone 1 Below -50°F
Zone 2 -50° to -40°F
Zone 3 -40° to -30°F
Zone 4 -30° to -20°F
Zone 5 -20° to -10°F
Zone 6 -10° to 0°F
Zone 7 0° to 10°F
Zone 8 10° to 20°F
Zone 9 20° to 30°F
Zone 10 30° to 40°F
Zone 11 above 40°F

Figure 2.1

Plant Hardiness Zone Map categories

Table 2.1

Each numbered rating is based on the average minimum winter temperature experienced at a site. Ratings are further divided into smaller categories designated by the letters "a" and "b."

°F	Zones	°C	*Example Cities*
70		21.1	
	13b		
65		18.3	
	13a		
60		15.6	
	12b		*Honolulu; Mayaguez, PR*
55		12.8	
	12a		
50		10.0	
	11b		
45		7.2	
	11a		*Miami, FL*
40		4.4	
	10b		*Los Angeles; San Francisco*
35		1.7	
	10a		*Naples, FL; Yuma, AZ*
30		-1.1	
	9b		*New Orleans; Tampa*
25		-3.9	
	9a		*Brunswick, GA; Houston*
20		-6.7	
	8b		*Seattle; Tallahassee*
15		-9.4	
	8a		*Atlanta; Dallas*
10		-12.2	
	7b		*Baltimore; Greensboro, NC*
5		-15.0	
	7a		*Boise; Nashville*
0		-17.8	
	6b		*Charleston; Santa Fe*
-5		-20.6	
	6a		*Chicago; Columbus, OH*
-10		-23.3	
	5b		*Milwaukee; Portland, ME*
-15		-26.1	
	5a		*North Platte, NE; Utica, NY*
-20		-28.9	
	4b		*Anchorage; St. Paul*
-25		-31.7	
	4a		*Fargo, ND; Sidney, MT*
-30		-34.4	
	3b		*Fort Kent, ME; Willow, AK*
-35		-37.2	
	3a		*International Falls, MN*
-40		-40.0	
	2b		
-45		-42.8	
	2a		*Fairbanks*
-50		-45.6	
	1b		*Fort Yukon, AK*
-55		-48.3	
	1a		
-60		-51.1	

Precipitation and Humidity

Arid locations with ample irrigation water are often ideal for orchards. The lack of rainfall during spring and summer greatly reduces apple scab, brown blight of stone fruits, fire blight, and other diseases compared to what you find in more humid areas. The higher precipitation levels and humidity east of the Rocky Mountains do not mean that you cannot grow orchard fruits organically, but it will be more challenging than for western growers. If you live in a rainy or otherwise humid climate, it is particularly important to select a site with excellent air drainage, improve the air movement on your site, and select disease-resistant crops and varieties. We'll discuss some approaches to modifying your orchard environment in later chapters.

Late spring and early summer rains can cause serious problems for cherry growers. The fruits are susceptible to cracking when rains come just as the fruits are ripening and into the harvest season. Some varieties are less prone to cracking than others. For rainy areas, it is important to select cracking-resistant cherry varieties.

How to Select Crops That Are Hardy for Your Site

1. Identify your hardiness zone using the hardiness zone map or table 2.1 on page 24. For example, Sandpoint, Idaho, is rated as Zone 6a (−5 to −10°F; −21 to −23°C).
2. Look up the extreme minimum winter data for your area using the NOAA Regional Climate Center website if in the United States, or the website of the National Climate Data and Information Archive if in Canada (see Resources for both websites). For Sandpoint, temperatures of −20°F (−29°C) are common, and the recorded low is −36°F (−38°C).
3. Compare the two sets of data to select a crop that will be hardy enough to withstand the winter temperature extremes that may occur during the life of your orchard. For Sandpoint, you may want to select a variety rated to Zone 5a or even 4a.

Important Temperatures

−15° to −40°F (−26 to −40°C): Depending on variety, tree fruit buds can tolerate some figure in this range during deep dormancy
28°F (−2°C): Blossoms are killed
32 to 55°F (0 to 13°C): Chilling units accumulate
45°F (7°C): Optimum temperature for chilling unit accumulation
Above 70°F (21°C): Chilling can be reversed if temperature maintained for 4 hours or more

Precipitation and humidity affect your irrigation practices. You want to provide enough water through precipitation and irrigation to meet the needs of the trees without forcing lush growth or encouraging root and collar rots. The rule is to keep the soil moist without waterlogging it.

The amount of precipitation your orchard receives will also have an impact on pests and diseases. If you grow tree fruits organically in a humid, high-rainfall area, be prepared to spend more for pest and disease management and have lower packouts of premium fruit than conventional growers in your area or organic growers in ideal climates. Before investing heavily in a commercial operation, be sure that your market will support the higher prices you will need to charge (called an organic premium) in order to earn a profit. Design your operation to market that portion of your crop that does not meet the high-quality standards for premium fresh markets; cider and processing are two options.

Winds

Wind damage is not a serious problem in many fruit-growing areas of North America, but there are exceptions. Frequent and prolonged wind can tatter leaves, dry out soil, and desiccate trees, the last of which is particularly a problem during subfreezing temperatures in winter. Winds can rub the fruit against branches and other fruits, causing bruises and scarring. Wind can also cause trees to lean, and the trees may require trellising to stay upright and properly trained.

If you are in an area known for frequent winds, it is a good strategy to select an orchard site on a bench (a more-or-less level area on a slope above a valley floor) or gentle slope facing away from the prevailing winds. Another option is to plant a windbreak of taller trees around your fruit trees to protect them. Windbreaks are common, for example, around orchards along both sides of the Columbia River in Washington and Oregon. Windbreaks can also be useful in the High Plains, the Prairie Provinces, and parts of the Midwestern United States.

Soils

SOILS ARE ONE OF THE MOST IMPORTANT CONSIDERATIONS in selecting and managing an orchard site. They are, quite literally, the foundation for your orchard. One of the most important steps in selecting or managing an orchard site is to test the soil. Soil test kits are available from garden centers and horticultural suppliers, but many do not provide adequate or accurate data. It is best to have your soil samples analyzed

ACME Soil Testing Laboratories, Inc 1827 Center Valley Spokane, WA 99201	Joe Smith 3151 Orchard Lane Anywhere, ID 83864
	County: Bonner
	Lab No. 11-0159
	Sample: apple block 1
	Crop: apples

Nutrient Availability		
Nutrient	Concentration ppm	Level
NO_3-N	17	Low
NH_4-N	6	Low
P	25	Low
K	230	High
Ca	4120	Adequate
Mg	0.5 meq/100 g	Low
SO4-S	15	Adequate
Zn	5.2	High
Fe	235	High
Cu	1.2	High
MN	7.2	High
B	0.9	Adequate

Soil properties		
Property	Value	Units
pH	5.7	---
pH buffered	6.5	---
EC	2.8	dS/m
CEC	15	Cmolc/kg
OM	4.3	%
% sand 20	% silt 60	% clay 20
Soil type: silt loam		

Recommendations:
Add 30 lb/acre actual N
Add 80 lb/acre P_2O_5
Add 1.5 tons/acre dolomitic limestone

Figure 2.2 · Sample soil analysis report

by a commercial or university laboratory that is familiar with the soils in your area. Because soils differ as you move from one region to another, different tests and interpretations are used.

In general, you will want to have your soil tested for pH, organic matter, buffering capacity, cation exchange capacity (CEC), phosphorus (P), potassium (K), calcium (Ca), magnesium (Mg), sulfur (S), and boron (B). If you are in an arid location or are planting on land that has been fertilized or irrigated, include a salinity (electrical conductivity or EC) test. Laboratories have standard tests, such as an extended fertility test, that may include additional nutrients. If you do not know what type of soil you have, a particle size distribution (PSD) analysis will be important in helping you develop soil management practices. Figure 2.2 shows a sample soil analysis report.

While much is made of soil fertility in crop production, soil nutrients are relatively easy to adjust, particularly when establishing a new orchard or replanting. The purpose of a preplant soil test is to establish baselines that will help you choose what amendments, if any, to add and how much of each. Chapters 4 and 8 address soil nutrition and fertilization.

Soil Types and Drainage

Even before we begin looking at soil pH and organic matter, consider soil drainage. With few exceptions, tree fruits perform best on deep, well-drained soils, and many problems encountered by orchardists can be traced to inadequate soil drainage. Provided that irrigation water is available, trees survive and produce much better on an excessively well-drained site than one that is too wet. Root rot is an obvious problem with wet soils, but poor drainage impedes the movement of oxygen to the roots and the movement of carbon dioxide out of the soil, and it interferes with nutrient uptake. Wet soils also warm slowly in the spring, interfering with tree growth and adversely affecting soil micro- and macroorganisms.

Classifying types. The type of soil, referred to as soil texture, has much to do with water drainage. We classify soils by their relative percentages of sand, silt, and clay. Sand particles are the largest and do little more than provide water drainage, gas exchange for the roots, and anchorage for tree roots. Sand does not hold water or nutrients well. Silt particles are the next smallest and hold water and greatly help form soil aggre-

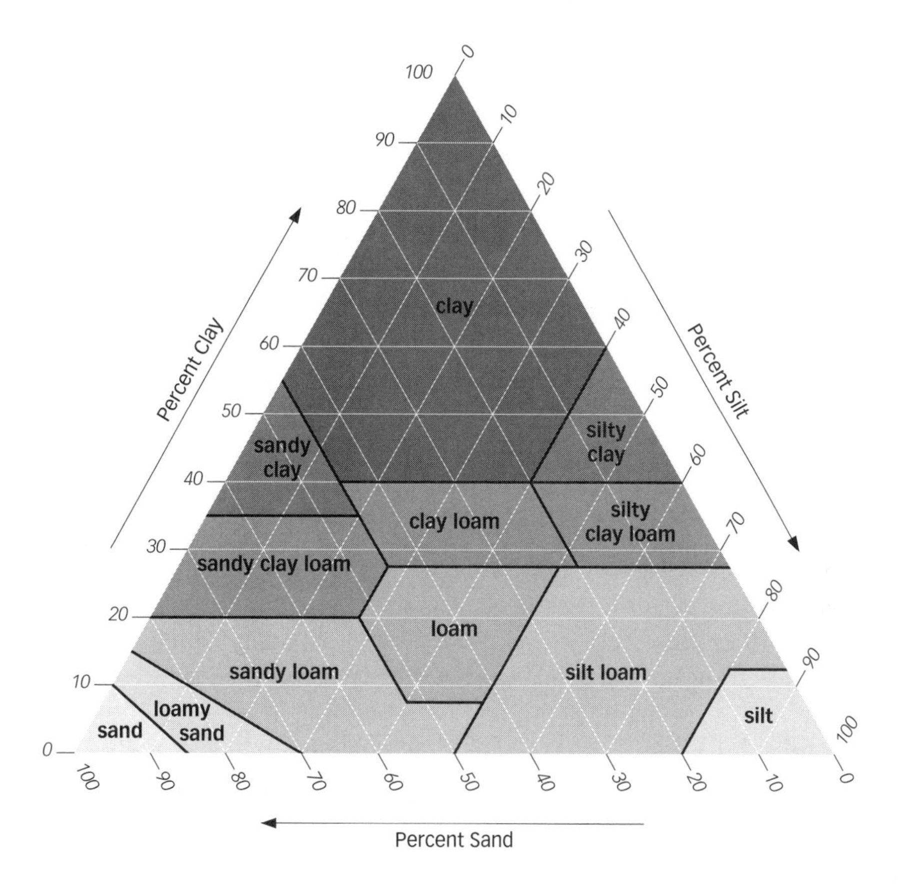

Figure 2.3 · The soil triangle shows soil classifications based on the percentages of sand, silt, and clay in a soil sample.

gates. Clay particles are the smallest soil particles. They help with aggregate formation and hold water and nutrients.

An effective blend of the three types of soil particles provides good water- and nutrient-holding capacity, but also good water drainage, gas exchange, and root penetration. Soils that are too sandy are droughty and require much irrigation. Heavy clay soils drain poorly and encourage root disorders.

Figure 2.3 shows how soils are classified, based on their percentages of sand, silt, and clay. To use the triangle, locate the percentage of sand, silt, and clay in your sample. Starting at those points, draw lines parallel to the adjacent counterclockwise side. Guidelines are shown in the figure. The point where the lines intersect is your soil type or texture. Say, for example, your soil particle size distribution test showed sand = 45 percent, silt = 35 percent, and clay = 20 percent. The classification for this sample would be loam — an excellent choice for orchard soil, provided that it is deep enough and well drained.

Soil types and crops. Loam and sandy loam soils are excellent for tree fruits. Given ample irrigation water, even loamy sands support orchards. With proper steps to ensure water drainage, silt loam soils can be effective orchard sites. Heavier-textured soils make tree fruit culture difficult. You can modify soils by adding organic matter to improve water-holding capacity, soil tilth, nutrition, and biological activity. Be aware, however, that simply mixing large quantities of organic matter into heavy soils can make a bad situation even worse by increasing water-holding capacity. Adding sand to improve soil drainage is usually prohibitively expensive on all but the smallest sites.

Soil drainage requirements differ among pome and stone fruits. In general, European pears tolerate heavy soils better than do other common tree fruit crops, followed by apples. Some apple and pear rootstocks perform much better on poorly drained soils than others. For example, M26 apple rootstocks produce trees about 8 feet tall and are generally popular. M26 rootstocks, however, perform very poorly on heavy or otherwise poorly drained soils. In such cases, a grower might choose a Bud9

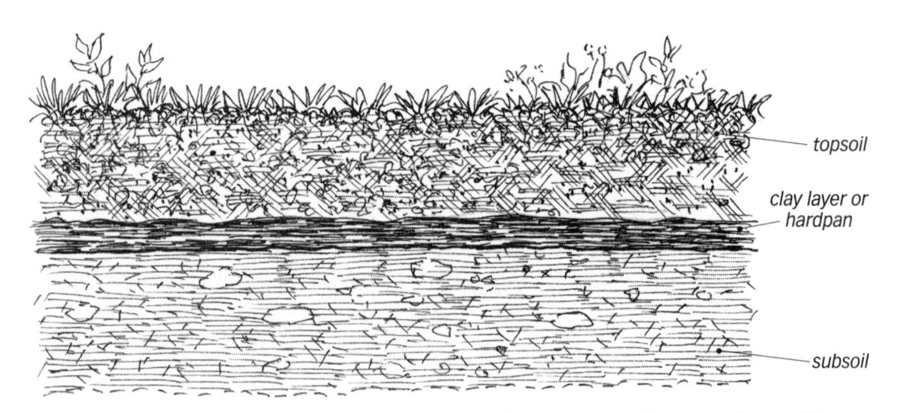

Figure 2.4 **· Sample soil profile**

Clay layers and hardpans can interfere with water drainage and root growth. Deep gravel layers can cause excessive drainage and droughty soils.

apple rootstock, which is more tolerant of heavy soils than M26. Stone fruits are less tolerant of heavy soils and poor drainage than apples and pears. Peaches, in particular, tend to grow and produce poorly when water drainage is inadequate.

I maintained a teaching orchard for 15 years, demonstrating different crops, varieties, rootstocks, and production practices. The orchard was located on a site with 8 to 12 inches of silt loam overlying 4 feet of fractured clay. Not an ideal site. European pears flourished. Apples on some rootstocks performed well, while other rootstock/variety combinations were severely stunted or died. Sweet cherry trees grew vigorously, while tart cherries suffered trunk splitting. The plum trees apparently grew well but had exceptionally poor anchorage. During a fruit growers' tour one spring, we discovered that some of my 15-foot-tall plum trees had literally fallen down. For me, it was embarrassing. For a commercial orchardist, the situation would have been far worse. Poorly drained sites are poor choices for orchards. If you must grow your fruit trees on poorly drained soil, you will need to take extra steps to improve water drainage in the root zone.

Examining your soil. If you are considering a commercial orchard site, an excellent practice is to dig several trenches, each 6 to 10 feet deep, with a backhoe. Examine the soil profile for clay lenses, hardpans, and deep gravel or sand layers. Be especially alert for dark grayish-blue to greenish-black, sometimes foul-smelling, layers that indicate waterlogged soils. This type of discoloration is called gleying.

Hardpans and clay layers that are not too deep can sometimes be broken up with a chisel or rip plows. These implements may be mounted on a tractor or bulldozer and consist of long shanks (and sometimes feet or blades) that extend deep into the soil. For shallow ripping on light-textured soils, a medium-sized tractor might suffice. For deep ripping on heavy-textured soils, you will need a large tractor or bulldozer.

Sandy and gravelly soils can be suitable for an orchard if you have adequate irrigation water and access to ample supplies of organic materials suitable for amending soils. Be very wary of waterlogged soils, however. Determine why they are waterlogged, and be sure you can develop adequate drainage before attempting to plant fruit trees on them. We'll cover specific ways to correct drainage problems in chapter 4.

By digging trenches, you can estimate rooting depth. Ideally, you should have 5 feet or more of rooting depth for fruit trees and a water table that remains at least 3 feet below the soil surface. You can certainly grow fruit trees on shallow soils, but doing so is more challenging than on deeper soils and may not be suitable for all fruit crops.

The types of vegetation growing on your orchard site will also give you information about your soil. Cattails, sedges, rushes, reeds, and certain grasses and other plants that are associated with wetlands or otherwise poorly drained soil are a giveaway that this location is not a good choice for an orchard.

Soil pH

Soil pH is usually manageable in most parts of the United States and Canada. The term pH refers to the soil's acidity or alkalinity, which has a profound influence on the availability of soil nutrients to plants. Soils with pH values less than 7.0 are acidic, while pH values above 7.0 indicate alkaline soils. Although an optimum soil pH for fruit trees is between 6.5 and 7.0, pH values ranging from 5.5 to 7.5 seldom create serious problems for most fruit tree crops.

Soils that are too acidic are seldom a serious problem for fruit trees because raising the soil pH with limestone or dolomite is generally straightforward and relatively inexpensive. Orchard management practices can be devised to help maintain an elevated soil pH.

Excessively alkaline soils are more challenging. As soil pH rises much above 7.5, fruit trees begin showing symptoms of iron chlorosis and other nutrient disorders. Yellow to yellowish-white leaves with green veins are the classic symptoms of iron chlorosis, which also includes stunted growth, poor yields, and even tree death. While iron chlorosis can be caused by a lack of iron in the soil, which sometimes occurs on acidic, organic soils, by far the most common cause is high soil pH. In these cases, iron may be abundant in the soil but is in a chemical form that is unavailable to the trees. Excessive concentrations of some other nutrients, such as phosphorus, can interfere with the uptake and utilization of iron. Iron chlorosis is particularly common on wet, cold soils.

Orchard soils can be acidified and iron can be added to the trees in foliar sprays, but these practices increase the work and costs of the establishment and maintenance of the orchard. By far the best orchard sites have well-drained soils with pH values roughly between 5.5 and 7.5. We'll discuss how to adjust soil pH in chapter 4. For now, be aware that growing fruit crops on soils with a pH above about 7.5 is more difficult than doing so on neutral or slightly acidic soils.

Soil Salinity

Salty (saline) soils can be a problem in some regions of North America, particularly on arid sites. Even in areas with relatively abundant precipitation, saline soils can develop due to excessive applications of fertilizers, including manures. Frequent, shallow irrigation dissolves salts and concentrates them near the soil surface as the water evaporates. Many horticultural crops are sensitive to high concentrations of salts and perform poorly on such soils.

Soil salinity is measured with electrical conductivity meters. Reporting units vary from one laboratory to another and include micromhos per centimeter (µmho/cm), millimhos per centimeter (mmho/cm), deciSiemens per meter (dS/m), and a few other units. Fortunately, it is easy to convert between units: 1000 µmho/cm = 1 mmho/cm = 1dS/m.

Soils with salinity levels of 0 to 2,000 µmho/cm (0–2 mmho/cm or 0–2 dS/m) are considered nonsaline and are suitable for all crops. Salinity levels between 2,000 and 4,000 µmho/cm (2–4 mmho/cm or 2–4 dS/m) indicate a very slightly saline soil and can create problems in sensitive plants, including blueberries, raspberries, and strawberries. Tree fruit crops tolerate very slightly saline soils better than their berry counterparts and can generally be grown on soils with salinity levels of 4,000 µmho/cm or less (4 mmho/cm or 4 dS/m), providing the soils are otherwise acceptable. For soils with salinity levels greater than 4,000 µmho/cm that cannot be reclaimed by leaching with large amounts of low-saline water, the production of horticulture crops is limited to a few vegetable crops, including beets, broccoli, squash, tomatoes, and asparagus.

Particularly troublesome are saline-sodic soils found in the American west. The abundant sodium and other ions in these soils interfere with the formation of soil aggregates that are necessary for water drainage. Such soils form slick, sometimes oil-colored, deposits that do not support plant life. Saline soils can be reclaimed, in some cases, by leaching with large amounts of water. Saline-sodic soils can be reclaimed using gypsum and leaching. A much better practice, however, is to avoid planting fruit trees on saline or saline-sodic soils.

Soil Organic Matter

The management of organic matter in soil is very important but sometimes misunderstood. Organic matter, as measured in soil tests, refers to chemically stable, invisible compounds, such as humic acids, that bind soil particles together. It does not refer to visible pieces of plant matter or manures.

As organic materials fall onto or are incorporated into the soil, they begin a complex process during which they break down and are changed physically and chemically. During this process, the organic matter provides food for micro- and macroorganisms in the soil, adds nutrients to the soil, and alters soil water-holding capacity and drainage. Chemicals are eventually formed from the organic residues and serve as glues that bind soil particles into larger clumps or aggregates. As the aggregates develop, open pore spaces also form in the soil. These pore spaces allow for the exchange of gases, facilitate root growth, and provide surfaces where roots can take up water and nutrients.

Sandy soils are inherently low in organic matter and respond slowly to programs designed to increase organic matter. In commercial Washington State orchards located on sandy soils, for example, soil organic matter may require 5 to 10 years to increase significantly after transition to organic practices. Forest soils and soils found in the American grasslands and Canadian Prairies often have naturally high concentrations of organic matter.

Measure organic matter before planting and every 2 years or so afterward to establish references or benchmarks. These benchmarks allow you to design soil-building programs and monitor the effects of your orchard management practices. Ideally, a new orchard site will have organic matter concentrations roughly between 3 and 5 percent, but do not be overly concerned if your site comes in higher or lower than that range.

Irrigation

Water is both a blessing and a curse for fruit growers. Trees obviously need water in order to survive, but excess water contributes to soil erosion, tree diseases, and generally poor fruit production.

The amount of irrigation water that you need depends on climate, soil, crop, tree size, and training practices. In the arid American West and parts of western Canada, frequent irrigation may be needed, particularly in orchards on light-textured (sandy) soils. In eastern North America, there is often less need for irrigation, and, indeed, you can often find domestic fruit trees growing wild with no irrigation. For optimum crop production and tree health, however, some irrigation is typically needed, even in high rainfall areas. Even short periods of drought stress during the growing season can reduce the number of fruits that develop and fruit size, as well as make trees more vulnerable to insect pests. Irrigation during the planting year is usually critical to getting your trees to survive and develop strong roots and tops.

Before investing in land or large amounts of planting stock and other establishment costs, estimate how much water you will need for irrigation during peak usage. It is difficult to make a general recommendation on the amount of irrigation water needed because it depends on many variables. If you are planning a large planting, your best strategy will be to consult local fruit growers, horticultural consultants, or university and governmental fruit specialists for their advice. In general, plan to replace the same amount of water that is lost due to evaporation as determined from evaporation pan measurements. In particularly arid regions you may need to increase irrigation for drought-sensitive crops.

Evaporation pans are simply round, metal pans. While at least two different designs exist, the U.S. National Weather Service uses Class A pans that are 47.5 inches in diameter and 10 inches deep. A hollow cylinder equipped with a metal needle in the center is used to mark a given water level. Each morning, enough water is added to restore the water level to that mark. The amount of water added is determined using a pitcher that is marked in hundredths of inches of water. That represents the amount of water lost by evaporation since the last filling. In agricultural areas, evaporation pan data is often available from the National Weather Service, a university, or various agricultural organizations.

Outside of these regions, evaporation pan data can be hard to locate. You might choose to install your own evaporation pan or to install any of several types of soil moisture monitoring devices, including vacuum tensiometers or soil moisture sensors and electronic meters. Inexpensive gypsum block sensors and handheld meters generally work well for orchard crops. Computerized systems are now available that monitor the soil moisture and transmit the data to your home or office computer. Some of these systems can be set to automatically turn on and off irrigation valves.

On my research farm, for example, we usually lost about 2 inches of water to evaporation each week during the warmest period of the summer, as measured with an evaporation pan. For the teaching orchard, we needed a source that provided 2 acre inches of water weekly during July and August.

Different types of irrigation systems use different amounts of water, so you will need to factor that in as well. Overhead sprinklers were once widely used in orchards, but they waste much water and increase diseases by wetting the foliage and fruit and increasing humidity within the canopy. If you intend to use sprinklers to control frost during bloom, plan to apply ¼ inch of water each hour to your entire orchard for as long as temperatures are below freezing (see chapter 3 for more information on overhead sprinklers). Drip or trickle irrigation systems are very efficient in applying water directly to the tree roots and also keep the foliage and fruit dry. Drip systems do not, however, allow you to irrigate an alley crop and do not necessarily work well for in-row cover crops. How you choose to manage your orchard floor will influence how much water you need for irrigation.

For organic orchardists, microsprinklers provide a good compromise between overhead sprinklers and drip systems. You can irrigate your trees, alley crops, and cover crops at one time while keeping the canopy dry. Specially designed microsprinklers can also be used to provide frost protection, depending on the size of your trees.

Once you have determined how much water you need, ensure that you can legally access that amount of water at the times that you will need it. Water rights vary greatly from one area to another. Do not assume that you can access the surface water on or adjacent to your property, or use well water drawn from your property for irrigating an orchard. Also, determine how much it will cost you to access the water.

Identify surface waters and wetlands on and around your property. While they can be assets, they can also create legal problems. Depending on your location, national, state, and provincial environmental laws relating to wetlands and surface waters can limit what crops you grow on your property and what management practices you use.

Topography

TOPOGRAPHY HAS A HUGE INFLUENCE on where tree fruit crops are grown commercially. You may have noticed that the best commercial orchards are often located on slopes and benches above valley floors. Such sites generally offer good water drainage and allow cold, frosty air to flow away from the orchards. Fruit trees produce best in sunny locations, and a good site will provide full sun throughout the day.

Low-lying sites are often poorly suited to orchards. Water settles into low-lying areas, making drainage difficult. Cold air settles into depressions, called frost pockets, increasing frost damage to blossoms and developing fruits. Humid air is heavy and settles into the depressions, increasing the incidence of fruit tree diseases. Always select an orchard site that provides the best possible airflow and water drainage away from the site.

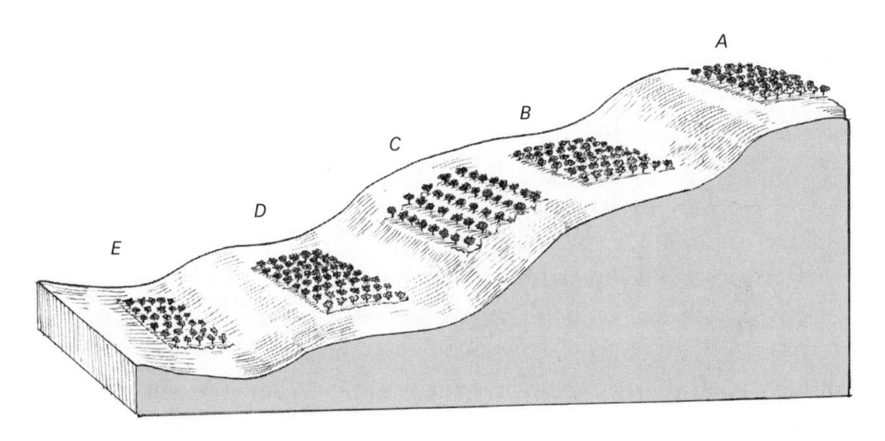

Figure 2.5 **· Geographic Profile of Good and Poor Orchard Locations**
The exposed hilltop (A) has excellent air and water drainage, but soil erosion can be a problem, as can winds. The bench area (B) is nearly level, lies above usual inversion layers, has good air and water drainage, and is an excellent site for an orchard. The slope (C) is too steep for orchard operations but can be transformed to an orchard site using terraces. The level valley floor (D) can be a fair orchard site but may suffer from frosts. Late-blooming crops and varieties might be best here. The low-lying area (E) is likely a frost pocket where heavy moist air collects, increasing both frost injury and diseases. This area is best not used for orchard crops.

Be careful, however, when planting on steep slopes. While you can find orchards on very steep slopes, orchard operations become more difficult and dangerous and the risk of soil erosion increases. If you must plant on a steep slope, consider building terraces to create level planting areas. Figure 2.5 illustrates good and poor orchard sites.

Site History and Neighborhood

BEFORE ESTABLISHING AN ORGANIC ORCHARD, find out as much as you can about the previous use of the site. This step is especially important if you plan to become a certified organic grower because some previous activities can interfere with certification. Arsenic, lead, and mercury, for example, are highly toxic elements that were once widely used in agricultural pesticides, and so might have been used on old orchard sites. Unfortunately, they are also highly stable in the soil and persist for decades. If in doubt, have the soil tested before you plant.

Another reason to be wary of old orchard sites is the risk of replant disease. This problem is particularly serious when following apples with apples. Commercially, apple orchards have traditionally been fumigated prior to replanting to help kill several fungal pathogens that inhabit the soil and severely stunt new trees.

Replant disease is generally thought to be caused by a complex of several pathogens that probably vary from one region to another. Throughout the life of an orchard, these pathogens build up in the soil, although the established trees often remain healthy. When the old trees are removed and new trees planted, the new trees can die or remain stunted and unproductive. While conventional growers fumigate the soil with highly toxic chemicals (methyl bromide was once very popular for this purpose), soil fumigation is not available for organic growers. If you are considering a site on which old fruit trees were grown within the past few years, replant could become a serious issue, especially if you follow apples with apples. Some certified organic apple growers prefer to fumigate new planting blocks and go through a 3-year recertification in order to avoid replant problems.

Try to determine if your orchard site was ever used for businesses that could have contaminated the soil with toxic chemicals, including such activities as metal plating, battery recycling, or leather tanning. Again, if in doubt, have the soil tested for heavy metals.

Present-day activities near your farm can also interfere with your ability to grow fruit organically. Look around for abandoned or poorly cared-for orchards, which are often reservoirs of pests and diseases. Highly mobile pests, such as codling moth, Oriental fruit moth, apple maggot, and oblique-banded leaf roller, can be extremely difficult to control when there is a dense source of them nearby. The same caution applies to sites near to or surrounded by woodlots, hedgerows, or windbreaks where wild pome and stone fruits are abundant. Both native fruit species and escaped domestic species serve as reservoirs of diseases and pests and can make organic production challenging. The level of difficulty increases in smaller orchards. In large orchards, management programs can often trap pests within a few fruit tree rows around the perimeter, leaving the centers relatively free of pests.

Beware of industries that might pollute your site through contaminated air or surface water runoff. The same caution applies to adjacent, nonorganic farms, especially if their operations include airblast sprayers or aerial applications of pesticides. Growing organic orchard fruit can be challenging if you are near large acreages of cereal grain, corn, soybeans, cotton, or other crops for which crop duster planes and helicopters are still used to apply pesticides and herbicides.

Pests and Diseases

We will go into detail on pests and diseases in chapters 10 and 11. Pest and disease pressures are far lower in the American West and parts of western Canada than they are in more easterly locations. In eastern North America, for example, plum curculio is a pest that attacks many tree fruit crops and has long been one of the most serious challenges for organic fruit growers. Apple scab, brown blight on stone fruits, and other fungal and bacterial pathogens are also more common and severe in the humid East.

As far as regions go, you should not let potential pest and disease problems be the deciding factor on whether to grow fruit organically.

New fruit varieties, more effective organic pesticides, and improved cultural practices allow you to grow organically, regardless of your location.

Access and Utilities

FOR COMMERCIAL GROWERS, transportation is a critical factor. You need to be able to get equipment, supplies, and workers into and out of your orchard easily, safely, and economically. Research also shows that much of the bruising and loss of quality to apples and other tree fruits occurs during transportation from the orchard to the packing house or other market outlet. Rough roads are a prime culprit, along with rough-riding trucks and other transportation equipment. Dust from dirt roads adjacent to orchards can contaminate fruits, requiring additional cleaning and handling to prepare them for market.

Wherever possible, paved roads offer a cleaner and smoother alternative to dirt or gravel roads. Paved roads may also mean being close to a population center or on a heavily traveled highway (think customers). If you must farm where paved roads are unavailable, try to select locations where the unpaved roads are well maintained and dust abatement programs are used.

You may also want to consider your proximity to other growers. While having your orchard be unique to an area may give you marketing advantages through lack of competition, you also face disadvantages. Those disadvantages increase with the size of your orchard. In established fruit growing regions, growers typically form cooperatives that enable them to collectively bargain with vendors for reduced rates on supplies and equipment. Cooperatives can also provide for sharing occasionally used, high-cost equipment, and for sharing farm workers. Food cooperatives can be excellent markets for your fruit. Such cooperatives may be hard to find in regions with few fruit growers.

Areas where organic fruit production is widespread also offer educational advantages. Organic orchardists in Washington and California, for example, enjoy abundant technical support from state universities, Cooperative Extension educators and specialists, private crop consultants, and analytical laboratories. In areas where organic fruit production has not become widely established, technical support will be more limited and not necessarily of the best quality.

Your orchard operation may or may not require utilities, including electrical power, natural gas, telephone, and Internet. Develop a production and marketing plan before you start looking for an orchard site. If you plan to transport all of the fruit to a packing house or other outlet, you may not need utilities. If you are planning to store your fruit on site or process it there, you are likely to need some utilities for refrigeration, lighting, heating, and other operations. Before investing in an orchard site, determine if the needed utilities are available and what their costs will be.

Labor and the Law

PRODUCING TREE FRUITS IS LABOR-INTENSIVE. Pruning, thinning, managing vegetation, controlling pests and diseases, and harvesting are the most labor-intensive operations. For a home or small market orchard, labor can usually be handled by the owner, the family, or a few employees. Large orchards, however, usually require much seasonal labor.

Before investing in an orchard site or establishment, be sure that you have a sufficient pool of laborers willing to work in your orchard at the times you need them and at wages you can afford. One advantage organic growers have over conventional growers is that potential employees often prefer to work in orchards free of toxic chemicals.

Depending on your area and goals, you may have to consider legal constraints to developing and setting up your orchard. For a home orchard, this is seldom an issue, although some subdivisions, even in rural areas, have covenants and restrictions that can interfere with establishing an orchard. Before starting a commercial enterprise, be sure that you will be allowed to grow and market your fruit where and how you want. Some counties, for example, severely limit where roadside stands can be placed and what can be sold. Zoning laws can help or hinder you. Before investing in a commercial orchard, thoroughly check out local and state laws and regulations that will pertain to organic fruit production.

CHAPTER 3

DESIGNING OR REDESIGNING YOUR ORCHARD

- Tree Size | 43
- Training Systems | 51
- Orchard Layout Design | 62
- Sample Orchard Design | 69

Here is where the fun starts. Whether you are planting one tree or 10,000, the same principles apply. For those of you with established orchards, this will be a good opportunity to evaluate your orchard and, perhaps, update your orchard layout and training systems. Take the time to get your design right before buying trees and equipment or changing management practices. Investing in trees and trellises, laying out fields and roads, and installing irrigation systems and fences are long-term decisions that can be expensive and difficult to change. For organic growers, it is very important to create a design that minimizes pest and disease problems and lends itself to organic management practices. And again, you also want to think about how you will market your crops, as discussed in chapter 14.

Creating and fine-tuning your orchard on paper is much easier and less expensive than correcting mistakes later on. Our goal is not to create a good orchard, but to create a great orchard!

Tree Size

One of the first decisions you need to make is the size of trees that you will grow. When I first started working in fruit culture, dwarf apples were just becoming commercially popular and dwarf sweet cherries were still a dream. Many orchards consisted of huge trees spaced far apart. Speaking from experience, pruning or harvesting a 20- to 30-foot-tall apple, pear, or sweet cherry tree is a highly overrated pastime. Worse still, yields and fruit quality are often poor, pest and disease management are difficult, and labor costs are high for large trees. If you are growing fruit in a home landscape, one or two large fruit trees can dominate your yard. By planting smaller trees, you not only enjoy greater diversity in your fruit production but are usually rewarded with better-quality fruit, higher usable yields, and less work.

The case for small trees. Modern orchards typically consist of relatively small trees that are sometimes supported on wires or poles. While an individual small tree produces less fruit than a large one, you make up the difference in yield by planting many more small trees in a given space. Although your initial establishment costs are greater because you need to buy more trees and may have to build supports for them, smaller trees offer many advantages.

First, large trees are less efficient fruit producers than small trees. If you examine a large, healthy fruit tree, you will find that most of the leaves and fruit are borne on a thin shell on the outside of the tree, rather like the skin of a balloon. According to research, few fruits form more than 18 to 24 inches from the outer surface of a dense canopy. The interior portion of a large tree is mostly branches and empty space that produce little fruit. The interior is also highly shaded. What few fruits are found there are usually small and poorly colored and may be misshapen. Much depends on the pruning and training, of course. Trees that are pruned to open shapes are more fruitful in the interior than are denser trees. Figure 3.1 shows the productive areas of large and small trees.

Labor needs are greater with large trees due to the need to use ladders for pruning and harvesting. Pruning also involves removing larger wood than is necessary with smaller trees and pruning as many as three times per year, instead of once during the dormant season. If you are operating a U-pick orchard, large trees greatly reduce the amount of fruit that is harvested because it is harder to pick and customers will not necessarily be willing to climb to the treetops to harvest fruit there. The risk of injury and potential liability also increase whenever your customers' feet leave the ground.

Large trees usually have more pest and disease problems than smaller trees. Large, dense trees trap humid air, restrict drying winds, and screen out sunlight. All of these factors favor bacterial and fungal infections. Dense trees can also tip the balance from beneficial insects and mites to pest species. Tall, dense trees make scouting for pests and diseases and spraying pesticides (organic, of course!) more difficult, and unharvested fruits in the tops of tall trees serve as nurseries for next year's pests and diseases. Large trees also require more pesticides and reduce the percentage of leaf and fruit surfaces that are effectively covered by the pesticides, compared with smaller trees.

Large trees also tend to begin bearing fruit crops slowly. Grafting a fruit variety onto a precocious, dwarfing rootstock can cut the time to production tremendously. In the Pacific Northwest, fruit specialists report that sweet cherry trees on Gisela dwarfing rootstocks begin bearing in the third leaf (third growing season from planting), are up to 50 percent smaller than trees on the older Mazzard rootstock, and begin turning profits in 8 years, as opposed to 15 years on Mazzard rootstock. The time needed to prune and harvest the smaller cherry trees is half that for larger trees. Depending on the training system and scion, sweet cherries can be maintained 8 to 10 feet tall, compared with 20-foot-tall trees in a traditional orchard.

Figure 3.1 · A) Much of the volume of large trees is shaded and unproductive.

B) Small, open trees are productive throughout the entire canopy.

While I advocate small trees, if you have a large, old fruit tree, do not feel the need to cut it down. The principles of organic fruit production still apply, although they will be rather more challenging and less effective with standard-sized apple, pear, and sweet cherry trees.

The right small size and number. It is possible to go too small. Extremely dwarfing rootstocks for apples create weak, unfruitful trees. Some nurseries promote genetic dwarf peaches, but these trees produce few or very small fruits. In cooler climates and on poor soils, a fruit tree's growth is naturally less than it would be in ideal climates; in such environments, it is best to plant moderately vigorous to vigorous trees.

Growing small trees can create financial advantages in the long run and financial disadvantages in the short run. High-density systems made up of many small trees often produce high early yields, but they cost more to establish than low-density orchards made up of fewer, larger trees. If you need to support the trees on trellises, the cost rises substantially. The advantage is that high-density plantings increase profitability by increasing yields. This concept became popular during the 1970s and 1980s. The extreme example was the "meadow orchard" system for peaches that contained 7,500 or more trees per acre.

Today's marketplace, in which new, high-value fruit varieties are being released frequently, has made commercial growers push to get as many trees as possible into bearing quickly while demand for a particular variety remains high. Early returns on investment, high yields, and premium prices for popular varieties can offset high establishment costs and make the strategy feasible. This strategy requires high-density plantings of small trees.

These designs often include support systems for the trees. The expected productive life of high-density, high-value orchards can be short — the strategy being to replace the trees with another hot new variety in 15 years or less. From a strictly economic perspective, quick-turn-around orchards make sense. From a long-term sustainability perspective, the answer is less clear and depends on your personal values and needs.

For many growers, particularly those new to fruit growing, small- to moderate-sized trees in the 6- to 14-foot height range work very well. For apples, pears, and occasionally other fruits, support systems are viable and can be economically rewarding. The trade-off is increased establishment costs for trees and trellises. Freestanding trees reduce establishment and maintenance costs, but you will probably have lower early yields, and it may take longer to achieve full production and earn a positive return on your investment.

Some organic orchardists remain leery of growing trees that are too small and prefer medium-sized trees. The rationale is usually that more vigorous trees are better able to cope with competition from alley cover crops, in-row cover crops, and/or weeds. This approach can be important to organic fruit growers, due to the lack of effective herbicides to manage the orchard floor. The larger trees also tend to fill the canopies more quickly than very dwarfing rootstocks. The trade-off here is that the more vigorous rootstocks can increase pruning labor and make maintaining abundant, small-diameter fruiting wood more difficult.

There is no perfect tree size and no perfect training system. Whether for a large commercial orchard, a small market orchard, or home fruit production, select trees and a growing system that fit your lifestyle and needs. Many options are available, from thousands of trees per acre hanging on wires, to a moderate number of freestanding trees, to one or a few gentle giants in your yard.

Controlling Size

Fruit trees are propagated in four ways: growing from seed, layering root suckers, budding or grafting pieces of a variety onto a rootstock, and inserting a short interstem between the rootstock and scion of a grafted tree. Depending on the crop and training system, trees that are 6 to 14 feet tall work well in modern orchards and home fruit plantings.

Trees that are grown from seed are produced from their own rootstocks. This method works well only for a handful of minor crops, such as Damson plum and Nanking cherries, as discussed in chapters 5 and 6. Alternatively, if you have found a seedling that you like, you can propagate it vegetatively by layering root suckers (creating roots on young shoots that grow from the roots or collar). As with seedling trees, you are growing the variety on its own roots. If you want to layer a fruit tree using suckers arising from the ground or base of the tree, be sure that the tree is not grafted. Otherwise, you will likely be propagating the rootstock (the bottom part of the tree), not the scion (the top part), and you will not get the type of fruit that you want. (See figure 3.2.)

Figure 3.2 · **Methods Used to Vegetatively Propagate Fruit Trees**

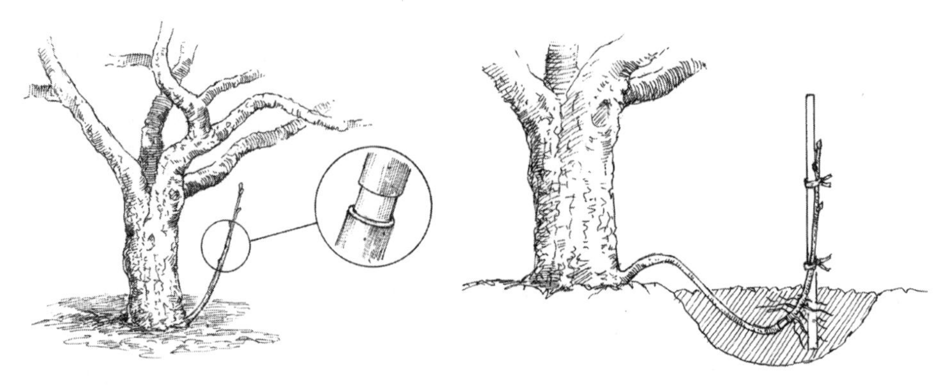

A) Layering using a root sucker

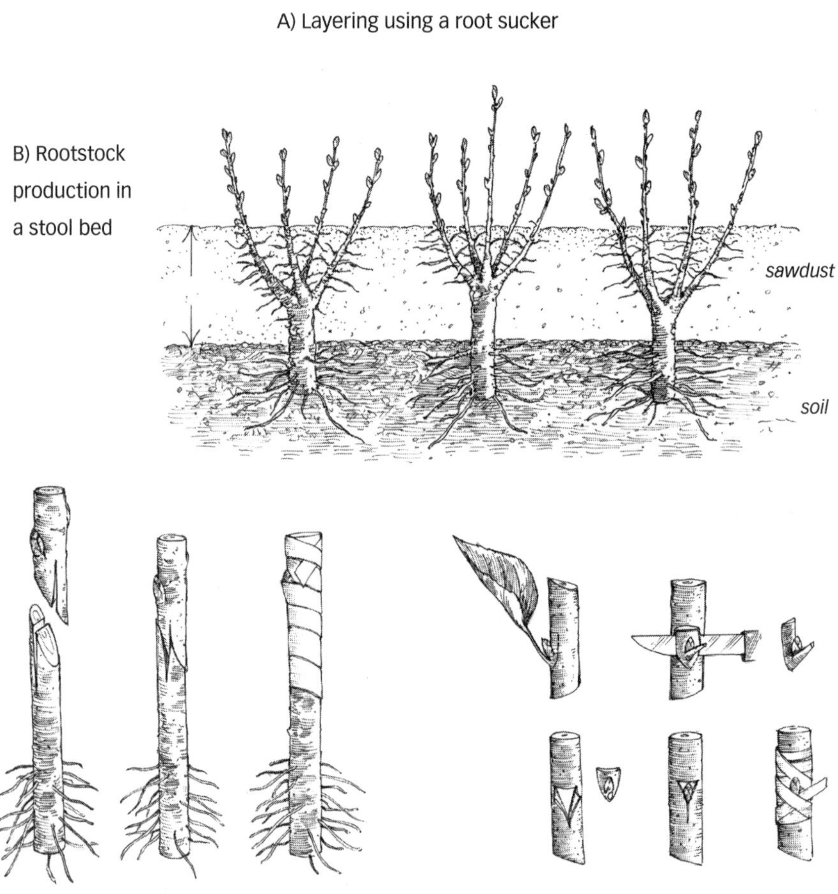

B) Rootstock production in a stool bed

C) Whip-and-tongue grafting (not for stone fruits!) onto a rootstock

D) T-budding onto a rootstock

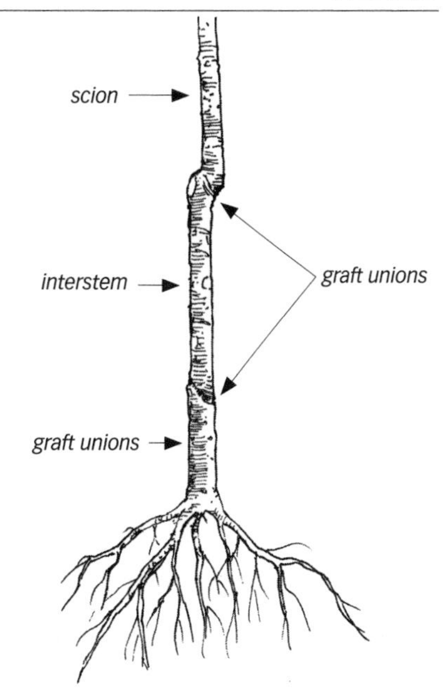

Figure 3.3 · A grafted tree using a root stock for soil adaptability, an interstem for size control, and a scion for fruit variety

Rather than layering fruit trees, we usually bud or graft pieces of a desirable variety onto a suitable rootstock. In a fourth method, we sometimes insert a short interstem between the rootstock and scion, usually to control the size of the tree when using a vigorous rootstock. The bottom part of the rootstock provides adaptability to the soil while the interstem controls the tree's size. The rootstock provides the foundation for the tree. The scion produces the variety of fruit that we want.

When selecting a rootstock, we consider its compatibility with the scion variety, its precocity (time to fruit bearing), its adaptability to different soil and climate conditions, and its resistance to pests and diseases. When choosing pear, apple, and sweet cherry rootstocks, size is often an important criterion.

Seedlings and trees grafted to seedling rootstocks generally grow to be about the same size as the seed tree. For apples, this might be 40 feet tall; for sweet cherries, 60 feet. For grafted or budded trees, the size of the tree varies according to the vigor of the rootstock, scion, and interstem (if used).

Fortunately, there are many good size-controlling and standard-sized rootstocks for apples, pears, and sweet cherries. These rootstocks allow you to grow small, medium-sized, or large trees. Many popular rootstock/variety combinations today produce apple and cherry trees that are 40 to 70 percent of the height of standard-sized trees. Some rootstocks produce even smaller trees. The scion variety also influences the final size of the tree, particularly with sweet cherries — more vigorous varieties produce somewhat taller trees than less vigorous varieties. Apricot, peach, nectarine, plum, and tart cherry trees can be produced efficiently on non-dwarfing rootstocks, although dwarfing rootstocks are sometimes available for these crops. (See chapters 5 and 6 for specific rootstocks.)

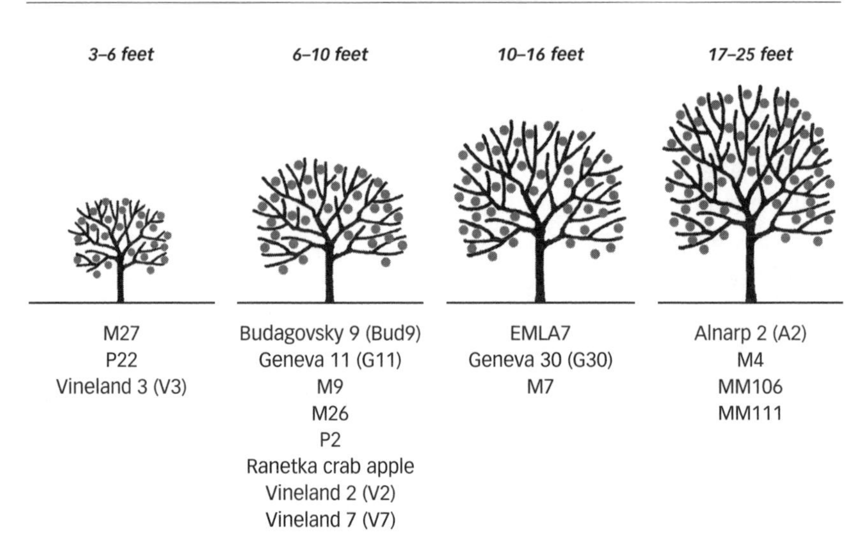

Figure 3.4 · **Effects of Rootstock on Apple Tree Heights**

Note that a particular rootstock/scion variety combination can vary substantially in height in different growing regions and is strongly affected by training and pruning practices.

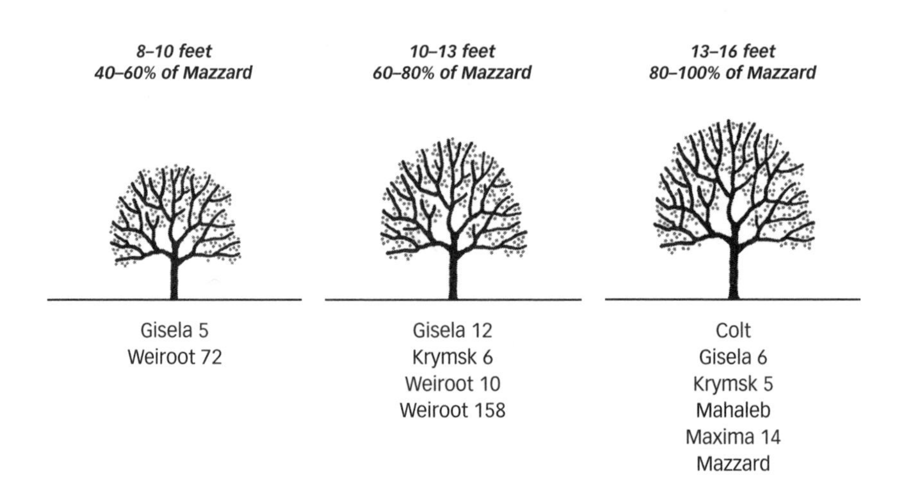

Figure 3.5 · **Effects of Rootstocks on Sweet Cherry Tree Heights**

Standard-sized cherry trees (Mazzard rootstock) can grow 40 feet tall if left unpruned but are normally kept about 16 feet tall in orchards. The relative heights in this figure are based on a 16-foot height for a full-sized tree. Note that rootstock/scion variety heights vary substantially in different regions and with different training and pruning practices.

Defining "Dwarf" and "Semidwarf"

You often hear the terms "dwarf" and "semidwarf" used to describe fruit trees on size-controlling rootstocks, particularly when talking about apples and sweet cherries. To be blunt, the terms are meaningless. A "standard" apple is about 25 feet tall or somewhat taller. You can reasonably argue that 20-foot-tall trees on a particular rootstock are dwarfs. Technically, that is correct. From a practical perspective, that amount of dwarfing is insignificant.

You can further reduce the size of trees by planting trees closer together, cropping the trees heavily, and pruning to remove excessively vigorous growth. The illustrations at left show typical sizes of apple and sweet cherry trees that develop on selected commercially available rootstocks.

Before you buy any trees, ask the salesperson what rootstocks the trees are grown on. If the response is a glazed look or gibberish about dwarfs and semidwarfs, find a nursery where the people know what they are doing. Again, select tree sizes that fit your needs. Don't be pressured into buying trees that will grow smaller or taller than you are comfortable with.

Training Systems

Regardless of the size of your trees, it is important to select a suitable training system to create an efficient and productive orchard. Choose a system that will optimize the fruit-bearing surface of the trees and the number of trees for your orchard. Note that I use the word *optimize*, not *maximize*. We want the system to be sustainable for the long term. We also want a system that works well for organic production and that fits the site and your lifestyle. Which training system you select determines the amount of labor needed to manage and harvest your trees, and it has a huge influence on pest and disease management strategies and effectiveness.

Training systems and rootstocks are introduced here to help readers begin planning the orchard layout. Detailed descriptions for various training systems and instructions for creating and maintaining them are found in chapter 12.

Pome and stone fruit trees can be grown freestanding or supported on trellis wires and/or poles. Until roughly the 1970s, freestanding trees were the rule, with the occasional exception of ornamental trellising of apples and pears in landscape designs. Since then, wire and pole systems have been developed for the high-density production of tree fruits. The goals of these systems are to produce fruit, rather than wood; bring the trees into production quickly; increase yields; and increase profits.

A word of caution: One of the most common mistakes new growers make is to plant their trees too closely together, which soon creates crowding and tree health problems. Rows that are too narrow and trees that are planted too closely together make orchard work difficult and reduce labor efficiency. Trees that are planted too far apart yield less fruit. The following sections provide suggested within- and between-row spacings for tree fruits.

Apple Training Systems

Of all the tree fruits, apples are the most complex crop because of the large numbers of varieties, rootstocks, and training systems. That being

Basic Apple Training Systems *Table 3.1*

System	Trees per Acre*	Tree Height†	Spacing† trees	Spacing† rows	Tree Supports
Central leader	272	8′–12′	10′	16′	none
Slender pyramid	443	12′–14′	7′	15′	none
Slender spindle	1,089	6′–8′	4′	10′	pole
Super spindle	2,178	6′–10′	2′	10′	pole or wires
Tall spindle	837–1,452	10′–12′	3′–4′	11′–13′	3+ wires
Vertical axis	518	10′–14′	6′	14′	wire & pole
Trellis	660–1,452	4′–10′	3′–6′	10′–11′	3–6 wires

*Trees per acre refers to solid blocks and does not take roads into account. These are average figures and can vary.
†Tree heights and spacings are given in feet. These are average figures and can vary. More details are in chapter 12.

said, apples are among the easiest tree fruits to grow and adapt to many different growing areas. Don't allow yourself to become overwhelmed by the sheer numbers of combinations and options available. Stay focused on basic fruit-growing principles and your needs. Table 3.1 reduces the many variations into seven basic systems and gives recommended planting distances within and between tree rows. Chapter 12 provides details on the training systems described below.

In the not-distant past, large apple trees were often planted 20 feet or more apart, allowing around 100 trees per acre. Plantings of 200 to 300 trees per acre were considered high-density. Today, apple orchards of 500 to 1,000 trees per acre are common, and some systems have more than 2,000 trees per acre.

In trials conducted at Cornell University, researchers evaluated various apple training systems. The lowest yields for the first 4 years of the trials came from freestanding trees trained to central leaders (218 trees/acre). For trees on supports, slender pyramid training (444 trees/acre) produced the lowest yields, followed by slender axis training (726 trees/acre). The tall spindle system (1,307 trees/acre) produced the earliest and heaviest yields. Figure 3.6 shows how tree planting density affects apple yields.

In related apple training system trials in New York, researchers estimated the earliest financial breakeven points for new orchards using each system (shown on the next page). Growing apples is not a get-rich-quick enterprise!

Yields			**Estab. Costs**	**Mgmt. Inputs**	**Uses**	**System**
2–5 yrs	***5–10 yrs***	***10+ yrs***				
L	L	M	L	L–M	hg	**Central leader**
L–M	M–H	H	L–M	M–H	all	**Slender pyramid**
L	M	M	L–H	L–M	hg, mg, up	**Slender spindle**
M	H	H	H–VH	M–H	mg, up, c	**Super spindle**
H	H	H	H	H–VH	c	**Tall spindle**
M	H	H	H–VH	H–VH	c	**Vertical axis**
L	L–M	M	M–VH	M	all	**Trellis**

Estab. costs = Establishment costs; Mgmt. inputs = Management inputs; L = low, M = moderate, H = high, VH = very high; Uses: hg = home orchards, mg = market orchards, up = U-pick, c = commercial grower pick

Slender axis: 12 years
Tall spindle: 13 years
Super spindle: 14 years
Vertical axis: 14 years
Slender pyramid: 17 years

Trials have been conducted in British Columbia, Pennsylvania, and other locations around the world with generally similar results. For new commercial apple orchards, most North American experts now generally recommend some variation of a vertical axis or tall spindle design.

Pear Training Systems

Pear orchards, particularly in North America, have traditionally been low- to medium-density plantings of freestanding trees or trees trained to palmette shapes and sometimes supported on trellis wires. Even today, that approach remains common in the United States. Internationally, research on high-density pear plantings has paralleled, but lagged behind, that on apples. Because the methods used to produce apples and pears are very similar, modern pear systems are largely adapted from apple systems. As with apples, the move toward higher densities

Basic Pear Training Systems

Table 3.2

System	Trees per Acre*	Tree Height†	Spacing†		Tree Supports
			trees	*rows*	
Central leader	151–453	10′–15′	8′–14′	12′–24′	none
Palmette	340–605	10′–15′	6′–8′	12′–16′	3+ wires
Spindle	453–1,320	6′–10′	3′–6′	11′–16′	2–3 wires
Slender spindle	726–1,210	6′–10′	3′–5′	12′	2–3 wires
Super spindle	>1,600	6′–10′	2.5′	11′	3 wires
Vertical axis	453–1,116	10′	3′–6′	13′–16′	3 wires
Double leader	990–1,320	10′–12′	3′–4′	11′	3 wires
V hedge	907	6′	4′	12′	2 wires & poles
Open Tatura trellis	968–2,234	6′–9′	1.5′–3′	13′–15′	6–8 wires

*Trees per acre refers to solid blocks and does not take roads into account. These are average figures and can vary.
†Tree heights and spacings are given in feet. These are average figures and can vary. More details are in chapter 12.

is largely driven by the desire to increase early yields and profitability. Unlike for apples, there are relatively few new pear varieties entering the market, so there is little need to establish short-term plantings to capture the market for hot new varieties. There are also far fewer pear rootstocks available than are available for apples.

Most existing commercial pear orchards in North America use either freestanding central leader trees or trees trained on wire trellises. Freestanding trees for home and commercial production are best trained to conic or pyramidal shapes, as for apples. High-density training systems that are being tested and used commercially worldwide include variations on the spindle, vertical axis, double leader, and split canopy designs (V, Y, and Tatura). Other than the double leader and split canopy systems, these designs are very similar to those we discussed for apples. Table 3.2 below lists basic pear training systems. In international and U.S. trials, split-canopy systems, particularly the Tatura, have often proven highly productive but increase the difficulty of managing pests and diseases. Perhaps the best choice for organic, high-density pears is the double leader system (see chapter 12 for more details).

Yields			Estab. Costs	Mgmt. Inputs	Uses	System
2–5 yrs	*5–10 yrs*	*10+ yrs*				
L	L	M	L	L	all	**Central leader**
L	M	M	M	M	all	**Palmette**
L–M	M–H	H	M–H	M	all	**Spindle**
M	H	H	H	M–H	all	**Slender spindle**
M–H	H	H	H	H	mg, up, c	**Super spindle**
M–H	H	H	H–VH	H	mg, up, c	**Vertical axis**
M–H	H	H	H	M–H	all	**Double leader**
M–H	M	H	H–VH	H	mg, up, c	**V hedge**
M–H	H	H	H–VH	H	mg, up, c	**Open Tatura trellis**

Estab. costs = Establishment costs; Mgmt. inputs = Management inputs; L = low, M = moderate, H = high, VH = very high; Uses: hg = home orchards, mg = market orchards, up = U-pick, c = commercial grower pick

Minor Pome Fruit Training Systems

In addition to apples and pears, loquat, mayhaw, medlar, quince, and saskatoon are pome fruits that are grown for personal and commercial use in North America. Being minor commercial crops, their production systems are usually very simple and involve freestanding trees or bushes (we'll cover these crops in more detail in chapter 5).

Loquats are a Chinese fruit long domesticated and grown in southern California, Florida, coastal Georgia, and Hawaii. These large evergreen shrubs or small trees can reach 30 feet tall but are generally kept about 10 feet tall, and quince rootstocks can be used to produce smaller plants. Loquats are generally grown freestanding and are planted 20 to 25 feet apart within and between rows.

Quince and medlar are very similar to pears in many respects. Table 3.3 shows typical spacings for freestanding trees. Medlar trees are spaced 7 to 20 feet apart in rows 15 to 20 feet apart, and quince trees are spaced about 15 feet apart in rows 20 to 25 feet apart. The newer training systems that work for apple and pear may be effective for quince and medlar (tables 3.1 and 3.2) and are worth experimenting with.

Mayhaws are hawthorns adapted to the southern and southeastern United States and have recently been domesticated. Although early recommendations were to plant mayhaws 15 to 18 feet apart, experience

Plant Spacing for Minor Pome Fruits *Table 3.3*

Spacings are given for plants grown as freestanding trees or bushes.

Crop	Plants* per Acre	Height (feet)	Plant Spacing** (feet)	Supports	Estab.	Mgt.	Uses
Loquat	69–108	10–30	20–25 × 25–35	none	L	L–M	all
Mayhaw	58–69	20–30	25 × 25–30	none	L	L–M	all
Medlar	108	8–25	7–20 × 15–20	none	L	L–M	all
Quince	116–145	8–15	15 × 20–25	none	L	M–H	all
Saskatoon	726–4,356	6–30	1–5 × 10–12	none	L–H	L–H	all

*Plants per acre refers to solid blocks and does not take roads into account. These are typical densities for the respective crops.

**Plant spacing refers to distance between plants × distance between rows.

Key: Estab. = establishment costs; Mgt. = management inputs

L = low, M = moderate, H = high, VH = very high

Uses: hg = home orchards, mg = market orchards, up = U-pick, c = commercial grower pick has shown that a spacing of about 25 feet apart within and between rows works better.

Saskatoons are bush fruits that resemble blueberries in size, shape, color, and flavor. Wild bushes can reach 30 feet, but the plants are usually kept to about 10 feet tall or less in cultivation. Very dense hedgerows designed for mechanical harvest can be created by planting the saskatoons 12 inches apart in rows 10 feet apart. A more typical spacing would be 3 to 5 feet apart in rows 10 to 12 feet apart.

Cherry Training Systems

As with apples, fruit growers around the world have made tremendous advances recently in developing new cherry varieties, rootstocks, and training systems. While the drive has primarily been due to the desire for increased yields and profits, market and noncommercial fruit growers are also benefiting. Not so long ago, cherry trees were planted on standard rootstocks and developed into large trees. Without intensive management, the trees often grew 25 to 40 feet tall, with little fruit within 8 to 10 feet of the ground.

With today's dwarfing rootstocks and improved training methods, we can keep cherry trees 8 to 16 feet tall. With some trees, 80 to 100 percent of the fruit can be harvested from the ground. Two genetic dwarf tart cherries grow only 6 to 8 feet tall and are ideal for home orchards.

That is the good news. On the downside, we are at the beginning of a very steep learning curve when it comes to dwarfing cherry rootstocks and high-density training systems. What we do not know is much greater than what we do, and growers must exercise caution when investing heavily in new varieties, rootstocks, and training methods. Further, combinations that work well in the Pacific Northwest may perform poorly in eastern North America and systems that are successful in Europe and Australia do not necessarily work well here.

Generally speaking, North American commercial cherry growers have been slow to adopt high-density cherry training systems. Research by growers and fruit specialists at universities and government agencies in the United States and Canada is underway and promising, but the present trend in North America is to continue with medium-density cherry orchards and rather simple training systems. One of the most serious problems with precocious, highly productive rootstocks is that, as the trees mature, they bear too many fruits and fruit size becomes too small to market. We are also finding that different cherry varieties respond quite differently to the same rootstocks (see chapter 6 for recommended rootstocks). Modern cherry training systems are generally similar to the spindle, axis, V-system, and Tatura systems described for apples and pears and are described in chapter 12.

Basic Cherry Training Systems

Table 3.4

System	Trees per Acre*	Tree Height (feet)	Tree Spacing** (feet)	Tree Supports	Estab.	Yields 2–5 years	Yields 5–10 years	Yields 10+ years	Mgt.	Uses
TART CHERRY										
Central-leader amarelle	110–156	8–14	14–18 × 20–22	none	L	L	M–H	H	M	all
Central-leader morello	194	8–14	15 × 15	none	L–M	L	M–H	H	M	all
Dwarf	202–272	6–8	10–12 × 16–18	none	M	L	L	L	L	hg, mg, up
SWEET CHERRY										
Steep ladder	91–151	12–14	16–20 × 18–24	none	L	L	M	H	M–H	all
Vogel central leader	202–363	10–12	8–12 × 15–18	none	M–H	M	H	H	M–H	all
Spanish bush	242–660	8	6–10 × 11–18	none	M–VH	M	H	H	M–H	all
Slender spindle	256–670	10–12	5–10 × 13–17	optional	H–VH	M	H	H	M–H	all

*Trees per acre refers to solid blocks and does not take roads into account. These are typical densities for the respective training systems and crops.

**Tree spacing refers to distance between trees × distance between rows.

Key: Estab. = establishment costs; Mgt. = management inputs

L = low, M = moderate, H = high, VH = very high

Uses: hg = home orchards, mg = market orchards, up = U-pick, c = commercial grower pick

Peach, Nectarine, Apricot, Plum, and Prune Training Systems

Peaches, nectarines, and apricots have traditionally been grown as free-standing trees trained to open centers. Despite efforts to develop high-density planting systems using the designs we discussed for apples, pears, and cherries, low- to medium-density orchards built on open center designs remain popular in leading peach and apricot growing regions. This is mainly due to three reasons:

- There are no suitable dwarfing rootstocks yet available for these crops.
- Although we can greatly increase yields with high-density plantings, fruit size is usually smaller than with lower-density systems.
- Depending on climate and soils, low-density orchards often match high-density orchards in yields beginning in about year 6 or 7, when the canopies fill in the spaces between trees.

The few genetic dwarf peach and nectarine varieties available are really novelty trees best suited for pots, patios, and ornamental gardens. The dwarfing is due to very short spaces between leaves (internodes) which create dense foliage and closely spaced fruits. Yields are very low, and you must thin excessively to avoid what has been described as "thumbnail-sized" fruits. The management of pests and diseases is complicated by the densely clustered foliage and fruits. Dwarfing in apple, pear, and cherry is accomplished by grafting vigorous varieties onto low-vigor rootstocks. The dwarfing is due to the genetics of the rootstock and does not result in the very short internodes found in genetic dwarf peaches.

Another problem with high-density, central leader training for peaches and nectarines is that the trees bear fruit on 1-year-old wood and must create a lot of new fruiting wood each year to bear the following year's crop. Fairly large, long-lived scaffold branches bear this abundant fruiting wood. Additionally, peach and nectarine trees have very upright growth habits that best lend themselves to open center training. Peaches trained to central leaders require extensive maintenance and are really going against their natural form.

Basic Peach, Nectarine, Apricot, Plum, and Prune Training Systems

Table 3.5

System	Trees per Acre*	Tree Height (feet)	Tree Spacing** (feet)	Tree Supports	Estab.	Yields 2–5 years	Yields 5–10 years	Yields 10+ years	Mgt.	Uses
PEACHES, NECTARINES, AND APRICOTS										
Open center	75–134	12–14	18–24 × 18–24	none	L	L	M–H	H	L	all
Perpen-dicular-V	311–544	12–14	5–7 × 16–20	none	H	M	H	H	M–H	all
Quad-V	268	12–14	9 × 18	none	M–H	M	H	H	H	all
APRICOTS ONLY										
Modified central leader	75–134	12–14	18–24 × 18–24	none	M	L–M	H	H	L	all
Open center	75–134	12–14	18–24 × 18–24	none	M	L–M	H	H	L	all
PLUMS AND PRUNES (EUROPEAN AND JAPANESE)										
Central leader or open center	108–272	12–14	10–20 × 16–20	none	M	L–M	M–H	H	L	all
PLUMS (AMERICAN)										
Free-standing bush	322–605	10–24	6–9 × 12–15	none	L–M	n/a	n/a	n/a	L–M	all
PLUMS (BEACH AND OTHER BUSH TYPES)										
Free-standing bush	726–871	3–12	5 × 10–12	none	L–M	n/a	n/a	n/a	L–M	all
PLUOTS, PLUMCOTS, AND OTHER PLUM × APRICOT HYBRIDS										
Central leader	193	12–14	15 × 15	none	M	L–M	M–H	H	L–M	all

*Trees per acre refers to solid blocks and does not take roads into account. These are typical densities for the respective training systems.

**Tree spacing refers to distance between trees × distance between rows.

Key: Estab. = establishment costs; Mgt. = management inputs

L = low, M = moderate, H = high, VH = very high, n/a = data not available

Uses: hg = home orchards, mg = market orchards, up = U-pick, c = commercial grower pick

Depending on where you live, however, high-density peach systems might be viable. In cold areas, where stem cankers and other diseases make for short-lived peach trees, high early yields can be important for profitability. In Ontario, peach growers have largely adopted central leader and spindle (Fusetto) training systems, despite difficulties with maintaining the tree shapes and increased stem canker diseases. Trials in New York, Georgia, and Arkansas support the use of high-density plantings using variations on open center training, including the perpendicular-V and quad-V designs (described in chapter 12). Research done in New York suggests that 500 to 600 peach trees per acre is the most profitable density for the northeastern United States and eastern Canada.

For warm, dry climates, such as in California and Texas, fruit specialists caution peach and nectarine growers against high-density systems. In these areas, trees have productive lives of more than 15 years and low- to moderate-density orchards work well.

Peach and nectarine orchardists have successfully used Tatura and other divided canopy systems to increase early yields. Trellises and other supports increase establishment costs and maintenance labor and the hard-to-reach crop row centers can complicate pest and disease control. I believe the perpendicular-V and quad-V systems are better alternatives for organic orchards.

Apricots can be trained to any of the open center designs, or to a modified central leader system, as for apples. In commercial orchards, they are usually trained the same way as peaches are trained, but somewhat less pruning is required.

You can train plums and prunes similarly to peaches — with an open center or modified central leader — but the trees are more upright. As with peaches, no proven dwarfing rootstocks are yet available.

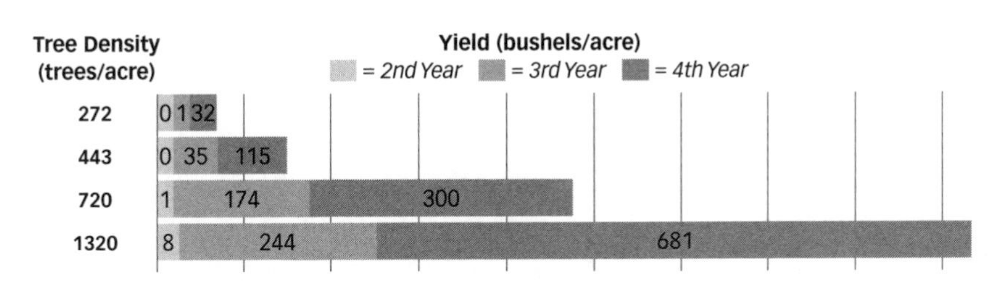

Figure 3.6 · **Average Apple Yields for a Newly Established Orchard**

As the numbers of trees per acre increase, yields, especially early yields, increase dramatically. Note that, even with trees at the same densities, yields differ substantially between rootstocks, varieties, training systems, management practices, and regions.

Across North America, plums and prune trees are typically spaced 10 to 20 feet apart in rows 16 to 20 feet apart. Only a very few high-density plum trials have been reported around the world, including experiments with tree densities greater than 2,000 trees per acre and spindle, axis, and V-shaped training systems. Although these high-density orchards had very high yields, the longest trials only lasted a few years, and we have little information on how plums respond to high-density training over a long period of time. In northeastern Europe, plum trees are planted around 6 to 7 feet apart in rows 13 to 14 feet apart. This closer spacing might work in particularly cold, short-season areas of the United States and Canada, but it will create a crowded orchard on good soils in favorable climates.

Orchard Layout Design

ONCE YOU HAVE DECIDED WHAT CROPS YOU WILL GROW and how they will be spaced and trained, you can begin designing the orchard. When laying out your orchard, avoid the temptation to crowd as many trees as possible into the space available. For commercial orchards, make sure there is quick, easy, and efficient access into and out of the orchard for workers and equipment. Crowding seldom results in higher marketable yields in the long run and it increases your labor and management costs by reducing efficiency. Narrow alleyways and headlands can lead to damage to trees from large equipment, and crowding greatly increases pest and disease problems.

Map It Out

Lay out your orchard on a piece of paper, and begin by mapping out the general crop and non-crop areas. Start with a fairly accurate outline of your property. If you do not have a property map, you might be able to find one at your local tax office. If you are purchasing a piece of property, be sure to have it accurately surveyed. The surveyor can provide you with a detailed map of the boundaries. Some online search engines now provide satellite maps of much of North America and enable you to identify and print off an aerial photo of your property. The great advantage of aerial maps is that they also show streams, bogs, rock outcrops, and other features of your site. Using these sources, create an outline map of your property, including features such as wetlands and low areas. Try to get the dimensions as accurate as you reasonably can.

The next step is to start designing your dream orchard. At this stage, do not worry about getting everything to scale. The important point is to develop a general idea of what you want to grow and how you want it laid out. Later on, you can refine the plan and drawing.

Non-crop areas. Begin by marking out crop and non-crop areas. Wetlands, low-lying frost pockets, and rocky outcroppings are best used for equipment, employee, and customer parking; equipment and supply storage buildings; refrigerated storage; sales stands; and any other structures your operation may need. Identify these non-crop areas, leaving the best parts of the site for your crops.

Other non-crop areas are buffer strips around the inside perimeter of your orchard. Certified organic orchard growers generally need to leave buffer strips between certified land and adjacent uncertified fields, rows, hedgerows, and roads. The width of the buffer strips will be determined by your certifying agency and the needs of your individual operation. As a general guideline, 30-foot-wide strips should meet most growers' needs. Exceptions to this would be if your orchard adjoins a nonorganic orchard where airblast sprayers or similar equipment are used to apply pesticides, or if it sits along roadways where nonorganic herbicides and other pesticides are sprayed. In these cases, you may need wider buffer strips and you should probably include a dense windbreak to reduce the amount of pesticide spray that drifts into your orchard. Even if you are not planning to become certified, creating a buffer strip to protect your fruit trees from pesticide drift is a good practice. The buffer strips typically serve as roads around the orchard and provide headlands for turning tractors and other equipment. Make the headlands wide enough to allow you to easily turn around the largest piece of equipment you will be using in the orchard.

Planting blocks. Once you have mapped out the non-crop areas, define the orchard blocks where each crop will be grown. Draw these blocks in relation to your already defined orchard boundaries and non-crop areas. These planting blocks will largely create themselves when you plot out buffer strips from boundary roads and fences. Every grower's site differs in topography and shape, and you will need to evaluate your own site to determine how best to lay out the orchard. For home orchards the trees may simply be spread out, more or less randomly, through the orchard space.

Lay out the entire planting area first, then divide that area into sections for different crops or varieties. Add crossroads to the plan to provide access to the different planting blocks. The road widths will depend on orchard size and the equipment you will be using. Twenty-foot-wide roads are about the minimum for small pickup trucks, with 30- to 40-foot-wide interblock roads being common in large orchards. For large orchards, include enough main and crossroads to allow bins of harvested fruit to be moved quickly from the fields with minimal rough handling.

Once you have the planting blocks drawn, lay out the tree rows. In general, run planting rows across slopes, not up and down the slopes. Determine the number of tree rows in each block. For each crop that you plan to grow, identify what training system you will use. The tables that appear earlier in chapter 3 give the distances between tree rows. These between-row spaces are called alleys.

Be sure to provide plenty of spacing between the trees within the rows in order to meet the needs of the individual crops. Again, use the tables that appear earlier in this chapter to determine tree spacing. Be cautious not to crowd the trees.

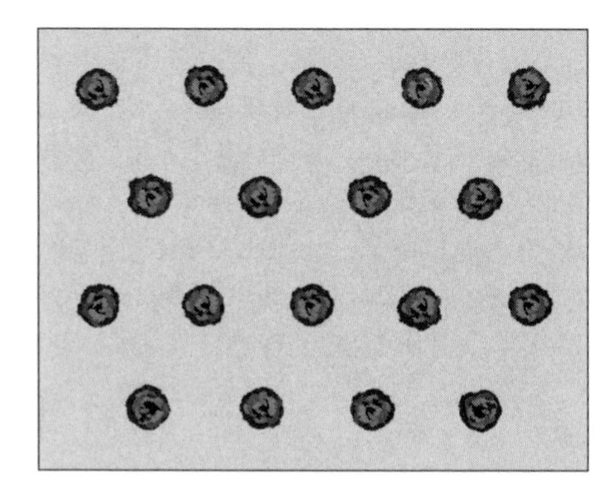

Figure 3.7 · For freestanding trees, particularly medium-sized to large trees, offsetting them in adjacent rows to form a triangular pattern maximizes the space for each tree. Given adequate spacing, offset plantings can be mown or cultivated along the alleys and on diagonals.

In laying out planting rows with freestanding trees, try to offset the trees in adjacent rows, forming a triangular pattern that maximizes the distance between trees. Figure 3.7 shows an offset or staggered planting.

While some of the spacings in the tables may seem too large, especially when trees are young and small, they are based on vast experience by growers around the world. Crowding does not increase long-term yields and greatly complicates orchard management.

As orchard size and mechanization increase, you will find straight rows desirable, although curved rows are often necessary due to curving property lines or slopes. Where planting rows must be laid out in curves, keep the curves gentle and uniform to allow easy access with orchard equipment, and easy installation and maintenance of trellises.

Irrigation Systems

Many orchard irrigation systems exist, from flooding to overhead sprinklers to high-efficiency drip systems. Which you choose will be influenced by the availability and cost of water, the cost of electricity for pumps and controller systems, the complexity and cost of the irrigation system (including installation, operation, and maintenance), and your personal preferences. A good irrigation system will do the following:

- Provide adequate water for the fruit crops when they need it
- Provide adequate water for alley and in-row cover crops
- Provide adequate water for frost control, if needed
- Keep foliage and fruit as dry as possible
- Prevent wet spots in the orchard that can increase root diseases and disorders
- Prevent soil erosion
- Be reliable
- Be easy to use

Although irrigation systems may seem simple, you need expertise and experience to design an effective system. For all but the smallest orchards, it is best to have a professional assist you with the design.

Make Safety a Priority

Personal safety is a concern in every orchard. When you have customers for U-pick or roadside stand orchards, safety becomes even more challenging. Design your operation to keep customers and visitors away from equipment, storage buildings, and potentially hazardous orchard operations. You want the orchard to be attractive and welcoming without exposing people to unnecessary hazards.

Consider installing fences and railings to keep people away from hazardous areas. Design planting blocks and access roads to confine U-pick customers to particular blocks while harvesting, rather than allowing them to roam throughout the orchard and "cherry pick" only perfect fruit. Figure 3.8 shows a sample orchard design.

Firms that sell irrigation equipment can often help you design a system that meets your needs.

Flooding. Years ago, orchards in regions where water was abundant were commonly flooded. Using a series of ditches and dams, the entire orchard floor would be flooded for up to several hours at a time before the dams were opened and the water allowed to flow out. While the trees received plenty of water, root diseases and micronutrient disorders were also common, as was soil erosion. The system required vast amounts of water and a perfectly level orchard. In most areas today, flood irrigation for orchard crops is rare.

Sprinklers. Sprinkler systems are an improvement over flood irrigation and can be good choices for home orchards, as well as for commercial orchards in areas where water is abundant and inexpensive. Sprinklers can be used in orchards on sloping ground, they work well when you have alley and in-row cover crops, and they are relatively simple to use. They can be set up to automatically turn on at regular intervals or when soil moisture drops to a specific level. Another plus is that you can easily spot a clogged sprinkler at a distance.

Overhead sprinklers are sometimes used in home and commercial orchards to provide frost control for a short period of time. If temperatures drop to around 34°F (1°C) and a damaging frost is expected, sprinklers are turned on and are run continuously during subfreezing temperatures until the temperature returns to above freezing. Although ice will form on the trees, as long as there is a film of liquid water on the ice, the temperature of the ice and tissues below it will remain at about 32°F (0°C), which will protect against frost damage. This approach to frost protection is based on the nature of water to release heat as it changes from liquid to solid.

If you use sprinklers for frost control, you will need to cover the trees completely with water and apply at least ¼ inch of water per hour for as long as there is a risk of frost. These systems work best for short-term frosts that last a few hours. They are not effective against masses of freezing air that settle into orchards for several days.

When sprinkler systems are not intended to provide protection against frosts, it is usually best to use sprinkler designs that keep the water close to the ground. This practice keeps the foliage and fruit dry and greatly reduces disease problems. When you use sprinkler irrigation, apply water early in the morning to allow the understory of the

Protecting against Frost

Most commercial orchards now use large fans mounted on towers to stir up the air in the orchard and help offset inversion layers during radiation frost events. Inversion layers commonly develop when cold air is trapped near the ground. You can see inversions when smoke from chimneys rises 100 feet or so in the air and then spreads out in a horizontal layer as if it has hit a glass ceiling. That ceiling is the boundary between cold and warm air layers. Mixing the cold layer with warmer air layers above it can help prevent frost damage in orchards.

Propane- or kerosene-fired heaters are sometimes used alone or in combination with fans to keep orchard temperatures above freezing during bloom. Specially designed orchard heaters called "salamanders" are available. On a small scale, one of my friends who has a home orchard places a portable barbecue grill under each blooming tree during frosty periods. The idea is to add just enough heat to the canopy area to keep temperatures at about 32 to 34°F (0 to 1°C) or above.

tree to dry before nightfall. This practice keeps the humidity level lower and helps reduce disease problems. Keeping the sprinkler pattern low also protects the kaolin clay film that is applied to the foliage, fruit, and branches as part of some pest and disease control programs. Sprinklers quickly wash the clay film off, leaving the crops susceptible to pests and diseases. We'll discuss kaolin use in chapters 10 and 11.

In addition to the potential for increasing disease problems, sprinkler systems have several major drawbacks, particularly for larger orchards. The first drawback is increased operation costs. A large commercial orchard requires large amounts of water and powerful pumps to lift the water from a well or other source and push it through the system, which can be quite expensive.

There are also several serious drawbacks with using sprinklers for frost protection: You need large amounts of water during and shortly after spring bloom, you need a system that will cover all parts of every tree, and, finally, the weight of the ice can break branches. As an alternative to sprinklers, you could use large fans or heaters to protect against frost (see box on page 67).

Drip irrigation and microsprinklers. Many orchards today are irrigated using drip systems that apply water directly to the soil at the base of the trees. These designs can be used on flat or sloping ground, use water very efficiently, and eliminate erosion problems associated with sprinkler and flood irrigation. One disadvantage is that it is harder to spot clogged emitters in a drip system.

Microsprinklers set in the crop rows to irrigate the fruit crops, alley crops, and in-row living mulches provide a compromise between a drip and a sprinkler system. It is easy to spot clogged microsprinklers, and this design uses much less water than older large-volume sprinklers. Microsprinklers have several advantages over drip systems. Microsprinklers cover relatively large areas. The wider distribution of irrigation water across the orchard floor can help encourage the development of larger, better-anchored root systems for fruit trees than is common with drip irrigation alone. The larger root systems can provide better drought tolerance for the trees than the limited root systems that sometimes develop under drip irrigation.

Microsprinklers and drip systems require either city water or a series of filters to trap sand and organic matter that would clog the emitters. They also require pressure regulators to ensure that each irrigation line receives about the same amount of water and at the correct pressure for the emitters you are using.

If you are using alley and in-row living mulch crops, microsprinklers are often the best choice. Set high-quality irrigation lines deeply in the tree rows before planting. Locate the risers and sprinkler heads in the centers of the rows and midway between adjacent trees.

For orchards that contain closely spaced trees and bushes and for those that are located in arid locations where water efficiency is very important, drip irrigation may be the best choice.

Fencing

Orchards often need fencing to protect crops from wildlife or livestock. Regardless of its size, if your orchard is located in an area inhabited by deer or moose, a well-designed and constructed orchard fence is a good investment.

When designing your fence system, keep convenience and efficiency in mind. While you can put a fence around each individual tree in a small orchard, caring for the trees becomes extremely inconvenient and inefficient. You will probably find that a fence enclosing a complete planting area will work better for you. For large orchards, enclose the entire orchard, rather than separate planting blocks. See pages 370–372 for several fencing options.

Sample Orchard Design

For our example, we will begin with a 5-acre property laid out in a square. Approximately 4.3 acres are devoted to fruit production, with 0.7 acre reserved for equipment, buildings, composting areas, and parking. Each fruit block is 178 feet long by 80 to 110 feet wide. This grower plans to sell fruit from an on-farm roadside stand, at a farmers' market, and through U-pick sales.

To reduce the risks associated with a single crop and to expand the market season, the design includes different tree fruits and berries. The berries mature quickly, with raspberries and blueberries bearing crops within 2 to 4 years. The berry crops also ripen early in the season, providing cash flow to help finance the tree fruit orchard. While multiple crops have marketing advantages, growing many different crops is much more challenging than focusing on a single crop. If you are new to fruit farming, start small and focus on one or a few fruit crops. As you gain experience, you can expand and diversify.

Our first step will be to complete an orchard planning worksheet. Much of this information you will develop while evaluating your site.

Sample Orchard Planning Worksheet

The following worksheet provides an example of how you might plan a new orchard or orchard upgrade. Filling in the different sections of the worksheet provides you with the information you need to use the site effectively and prepare a diagram or blueprint for your orchard.

Site description: A square parcel consisting of 5 acres lying 4 miles outside of a town of 6,000 people. The site is adjacent to the main highway leading into town. Four other towns with 500 to 2,000 people are within 10 miles of the orchard and a city of 50,000 people is 45 miles away. (This information is critical for evaluating your marketing plan.)

Slopes: Approximately 2 acres next to the highway are mostly level, with a few low-lying areas. The remaining 3 acres slope upward and away from the highway. Slopes average 6 to 10 degrees, and the surface is generally flat.

Surface water drainage: The slopes are grass-covered pasture and show no signs of erosion. Drainage appears good; there is no standing water on the slopes, but several areas near the highway retain standing water for 1 to 2 days after a rain.

Soil type and drainage: The site is covered with 24 to 30 inches of loam overlying 48 inches of sandy clay loam lying over gravel. Percolation tests show adequate drainage on the 4 acres away from the road. Several clay lenses near the highway are causing poor drainage and some standing water.

Cold air drainage: Cold air drainage (the ability of cold, heavy air to flow out of your orchard) on the 3 acres away from the highway is good. Next to the highway, the land is more or less level for several miles around. Late spring frosts are common on the lower 2 acres.

Wind protection needed: The area experiences occasional summer and winter winds with steady speeds of 20 to 30 miles per hour and gusts of 50 to 60 mph. The winds are usually from the north and east. Windbreaks will be used on the north, east, and west sides, leaving the highway side of the orchard open. The windbreak will consist of the Penn State University clone of hybrid poplar with trees spaced 10 feet apart.

Low spots or other problem areas: There are several low spots with occasional standing water next to the highway. A few areas between the base of the slope and the highway need leveling before fruit crops are planted.

Fencing required: Whitetail deer are abundant in the area. To exclude deer and reduce vandalism to farm equipment, an 8-foot-tall fence will be constructed around the entire property, with the exception of the roadside stand area. The fence will be 4-inch-square wire orchard mesh to a height of 60 inches, and above that individual high-tensile wires will be spaced every 10 inches to the tops of the posts. Two 10-foot-wide

Orchard-Layout Planning Worksheet

Table 3.6

Total orchard size: 5 acres

Item	Purpose	Area (feet)
Pole barn	farm equipment and supply storage	40 × 80
Roadside stand		40 × 40
Parking	customers and employees	65 × 100
Parking and staging	farm equipment, harvest supplies, fruit	65 × 200
Buffer strips	buffer zone around perimeter and headlands at ends of fruit blocks	40 wide
Roads	access between fruit blocks and headlands between blocks	30 wide
Planting block: trees	apple, pear, cherry, peach	178 × 110
Planting block: berries	blueberry, raspberry	178 × 80

Crop	Training	Between-Row Spacing (feet)	Within-Row Spacing (feet)	Plants Needed
Apple	tall spindle	11	3	593
Pear	double leader	11	4	445
Cherry	slender spindle	15	6	217
Peach	perpendicular-V	15	6	217
Blueberry	freestanding	10	4	356
Raspberry	"I" trellis	10	2.5	569

rolling gates will be installed to allow for a 20-foot-wide entrance. The fence will go on the outside of the windbreak.

Access to county road or highway: A gravel driveway and parking area will connect directly to the highway.

Utilities available and needed: Electricity and telephone will be needed. Both are available at the property edge along the highway.

Irrigation system: Microsprinklers will be used in all tree fruit blocks. Risers are spaced 5 feet apart, with enough spread to cover the crop rows and alleys at the rate of 10 gallons per hour per emitter. For the blueberries and raspberries, rigid polyethylene tubing will run the length of the rows with 1-gallon-per-hour emitters. The tubing will be supported on a trellis wire 18 inches above the ground for raspberries. Schedule 40 PVC pipes 6 inches in diameter will be used for main lines; for secondary lines, 4-inch pipe will be used with the tree fruits and 3-inch pipe will be used in the berry field. Pipes will be buried 48 inches below the soil surface with suitable drainage outlets to remove all water from the system for winterizing. A hybrid sand and cartridge filter system will be used to clean the water before distributing it to the microsprinklers or drip lines.

Source of water: One or more wells will be needed to produce at least 60 gallons per minute.

Pump requirements: An electrical pump capable of sustained rates of up to 75 gallons per minute.

Holding or irrigation pond needed: None.

Source of potable water and water for pesticide mixing: Well.

Analyzing the Sample Plan

Slope management. The site slopes upward from south to north. Cherries and peaches bloom earlier than apples and are more susceptible to spring frost damage. By locating them as high as possible on the slope, heavy cold air drains away from the trees, reducing the danger of frost damage to the blossoms and developing fruit.

Crop row orientation. Crop rows are oriented east–west. The orientation of fruit tree rows is relatively unimportant for most growers. For rows of very large trees, a north–south orientation allows somewhat more light to reach the orchard floor than an east–west orientation, but not much from a practical standpoint. With smaller trees, you should obtain adequate sunlight penetration regardless of the planting direction. In this example, the greater consideration is preventing soil erosion. Running the tree rows north–south on this site would create serious erosion problems. If the slope is too steep to allow vehicles and tractors to move safely across it, one possibility is to create a pattern of narrow, level terraces along the tree rows that would be wide enough to handle farm equipment, although terraces will probably not be needed for this orchard. The alleys are maintained in permanent vegetation, which will help reduce erosion and improve access in wet weather while supporting beneficial organisms.

Crop spacing. The space between crop rows varies from 10 feet for berries to 11 feet for apples trained to slender wire-trained spindles to 15 feet for freestanding, perpendicular-V-trained peaches. Keeping the row spacing the same and the alleys aligned with adjacent blocks makes it easy to move equipment between the blocks and reduces the need for additional headlands. Spacing between trees is 3 feet for apples, 4 feet for pears, 6 feet for peaches, and 6 feet for sweet cherries. The height of

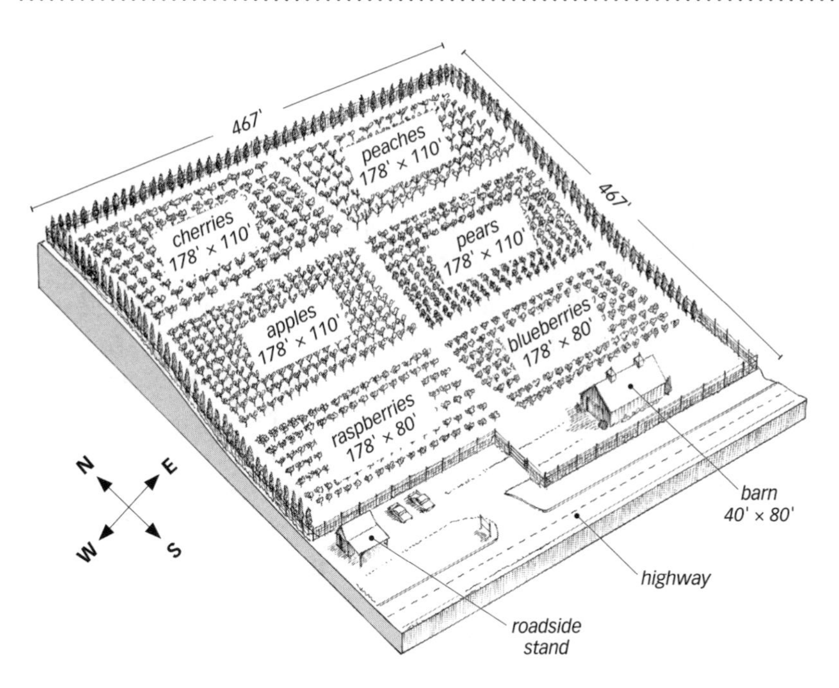

Figure 3.8 · Sample orchard design for a square 5-acre orchard. The design allows for U-pick, a roadside stand, and farmers' market sales.

the trees will be limited to 10 feet. Remember that these dimensions will vary from one orchard to another, depending on crops, varieties, rootstocks, and management practices.

Windbreaks. Hybrid poplar trees are placed just inside the fence, with the trees spaced 10 feet apart. Hybrid poplar trees grow rapidly, have straight trunks, and can easily be trained to single trunks. The Penn State University clone is shorter than many other hybrid poplar clones and produces a denser canopy, which improves its ability to block wind. Further, the poplars are not hosts for most serious fruit crop pests and diseases. No trees are planted on the southern perimeter, so that cold air is allowed to drain out of the orchard.

Fencing. The 8-foot-tall fence, with 5 feet of 4-inch mesh and high-tensile wires, will keep out deer and other browsers. This will work well for this orchard, although there are many fence designs, some of which were developed specifically for orchards, as we discuss in chapter 11. While electrified fences can be used for orchards, they are not well-suited to populated areas and U-pick farms, where visitors and customers could be hurt.

Buffer strips. The buffer strips around the orchard's perimeter serve as roads and headlands. In this example, they are kept in permanent sod, which helps lower dust contamination on the fruit, reduces erosion, and provides improved access during wet weather. On the downside, the sod roads require mowing. Twenty- or thirty-foot-wide roads between orchard blocks provide room for equipment operations, as well as staging and transport of fruit bins during harvest.

Access roads. A 30-foot-wide road running north–south between the blocks facilitates harvesting by U-pick customers. In a strictly grower-pick orchard, eliminating the central road would increase the amount of space available for crops, but it would also make moving equipment and fruit bins through the orchard more difficult. In this design, the grower has chosen to emphasize labor efficiency, high fruit quality, and open canopies, rather than maximum yields. For your design, you will need to find a balance between yields and efficiency that works for you.

Irrigation. These trees will be irrigated using microsprinklers set into the tree rows. The microsprinklers will irrigate the trees, alley crops, and living mulches within tree rows, while still keeping the tree foliage and fruit dry. The grower will use the Swiss sandwich system for in-row living mulches (see chapter 9). The Swiss sandwich system uses perennial crops in the alleys, typically grass or alfalfa. Underneath the trees and centered on the tree row is a 2- to 3-foot-wide strip of herbaceous perennials, such as sweet woodruff or grass. Between the in-row and alley cover crops are narrow strips of bare earth along each side of the tree row.

Running irrigation lines up and down a slope would make uniform irrigation difficult. Water pressure would be greatest at the lowest emitters, causing them to apply more water to the trees at the bottom of the slope than to those higher up. When the irrigation pump is turned off, water from the irrigation lines would drain out of the emitters near the bottom of the slope, adding to the water already applied to those trees. Since cherry and peach trees are quite susceptible to root diseases encouraged by wet soils, trees at the lower edges of the planting blocks would likely suffer more disease problems than those higher up. Additionally, cherries and peaches are particularly susceptible to iron chlorosis on alkaline soils (pH above 7.0), which is aggravated by wet soils. Running the tree rows and irrigation lines across the slope should reduce or eliminate these problems.

Non-crop areas. The roadside stand is next to the highway, making it highly visible and easy for customers to approach. It also provides a large, graveled parking and turnoff area. The barn and equipment parking areas are inside the orchard fence and away from customers. The design provides a safe, efficient traffic flow for farm operations, employees, and customers. Whenever possible, keep customers and visitors away from equipment and storage areas to reduce the risk of injury and liability.

PREPARING OR UPGRADING YOUR ORCHARD SITE

- Drainage | 78
- Staking the Orchard | 85
- Setting Your Soil Standards | 88
- Mineral Nutrients | 97
- Organic Amendments | 108
- Microorganisms | 111
- Other Soil Amendments | 112
- Weed and Pest Control | 113

Early organic horticulturists focused heavily on soil health, declaring that healthy soils were essential for healthy crops. Although organic advocates have continued to use that mantra for more than a century, the concept was widely ridiculed or minimized by many commercial fruit growers and agricultural scientists. By adding enough fertilizers and using herbicides, orchardists were able to produce abundant crops with little regard to the soil, although root and collar diseases and replant disorders were common. We are finding, however, that orchard soils are far more complex than we once thought and do, indeed, play a key role in tree health and productivity.

Soils are critically important in organic systems because they impact not only tree health and productivity; they also affect beneficial organisms that are keystones in our pest and disease control programs. The soil is, quite literally, the foundation of your orchard. If you don't get the soil right, not much else you do will matter. All organic systems emphasize healthy, biologically active soils that:

- Are moist but well drained
- Have good soil structure (tilth)
- Contain adequate organic matter
- Have appropriate pH for the crop
- Contain optimum amounts of nutrients
- Maintain optimum temperature

To prepare or upgrade your orchard, you will need to develop healthy soils. This is accomplished by identifying and correcting any drainage problems, setting soil standards, establishing the right pH and salinity levels for your soil, adding the appropriate amendments to your soil, and controlling weeds and pests. In chapter 9, we will go into detail about how to maintain healthy soils after your trees are established.

The following sections provide guidance on preparing a new orchard site for planting or upgrading an established orchard site. The steps are:

1. Correct drainage problems across the orchard site.
2. Stake out the orchard planting blocks, roads, buffer strips, and non-crop areas.
3. Test your soil, and set soil standards.
4. Adjust mineral nutrients.
5. Add organic and other amendments, if necessary.
6. Control weeds and pests.

Drainage

REGARDLESS OF THE CROP, fruit trees perform best on deep, well-drained soils. Poor soil drainage creates many diseases and physiological disorders in fruit crops. Unfortunately, not everyone has ideal soils, and extra work can be needed to prepare a planting site. Correcting drainage problems may involve grading down high spots and filling in low spots, installing drainage tiles, and installing drainage ditches and culverts. These steps are best done before staking out planting blocks and roads. It will be enough to have identified your general orchard outline at this point.

Telltale signs of poor drainage are standing water; very dark soils (sometimes with a foul or swamp odor); low-lying areas with sparse vegetation; or the presence of rushes, sedges, reeds, cattails, and other wetland plants. As mentioned in chapter 2, some wetlands are regulated by federal, state, or provincial laws and you may be limited on what you do with such lands. If in doubt as to the status of your land, check with your regulatory agency before making any modifications to wetlands.

For commercial orchards or a large home orchard, take the time to dig several pits or trenches about 6 feet deep. Look for hardpans, other impermeable layers, and dark bluish or greenish streaks and layers in the soil (see chapter 2). Map out areas of the orchard where drainage appears to be poor. They will require remedial treatment before planting and possibly special management after planting. Poorly drained spots may never be suitable for fruit trees and may best be used for non-tree crops or staging equipment.

To improve poor soil drainage, you can employ the following strategies:

- Modify the soil texture with sand or organic materials.
- Break up compacted soil and hardpans.
- Install drain tiles.
- Grade the site to allow surface runoff.
- Grow your crops on raised beds or berms.

Adding sand or organic materials. In chapter 2, we discussed how soil texture or type affects drainage. Unfortunately, there is little you can do to change soil texture in any but the smallest plantings. Hauling in enough sand or clay to modify the soil on a large scale is generally prohibitively expensive. What is more, adding sand or organic matter to heavy, low-lying soils seldom improves the drainage because the water still has no place to drain to. For example, say you have a slight depression where water stands for a day or more after rain. Filling the depression with well-drained soil will not improve the drainage because the water will still perch on top of the heavier soil at the bottom. This practice simply creates a pot without a hole in the bottom and can actually make a poorly drained soil even worse.

If you choose to amend your soil with sand and/or organic matter, spread the amendments across the entire planting area, including alleys, and till them into the soil. Then lay out your tree rows and plant the trees. Do not simply dig a trench along a tree row and fill it with amended soil or add amendments to the planting holes. Doing so creates a boundary layer between the amended and native soils that impedes root growth and the movement of water. Again, you have created a pot without a hole in the bottom.

Sand and organic amendments can effectively lighten soils and improve drainage when combined with raised beds, and such practices can be effective for small orchards.

Identifying and breaking up hardpans. Hardpans are impermeable layers of soil that prevent water from draining through the soil. This limits the movement of oxygen into the root zone, reduces micro- and macrobiological activity, increases root and collar diseases, and makes certain nutrients less available to plants. Hardpans also limit how deeply roots grow and can create weakly anchored trees that experience further root damage as they sway excessively in the wind.

Hardpans can be detected by digging test pits in your planting site (see chapter 2). Puddles that remain more than a few hours after a heavy rain or snow melt are also indications of hardpans. In many cases, impermeable layers result from poor soil management practices, such as excessive tilling (see box on page 81) or compaction from equipment and livestock.

Hardpans also occur naturally due to clay layers (sometimes called lenses) or mineral deposits (caliche). The mineral deposits are usually some form of carbonate (lime) and occur most often in arid regions. Frequent, shallow irrigation that draws soil minerals to the surface and deposits them when the water evaporates can also create drainage problems. Less frequent and deeper irrigation will help flush excess mineral salts out of the root zone.

If a hardpan is not too deep, you can often break it up using a long, narrow chisel plow or shank in a process called "deep ripping." Shallow tiller- or plow-pans can usually be broken up by ripping using a medium-sized tractor. Deeper and harder pans usually require a large tractor or bulldozer, often pulling two to several deep-ripping shanks. Ripping is

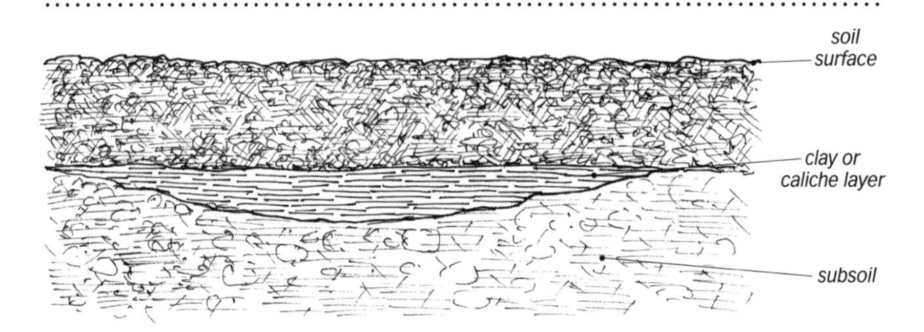

Figure 4.1 · In this soil profile, a clay or caliche layer impedes water drainage and root growth. Deep-ripping the soil can sometimes break up the impermeable layers and improve drainage.

more effective with mechanical or mineral hardpans than with clay layers because the clay particles tend to settle back into their original positions when the soils are wet.

For new or replanted commercial orchards, growers commonly cross-rip the fields before planting, first deep-ripping in one direction and then ripping at right angles to the first plowed rows. Cross ripping before planting can significantly reduce problems due to compacted soils and hardpans. A small investment now often reaps large benefits later.

Rototillers and Hardpans

Rototillers are especially likely to create hardpans for two reasons:

1) When used excessively, their stirring and chopping action destroys soil structure by breaking down large aggregates. This allows soil particles to settle more densely, creating smaller and fewer macropore spaces for water to drain through.

2) The flat, horizontal tiller blades also pound on and compact the soil at whatever depth you have set the tiller to be. This is especially likely on wet soils and on those with high clay percentages.

Mechanical tillers with vertical tines that stir the soil, rather like an eggbeater, are less likely to create a tiller pan than tillers with traditional flat, revolving blades. Disk and plow blades are less likely to damage soil structure than tillers, but they can create mechanical compaction when used excessively. While rototillers, disks, and plows are useful farm implements, use them carefully and sparingly.

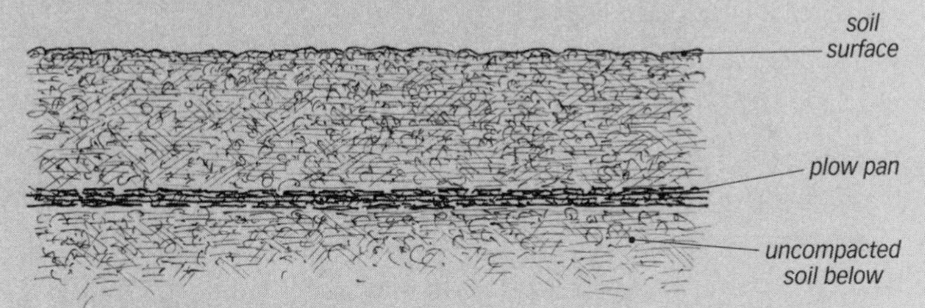

This soil profile shows a compacted layer caused by plow or rototiller blades. Deep ripping can be used to break up the compacted layer and improve water drainage.

Creating ditches, installing drain tiles, and grading. These practices can all be effective in improving drainage, provided you have a lower area to drain the water into.

Simply creating drainage ditches through your orchard may help improve soil drainage. Keep the ditches on more-or-less level ground to reduce problems with erosion. Install and cover culverts where equipment and people need to cross over the ditches. Due to safety concerns, open ditches are best not used in orchards where young children may be present.

Drain tiles or perforated pipe can be buried in the orchard to accumulate excess soil water and carry it off site or to a storage pond. The great advantage tiles and perforated pipes have is that they are entirely underground, require little or no maintenance, and do not interfere with orchard operations or movement of people and equipment. Be sure that the tiles and pipes are surrounded with weed barrier fabric to allow water to enter the pipes but prevent tree roots from entering into and clogging them.

Orchard sites can be graded to provide level or at least uniform sloping planting areas. Grading can also be used to remove high spots and fill in low-lying areas. Be especially cautious with grading, however, and ensure that you do not reduce portions of your orchard to infertile subsoil by scraping away all of the topsoil.

Grading fields, establishing drainage ditches, and installing drain tiles often require specialized equipment and expertise. For commercial orchards, these activities are usually best left to professionals. Correcting drainage problems is much more difficult once the trees are planted, so be sure to get this process right from the start.

Using raised beds. Raised beds are simply mounds of soil that lift crop plants above the surrounding soil and provide increased rooting depth. In gardens, raised beds are often used and quite typically are enclosed on the sides with stone, brick, wood, or composite materials. In orchards, raised beds are most often simply flattened mounds of soil. These beds will be formed after the orchard is laid out, usually just before planting the trees. This section is included here so that you can consider whether raised beds will meet your needs later or if you also need to take other remedial action to correct drainage problems before laying out the planting blocks.

Figure 4.2 · **Raised Beds**

A) A raised bed along the center of the fruit crop row can be used in home and commercial orchards.

B) Raised planting areas work well for fruit crops grown in a landscape setting.

Planting on raised beds is rapidly becoming a standard practice in commercial berry production in many parts of the world. For very little investment, you can ensure excellent soil drainage. Saskatoons and bush cherries and plums, for example, can be grown in raised beds that are 10 to 12 inches high and 3 to 4 feet wide. Even for the larger tree fruits, more fruit specialists are now recommending the use of raised beds. Because fruit trees have larger root systems than berries, the beds must be somewhat wider. While the use of raised beds in commercial tree fruit orchards is still largely novel and untested, it appears likely that the practice would work best with relatively small trees on low-vigor rootstocks.

Raised beds are useful on poorly drained home orchard sites where you can create a berm or raised ridge to plant your trees on. Berms are generally higher and wider than the raised beds used in orchards and gardens. Likewise, raised landscape beds perhaps 10 feet or more in diameter can be used to improve soil drainage. If your native soil has high amounts of clay and silt, combine it with sand and organic materials to create an amended soil that will improve drainage and biological activity in the berms or raised beds. Figure 4.2 shows two raised bed designs.

Excessive Drainage

Excessive drainage usually occurs on soils high in sand and/or gravel. They have little water-holding capacity, and without irrigation, plant growth is sparse. On a small scale, you can improve the water-holding capacity of droughty soil by amending it with clay or silt soils, peat soils, compost, wood chips, bark, or similar materials. On a large scale, adding

A Case Study on Preparing an Orchard Site

With a few improvements, some orchard owners that I consulted with turned a poorly drained site into a productive berry orchard. This family-run farm was located on level to rolling pasture land with a low-lying strip next to the adjacent county road. I examined the site during late winter when areas of standing water from rain and melting snow were clearly visible in some of the low-lying areas of the field.

To correct the problem, the owners devised a series of ditches between the planting blocks, providing drainage into an inlet from a nearby lake. They used grading to level the fields and to fill in some low-lying areas in the main body of the orchard, and they installed drain tiles leading into the ditches. The strip along the county road could not be improved enough to grow fruit on without great expense, so it was kept as a buffer strip for a future roadside stand and parking area. The result was that most of the 15-acre property was converted from rather marginal pasture and forage fields into manageable orchard land.

large amounts of off-site soils or organic amendments is usually prohibitively expensive.

In all cases, tilling under one to several green manure crops before planting your trees will start the process of building organic matter. After the trees are established, annual alley crops, such as barley, can be tilled into the soil. If you use permanent alley cover crops, blow the clippings into the tree rows during mowing. Increasing soil organic matter concentrations, however, is a slow process, and you may need 5 to 10 years of careful management (discussed in chapter 9) to achieve significant increases. If you plan to establish an orchard on a droughty site, ensure that you have ample irrigation water and install an effective irrigation system before you plant any trees.

Staking the Orchard

BEFORE YOU STAKE OUT PLANTING BLOCKS and other parts of your orchard, take the time to carefully examine the site and identify problem areas, as we discussed in chapters 2 and 3. Of particular importance, identify areas that are poorly or excessively drained, rocky outcrops unsuitable for fruit trees, and other portions of your site that require special management. Then adjust your orchard design, if necessary, to take these areas into account.

Once you have corrected drainage problems, lay out your orchard blocks and roads according to the design you created in chapter 3 and modified after a close inspection of the site. For this process, you simply need stakes, brightly colored flagging tape, and a measuring tape. Bamboo plant stakes from the garden center and 1- or 2-inch-square tomato stakes 3 to 4 feet long work well.

You can purchase marking flags on plastic or metal stakes from Internet retailers. You will find that these are much easier to carry and insert into the ground than are wooden stakes. You might find it convenient to make several metal stakes from ¼- or ⁵⁄₁₆-inch diameter wire rod to hold the end of your measuring tape in place as you lay out and measure reference lines. These stout, metal stakes are different from the marking stakes that you leave in place to identify corners and reference lines. The rod should slip easily through the end grommet on your measuring tape.

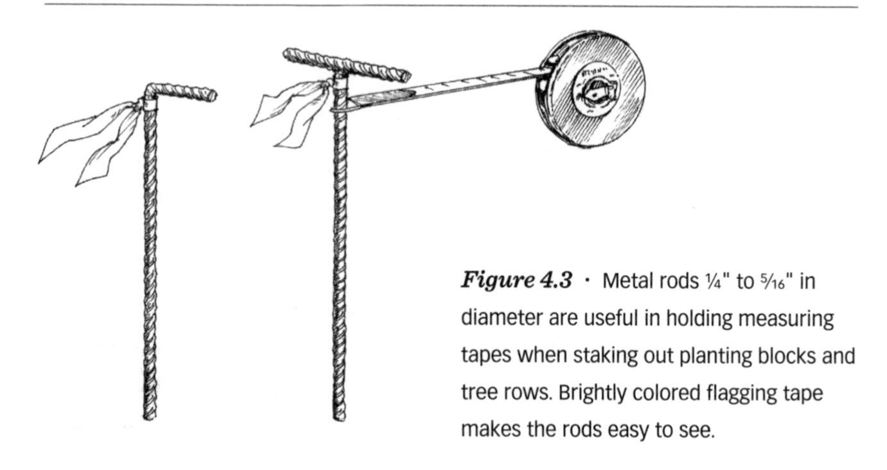

Figure 4.3 · Metal rods ¼" to 5⁄16" in diameter are useful in holding measuring tapes when staking out planting blocks and tree rows. Brightly colored flagging tape makes the rods easy to see.

Bend the rod to form an L with the main shank about 24 inches long and the handle about 6 inches long, or weld a cross piece onto the main shank to form a T-shaped stake. Tie flagging tape to the tops of the rods to make them easy to see. Fiberglass measuring tapes 200 to 300 feet long are inexpensive and ideal for both measuring lines and creating accurate right angles.

Mark buffer strips and roads. Fences and roads provide good reference lines for the edges of your orchard. If roads or fences are not available, stretch string or baling twine along the property line. Simply mark the width of the buffer strip, perhaps 30 feet, by placing a flagged stake at a right angle to your reference line. Place stakes every 50 to 100 feet around the entire perimeter of where your trees will be planted. If using a windbreak, be sure to allow room for it outside the buffer strip.

The buffer strips also serve as roads and headlands, so before going any further, test to make sure they are wide enough for your equipment. Try pulling your largest equipment from the orchard block into the headlands, turning around, and returning into the orchard block as if you were entering the next alleyway or every other alleyway. If you cannot easily make the turn, increase the width of the headlands. Cramped headlands waste time and cause damage to trees and equipment.

Create right angles for planting blocks. If your design includes right angles for the corners of the orchard blocks, you will need a reference line along the orchard block, a starting corner, three stout stakes, and a measuring tape 200 to 300 feet long. The idea is to create a triangle that has side lengths in a ratio of 3:4:5. This is easier to do than it sounds.

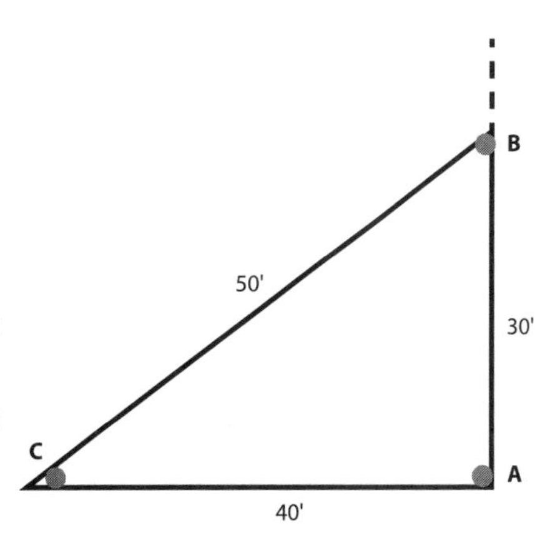

Figure 4.4 · **Laying Out a Square Corner**
Place stake (A) at the corner of the planting block and on the reference line. Fasten the end of a 200-foot-long tape measure and the 120-foot mark of the tape to stake (A), forming a loop. Place stake (B) on the reference line and 30 feet from (A). Bring the tape to the outside of stake (B) and walk in the direction of (C) holding onto the 80-foot mark on the tape. When both legs of the tape from stakes (A) and (B) are tight, you have reached (C) and formed a 90-degree angle. Place stake (C) on your new reference line.

Place one stake (A) on the reference line where you want the corner and slip the end grommet of the measuring tape over the metal stake (see figure 4.3). Wrap the 120-foot mark on the tape around stake (A) and fasten the tape in place so that it cannot slip. Stake (A) forms the first corner of the triangle. Measure 30 feet along the reference line that you laid out earlier and place a second stake (B) at this point. Run the measuring tape on the outside of the stake. This is the second corner of the triangle. Holding onto the 80-foot mark on the measuring tape, walk at approximately right angles to your reference line from stake (A), stretching the tape until both segments from stakes (A) and (B) are tight. Place a stake at (C) to mark the line. You have now formed a triangle at a right angle to your reference line. The stakes serve as the corners and the measuring tape shows the sides of the triangle.

For greater accuracy, use a 300-foot-long tape and make the sides of the triangle 60, 80, and 100 feet long. With a 300-foot tape, place the end grommet and the 240-foot mark on the measuring tape at stake (A). Place stake (B) 60 feet along the reference line and run the measuring tape on the outside of the stake. Holding onto the 160-foot mark on the measuring tape, walk in the general direction of (C), as described above. When both segments of the tape from stakes (A) and (B) are tight, place stake (C).

Extend your right angle line for the width of the block by stretching your measuring tape or twine from the corner stake (A) and just touching the right angle corner stake (C). Repeat the procedure until all of the planting blocks and roads are marked.

Later, you will need to lay out the tree planting rows (see chapter 7). For now, it is enough to have the planting areas, buffer strips, roads, and headlands clearly marked.

Setting Your Soil Standards

ALL TOO OFTEN, ORGANIC GROWERS apply soil conditioners and fertilizers blindly in the hope that something good will happen. Unless you know what you are starting with, however, adding materials to the soil is a gamble, at best, and can actually create problems in an orchard. Make the investment to have your soil tested before planting a new orchard or changing management practices in an established orchard. That small investment will reap large benefits.

Test your soil. If you are not already sure what your soil type is, have your soils tested for particle size density (texture or soil type). Also test for pH, liming requirement, organic matter concentrations, cation exchange capacity (CEC), nutrient concentrations, and salinity (see chapter 2 for specific tests that are needed). Remember that soils can change dramatically over very short distances. If necessary, divide your orchard into separate management areas based on topography and soils. The topography of your site often provides clues as to where different soils might be. Given a site with a ridge, level area, and low-lying area, you might consider testing the soils in each of those areas. Patches of different types of vegetation on your site may indicate corresponding areas of different soils.

Even in a small home orchard where the site is not uniform, establishing management zones is important. For example, an organic home orchard I visited covered less than an acre, but their planting sites ranged from a low-lying creek side with rather gravelly soils to a level area with silt loam to a low-lying, poorly drained meadow on silty-clay soil. Each of the areas required its own preparation, planting, and management strategies. If you have two or more different management zones in your orchard, have separate soil samples tested for each of the zones.

Strategize for amendments. Consider how you will apply amendments in your fruit planting. In berry and grape culture, growers often amend only the soil within the crop rows. For crops with small root systems, this practice works well, but for crops like fruit trees that develop extensive root systems that spread far from the trunks, it is not effective. Apple tree roots can spread over a diameter two to three times as far as a tree is tall and extend downward until they hit a permanent water table or impervious layer of subsoil. For that reason, orchardists usually amend all of the soil in their planting blocks, rather than only those soils within the tree rows. In a home orchard with a few trees, estimate the final height of the trees and draw circles around the planting spots and extending out 1 to 1.5 times that height. Apply amendments, as described below, throughout the entire circles. Figure 4.5 shows the above- and below-ground structure of a fruit tree.

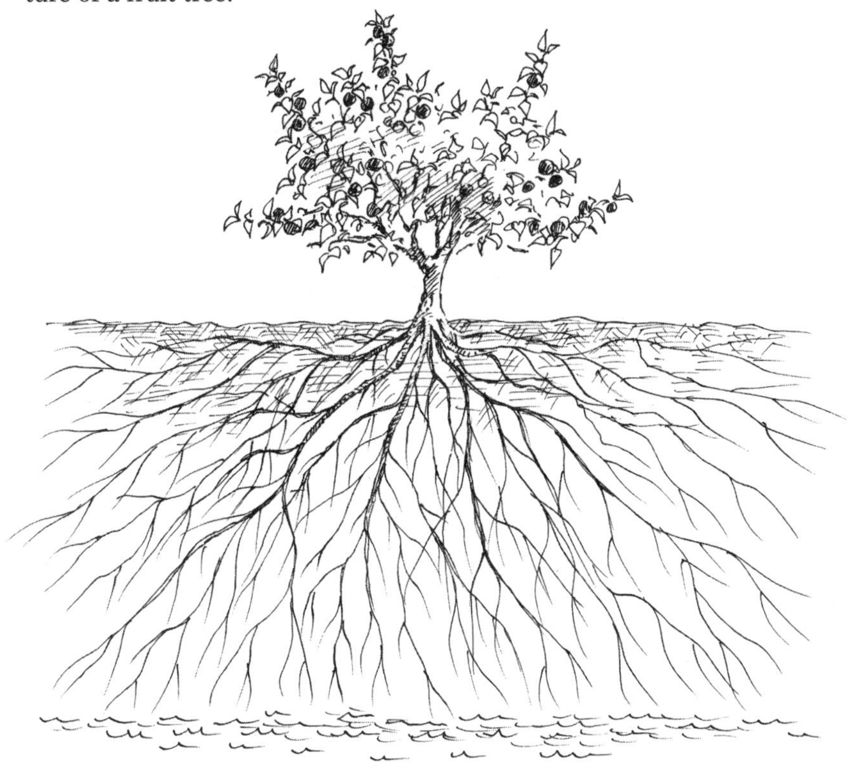

Figure 4.5 · Fruit tree roots can spread up to three times the width of the canopy and downward until they are stopped by a water table, impervious soil layer, or other unfavorable soil environments.

Unless you know what the soil pH, nutrient concentrations, and organic matter should be, even the best analytical results will mean little. Now is the time to begin developing standards. Soil standards are simply minimum benchmarks or ranges in pH, soil nutrient concentrations, and organic matter. We have similar standards for foliar nutrient concentrations, as we will discuss in chapter 8. These standards help you decide which amendments to apply and help you monitor changes in your orchard over time. The goal is to create and maintain a stable, pH-appropriate, nutrient-rich, biologically active soil.

Soil pH

A soil's pH affects the availability of mineral nutrients to plants, so it's important to get it right. Remember that a pH of 7.0 is neutral. Values above that are alkaline (basic) and those below 7.0 are acidic. A laboratory will provide the most accurate analysis, although there are several types of kits you can purchase to measure pH yourself. A handheld pH meter purchased from an agricultural or laboratory supply company will provide accurate results when used properly and can be a good investment for growers who wish to test soil pH frequently. You can also purchase one of several "color kits" for measuring soil pH and nutrient concentrations. Some are extremely accurate (and expensive!) while others provide poor accuracy. For serious fruit growers, it is best to send a sample to the laboratory or purchase a good-quality pH meter. Simple pH "meters" with metal probes are readily available in garden centers, but these often provide very poor accuracy and should be avoided.

As we discussed in chapter 2, aside from blueberries and a few other acid-loving crops, temperate zone fruit crops grow and produce best on soils that are slightly to moderately acidic. As the soil becomes more strongly acidic (lower pH), the macronutrients nitrogen, phosphorus, potassium, magnesium, calcium, and sulfur become less available to plants, as does the micronutrient molybdenum.

As soils become alkaline, the micronutrients boron, cobalt, copper, iron, manganese, and zinc become less available. At pH values above roughly 7.5, iron chlorosis often develops in apples, peaches, cherries, and other fruit tree crops. This disorder is characterized by yellow to white leaf blades with dark green veins. Unless corrected, affected trees become weakened and stunted with poor fruit production. Under severe conditions, trees may eventually die. Similar micronutrient deficiency disorders are sometimes seen with a lack of available boron, manganese, and zinc, even when the nutrients are relatively abundant in the soil. Figure 4.6 shows how soil pH affects nutrient availability to plants.

Most tree fruits tolerate pH values between 5.5 and 7.5. For a productive, healthy, long-lived orchard, a pH of 6.0 to 7.0 should produce excellent results. Bear in mind that the soil pH fluctuates during the year due to wetting and drying, microbial activity, and chemicals added to and removed from the soil by plants. In western Oregon orchards, for example, the soil pH typically fluctuates up to 0.5 pH unit and can fluctuate twice that amount on sandy soils. There, the highest soil pH is usually recorded in late winter. Soil drying, increased microbial and plant activity, and fertilizer applications cause the pH to drop during the growing season, reaching an annual low in late summer or early fall.

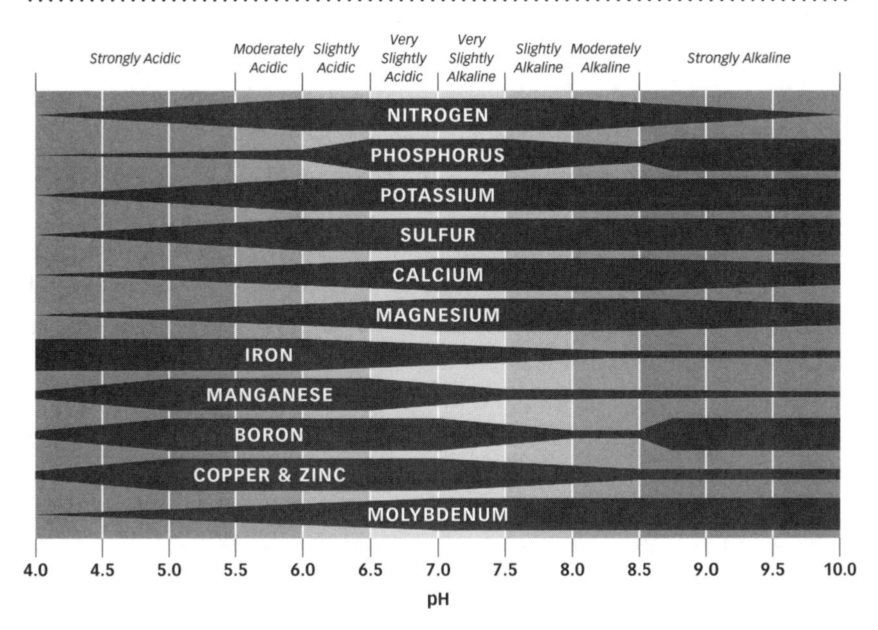

Figure 4.6 · Soil pH affects the availability of mineral nutrients to plants. In this chart, the wider the bar, the more available that nutrient is to plants. As the bars become narrower, nutrient availability decreases. For pome and stone fruit crops, both macro- and micronutrients are readily available at pH 6–7.

Raising pH

For organic production, some form of limestone or dolomite is normally used to adjust soil pH.

Limestone. It is quite easy to raise pH on acidic soils by using limestone (calcium carbonate). If your soil test shows both low pH and low magnesium, apply dolomitic limestone (also known as dolomite), which is primarily calcium-magnesium carbonate. Do not use dolomitic liming products if your soil calcium is low (see table 4.6) unless soil tests show soil magnesium concentrations are also low. This is particularly important for apples. Magnesium can interfere with the uptake and utilization of calcium in fruit trees, increasing problems with bitter pit disorder in stored apples.

Wood ashes. Strongly alkaline, wood ashes can be used to raise soil pH while also supplying some nutrients, as shown in table 8.4. They are allowed in organic production, although they can be problematic. Use only ashes from untreated and unpainted wood. Ashes created from treated woods, colored paper, plastic, and other materials are prohibited under organic certification guidelines. Never use coal ash. Apply ashes sparingly because excessive amounts can cause pH and nutrient imbalances. Wood ash is highly alkaline and reacts very quickly in the soil to raise the soil pH.

Unfortunately, the liming equivalent of wood ashes varies greatly depending on the source and handling of the ashes. In general, wood ashes have a calcium carbonate equivalent (CCE) of 25 to 60 percent of calcium carbonate, and you need to add two to four times as much wood ash as limestone, by weight, to create the same change in soil pH.

Do not apply wood ashes to soils that are already at or above the desired pH, and do not add more than 20 pounds per 1,000 square feet per year. At this rate, you would be adding the equivalent of about 220 to 520 pounds of 100 percent calcium carbonate per acre per year.

Sugar beet lime. Also called spent lime, sugar beet lime can be used for raising the soil pH in some organic orchards (check with your certifying agency). This material is a by-product of sugar beet processing in which a suspension of finely ground calcium hydroxide (slaked lime or hydrated lime) is mixed with carbon dioxide and the sugar extracted from sugar beets. The lime does not come from the sugar beets themselves or any other plant material.

Sugar beet lime can be problematic. Its high water content increases transportation costs and makes uniform application difficult. The uneven application of lime in your orchard can create serious production problems. Unless you have a particular reason for using sugar beet lime, traditional liming materials will probably provide better results.

Hydrated lime is allowed for the preparation of Bordeaux fungicide and can be used in organic livestock production but is not allowed as a soil amendment.

Quicklime (burnt lime) is more potent than limestone. It is very reactive and is difficult and hazardous to handle, however, as well as damaging to plants. The U.S. National Organic Program makes no mention of quicklime, and it is best avoided.

Adding the liming material. The amount of liming material you should add depends on your soil's pH, texture, and buffering capacity. The pH value alone tells you only whether your soil is acidic or alkaline — it will not tell you how much liming material to add. Your soil analyses should include a "liming requirement" test, which will tell you how much standard liming material to add to raise soil pH by a given amount. You will then need to calculate how much material to apply, based on that material's CCE. Pure calcium carbonate has a CCE of 100 percent. Hydrated lime has a CCE of 135 percent. Most commonly available liming materials have CCE values less than 100 percent.

If the material you plan to use has a CCE of 50 percent, you would need to add twice as much as you would for pure calcium carbonate. In reality, determining how much material to apply is very easy once you receive your liming requirement results and know the CCE of your liming material. Your soil test report should tell you how much 100 percent CCE material to add.

Allow time for the liming materials to react in the soil and for the soil pH to stabilize. If you are making substantial changes to the soil pH, apply liming materials at least one full year before planting your trees. If the soil pH in an established orchard is 5.0 or below, raise it gradually by adding liming materials over 2 or 3 years to avoid damaging the trees and to allow the soil pH to stabilize. Be sure to monitor the soil pH at least once yearly before liming so as not to add too much liming material.

Lowering pH

While raising soil pH is relatively easy and inexpensive, lowering soil pH can be both difficult and expensive, particularly on a large scale. In order to significantly lower pH in an organic orchard soil, you usually need to add elemental sulfur.

Lowering soil pH is more technically difficult than raising pH with liming materials. Liming materials undergo simple chemical reactions with the soil particles. Elemental sulfur, however, must be metabolized by soil bacteria and converted into the acidic SO_4^- ion, which then reacts with basic (alkaline) compounds in the soil. Although simple in theory, sulfur oxidation depends on which types and how many soil microorganisms are present, the soil type and moisture level, the carbon nutrient sources for the microbes, soil temperature, and more.

Note also that sulfates, such as calcium sulfate and magnesium sulfate, are salts formed by the reaction of SO_4^- ions and do not alter soil pH. This is because the SO_4^- has already reacted and is no longer acidic. In conventional farming, ammonium sulfate is sometimes used to lower soil pH. In these cases, it is the ammonium, not the sulfate, that reacts to acidify the soil. Ammonium sulfate is sometimes recommended for acidifying soils, but it is not allowed in organic production. Aluminum sulfate is sometimes recommended for acidifying soils, but it is not mentioned in the NOP. Although some aluminum sulfate products are marketed as "organic," nearly all aluminum sulfate used commercially is manufac-

Elemental Sulfur

Elemental sulfur is a yellow solid and is usually sold as fine powders and small pellets. It is readily available at many garden centers and through farm supply stores. The fine powder is generally used as a fungicide to control diseases. Pellets are more commonly used to amend soil. Costs vary according to the form of the sulfur and the amount purchased. While various organic materials, such as pine needles and cottonseed meal, are somewhat acidifying, it is better to use organic acidifying amendments to help maintain a desirable pH once you have it within your desired range.

Elemental Sulfur Needed to Acidify Soil

Table 4.1

The approximate amount of elemental sulfur, in pounds, to add per acre to acidify the soil pH to 6.5.

	SOIL TYPE					
	SAND		LOAM		CLAY	
Original Soil pH	pounds per acre	pounds per 1,000 sq.ft.	pounds per acre	pounds per 1,000 sq.ft.	pounds per acre	pounds per 1,000 sq.ft.
6.5	0	0	0	0	0	0
7	174	4	523	12	789	18
7.5	348	8	1045	24	1577	36
8	522	12	1568	36	2367	54
8.5	697	16	2091	48	3156	72

Adapted from Cornell University and Oregon State University guidelines. Do not add more than 3,000 pounds of sulfur per acre per year.

tured synthetically. The naturally occurring form is rare and typically found in burning coal mining waste dumps and volcanoes. The improper use of aluminum sulfate can cause aluminum toxicity in sensitive plants. It should not be used in an organic orchard.

Adding the sulfur. The best approach to lowering soil pH is to move slowly and conservatively. Deciding how much sulfur to add is often difficult because of the factors mentioned above and the strong influence of soil type and original soil pH. Table 4.1 provides estimates of how much elemental sulfur you will need to lower the pH of alkaline soils to between pH 6 and 7. The more finely ground the sulfur, the more quickly it will react in the soil to lower pH.

Avoid adding excessive sulfur and acidifying the soil too much. In no case should you add more than 3,000 pounds of elemental sulfur per acre at one time, regardless of the amounts listed in this or any other table. If you need to add more than 3,000 pounds of sulfur per acre, you should do so in stages over a period of several years. Allow time for the sulfur to be metabolized and for the soil pH to stabilize, processes that can take 2 years or more from the time you apply the sulfur. The acidification process can be hard on plants when making dramatic changes, and major adjustments to the soil pH are best done before planting your orchard crops.

For new orchards, apply the sulfur uniformly across your planting block at least a year before planting your trees and mix it thoroughly into the top 6 to 8 inches of soil. Remember that the soil must be moist, but not saturated, in order for the microorganisms to metabolize the sulfur. Six months after application, measure the soil pH from several locations across your planting block. Follow up with additional pH tests 12 months after applying the sulfur to see how the process is progressing. It is best to wait for 2 years from the first application before applying more sulfur.

For established orchards, broadcast the sulfur throughout the orchard and cultivate it into the soil no more than 2 inches deep to avoid damaging the tree roots. If the pH is 7.5 to 8.0, apply half of the needed sulfur one year and half the next year. If the pH is above 8.0, apply one-third of the needed sulfur one year and one-third during each of the two following years. Test the soil pH each year, and if more sulfur is needed, adjust the amount accordingly.

Soil Salinity

As we discussed in chapter 2, soil salinity, or soluble salts, refers to electrolytes, such as sodium chloride, that dissolve in the soil water. When soil salinity levels become excessive, plants take up less water and grow roots more slowly. The tips of leaves often die and become brown, sometimes referred to as "burning" or "scorching." Shoots grow more slowly, and plants produce fewer and poorer-quality fruits. At still higher salt concentrations, most plants die. Tree fruits tolerate somewhat higher salinity than do most berry crops and should perform reasonably well with salinity levels of 2 to 4 millisiemens per centimeter ($mS \cdot cm^{-1}$), although lower levels are generally preferred (see page 33 for more information on salinity measures). If your orchard includes blueberries, raspberries, or strawberries, keep your soil salinity levels at 2 $mS \cdot cm^{-1}$ or less.

Provided that you have a sufficient amount of low-salt irrigation water available, you can sometimes reclaim an orchard site that has excessively salty soils by applying large amounts of water to leach the salts from the soil. You will need a sprinkler irrigation system that provides 100 percent coverage of the orchard blocks. Reclaiming a site this way works well when the excess salts accumulate because of applying too much fertilizer or manure, or if frequent, shallow irrigation concentrated the salts near the soil surface. Depending on calcium, sulfur, and magnesium levels (discussed below), you can help displace some of the unwanted salts from the soil particles by applying gypsum to the soil before leaching.

Unfortunately, some saline soils in arid regions of North America are associated with irrigation water that is also high in salts and, therefore, unsuitable for leaching purposes. On sites with extreme soil salinity, producing orchard crops can be difficult or impossible.

Mineral Nutrients

SOIL TESTS PROVIDE A STARTING POINT for amending your orchard soils before planting trees (see tables 4.2 through 4.5 for suggested preplant soil concentrations and application rates for selected nutrients).

Our goal at this stage is to make sure ample supplies of nutrients will be available to the fruit trees, particularly those nutrients that move slowly in the soil. Once your trees are planted, it can be difficult to amend the soil with certain nutrients. On low-pH (strongly acidic) soils rich in aluminum and iron, for example, phosphorus is very immobile, and applying it after the trees are planted is rather ineffective. When phosphorus is applied to the surface of such soils, it tends to become tied up near the soil surface and remains unavailable to the plant roots. Calcium, potassium, and sometimes boron can also be relatively immobile in soil and are best applied before you plant your orchard. Most other amendments can be applied to the soil up to the time of planting the trees.

Bear in mind that soil tests do not necessarily show the actual amount of nutrients in the soil. Instead, they show the approximate amounts that should be available to plants, taking the soil pH and chemical forms of the nutrients into account. Nitrogen availability is shown as equivalent amounts of ammonium (NH_4^+) and nitrate (NO_3^-) nitrogen. Phosphorus, being chemically reactive, is found in many different chemical compounds in the soil and is defined as the equivalent of P_2O_5 (phosphorus pentoxide), regardless of the chemical forms actually present. Potassium is reported as equivalents of K_2O (potassium oxide or potash).

Preplant Phosphorus Fertilizer Rates

Table 4.2

Rates are in P_2O_5, based on soil tests. Soil test P can be determined by several procedures, three of which are shown here: sodium acetate (NaOAc), Bray I method, or by sodium bicarbonate ($NaHCO_3$). Sodium bicarbonate should not be used on soils with pH values less than 6.2. Use the column indicated by your soil test report.

SOIL TEST P_2O_5 (PPM)			
NaOAc	Bray I	$NaHCO_3$	Pounds per Acre (pounds per 1,000 feet)
0 to 1	0 to 10	0 to 4	135 (3)
1 to 2	10 to 20	4 to 8	100 (2.3)
2 to 3	20 to 30	9 to 11	80 (1.8)
3 to 4	30 to 40	12 to 14	70 (1.5)
4 to 5	40 to 50	15 to 17	50 (1.13)
5 to 10	50 to 100	18 to 25	40 (0.88)
> 10	> 100	> 25	none needed

P × 2.29 = P_2O_5, or P_2O_5 × 0.44 = P, where P refers to the actual amount of pure phosphorus.
Adapted from University of Idaho guidelines.

Preplant Potassium Fertilizer Rates

Table 4.3

Rates are in K_2O, based on soil tests using sodium acetate as the extractant.

Soil Test K_2O in ppm	Pounds per Acre (pounds per 1,000 square feet)
0 to 50	90 (2)
50 to 75	60 (1.5)
75 to 100	40 (0.94)
more than 100	none needed

K × 1.20 = K2O or K2O × 0.83 = K, where K refers to the actual amount of pure potassium.

Preplant Sulfur Fertilizer Rates

Table 4.4

Application rates are in actual S, based on soil tests.

Soil Test S in ppm	Pounds per Acre (pounds per 1,000 square feet)
< 10	30 (0.69)
10 to 20	none
> 20	none

Adapted from University of Idaho guidelines

Preplant Boron Fertilizer Rates

Table 4.5

Broadcast boron-containing fertilizers throughout the planting site or orchard. Never band them in rows. Boron-containing sprays can be applied to the fruit crop foliage in established orchards. Boron becomes toxic to plants at very low concentrations. Do not exceed the rates in table 4.5. Rates are in actual B, based on soil tests.

Soil Test B in ppm	Pounds per Acre	Ounces per 1,000 Square Feet
< 0.5	1	0.35
0.5 to 1.0	none	none
> 1.0	none	none

Adapted from University of Idaho guidelines

Nitrogen

Nitrogen is essential to plants and is the nutrient most often deficient in orchards. In conventional orchards, nitrogen is easy to add in the form of rapidly available industrial fertilizers, and it is seldom added before planting trees unless soil concentrations are very low. The situation for organic orchardists is less clear because many organic nitrogen fertilizers become available to plants relatively slowly. In some organic orchard trials, the establishment and growth of trees in organic plots treated with organic nitrogen fertilizers was very poor when compared with adjacent conventional treatments. The problem has likely been due to low amounts of available soil nitrogen and the inability of growers to rapidly increase available nitrogen to the newly planted trees using slow-release, organic materials.

Compounding the problem is that the nitrogen cycle in the soil is very complex, and nitrogen is constantly shifting from one chemical form to another as it moves through micro- and macroorganisms and plants. Rather than simply looking at nitrogen (NH_4^+ and NO_3^-) concentrations in a soil test report, a better guide to available nitrogen for a planned orchard is to look at the soil test results for organic matter. As organic matter decomposes in the soil, it adds nitrogen that is available to the plants.

Each 1 percent of soil organic matter adds up to about 20 pounds of plant-available nitrogen per year. Soil temperatures and moisture affect the actual amount of nitrogen released, which will be less in cold, wet soils. In reasonably well-drained soils, organic matter concentrations of 5 percent or more should provide sufficient available nitrogen for planting tree fruit crops. Additional nitrogen may be needed later, as the trees mature and begin bearing fruit.

Physical signs of low soil nitrogen are stunted and yellowish crop plants and weeds on the planting site. If crop and weed growth are lush and vigorous, and leaf colors are normally green, nitrogen deficiency should not be a problem for your newly planted trees.

Ways to Add Nitrogen

Composts, manures, alfalfa pellets, alfalfa meal, soybean meal, feather meal, fish emulsions and meal, and cottonseed meal provide moderate to slow nitrogen release into the soil. These products, their mineral contents, and their uses in organic orchards are described in chapter 8. If you need a quick fix of nitrogen just before or at the time you plant your trees, nitrogen from dried blood (blood meal) (12 percent N) is available almost immediately. Blood meal must be used with care as it can damage plants when too much is applied.

Sodium nitrate (also known as Chilean nitrate) (16 percent N) provides nitrogen rapidly, but it is discouraged in organic crop production because of its high sodium content (26 percent Na), which can damage sensitive plants and contaminate groundwater. Certified organic growers should check with their certifying organization before using sodium nitrate, as some organizations limit sodium nitrate to no more than 20 percent of the total nitrogen applied, and some certifying organizations prohibit its use altogether.

Nitrogen-fixing green manure crops can help get your trees off to a good start. Plow them under the year before planting. Clovers, beans, and peas are typically used to fix nitrogen as green manure crops. Alfalfa establishes more slowly than these crops and, depending on your location, is best left in place for two or more growing seasons before tilling it into the soil. Inoculate your legume seeds with a *Rhizobium* preparation to ensure that your crops fix nitrogen. Check the inoculant label carefully to be sure that it is organically certified and not produced using genetically modified organisms.

Depending on the materials that went into its making, compost is also a good source of nitrogen and can be a cornerstone of nutrition programs in small orchards. Composting has the added benefit of recycling discarded fruit, prunings, and other waste products from the orchard.

Applying Nitrogen

Apply nitrogen fertilizers directly to the soil. Although some people advocate "foliar feeding" by spraying fertilizers onto the leaves of crop plants, nitrogen and other macronutrients are generally not taken up by foliage in amounts large enough to benefit the plants. Foliar feeding can be used effectively for some micronutrients.

If the level of organic matter in your soil is below about 4 to 5 percent and nitrogen concentrations are also low, as shown in a soil test, you may want to begin incorporating slow-release forms of nitrogen before planting your trees. Your soil test should recommend suggested amounts of nitrogen to add. If you are using slow-release forms of nitrogen, such as compost, add the amendment at least by the summer or fall before planting your trees. For fertilizers that release nitrogen more rapidly, such as blood meal, apply one-half of the needed amount 1 to 4 months before planting the orchard. Apply the other half shortly after planting your trees.

Be careful not to apply too much nitrogen before or after planting. In young trees, excessive nitrogen can cause too much shoot growth at the expense of root growth. In established orchards, excessive nitrogen causes poor fruit set; soft, poorly colored fruit; and excessive amounts of lush, pest- and disease-prone foliage and shoots that require much labor to manage. Remember that organic soil systems move more slowly than those in conventional orchards. If you have incorporated a green manure crop or applied composts or manures, the amount of nitrogen available to the plants will increase for several years after each application. Unlike a conventional orchard where soil nutrient concentrations can be changed quickly, we want to change organic soils gradually, nudging them into the desired condition and not overshooting the mark.

Phosphorus

Phosphorus is rather a puzzle when it comes to orchard crops. Although it is required by plants and soils can be deficient in available phosphorus, we seldom see shoot growth or fruit yield increases following phosphorus fertilization of fruit trees.

Ways to Add Phosphorus

Rock phosphate, also called hard rock phosphate, is mined from ancient ocean deposits and contains only 1 to 2 percent available phosphate in a form that is available very slowly to most plants, particularly in alkaline soils. It typically becomes available over a period of years to centuries. From economic and horticultural perspectives, this is an inefficient and expensive form of phosphorus.

Colloidal phosphate, also known as soft rock phosphate, consists of clay particles surrounded by naturally derived phosphorus. Although it contains around 20 percent phosphorus, only about 2 to 3 percent is available phosphate. Its phosphorus is also available to plants slowly, but it is more readily available than the phosphorus from hard rock phosphate.

Bonemeal is a rich source of phosphorus that is readily available to plants. Use steamed bonemeal to reduce the risk of pathogens that can infect humans. Other organic sources of phosphorus are available, although most have low concentrations of phosphorus.

Wood ashes can be added to supply phosphorus if potassium is low and the soil pH is below 6.5 (see page 92).

Dried blood, fish meal, shrimp waste, and oyster shell products also provide phosphorus, but usually at low rates of available phosphate. The downside of commercial organic phosphorus fertilizers is their cost, which can be prohibitive for large orchards.

Phosphates and Manure

Another effective use of colloidal phosphate and rock phosphate is to include them with animal manures during the composting process. Acids in the manures dissolve the available phosphate, which then helps to stabilize the nitrogen in the manures.

Applying Phosphorus

Phosphorus is relatively immobile in soils under most conditions and is best applied before planting your trees. Unfortunately, phosphorus can react with other minerals in the soil and become unavailable to plants. On moderately to strongly acidic soils, phosphorus combines with aluminum and iron. Under alkaline conditions, phosphorus reacts with calcium to become unavailable. For instance, when applied directly to the orchard floor, rock phosphates are most available and effective on acidic soils and remain generally unavailable on neutral to alkaline soils. A key way to effectively manage phosphorus is to keep your soil pH between 6.0 and 7.0 and, ideally, 6.5 to 7.0.

Base your phosphorus applications on "available phosphate" or P_2O_5. Many materials contain high concentrations of phosphorus, but they are in forms that plants cannot use. The given amount of phosphorus in all fertilizers refers to the amount of available phosphate that is released from fertilizer materials in a weakly acidic solution. This solution mimics chemicals that are released by roots to make certain mineral nutrients more available and assist in their uptake.

Because of its tendency to become immobilized in the soil, you may find it beneficial to band phosphorus materials along your tree rows, rather than broadcasting them throughout the orchard. Do not band materials that contain boron, nitrogen, or potassium.

To make phosphorus available more quickly to your fruit trees, add the phosphate materials to the soil 1 to 2 years before planting your trees and then grow one or more green manure crops of buckwheat. Cornell University reports that buckwheat is very effective at using these sources of phosphorus. When the buckwheat is tilled into the soil, the phosphorus in its tissues becomes available to other plants. Be sure to till under the buckwheat before it sets seeds because it can become a weed problem.

Potassium

Potassium is abundant in many North American soils and often does not need to be added before planting an orchard. Check your soil analysis results to see if potassium is deficient before adding this nutrient to your planting site.

Ways to Add Potassium

Granite dust contains about 5 percent potassium, but little of it is available for plant growth, and this material has little value for organic fruit growers. If used, the potassium is most rapidly available on acidic soils. Granite dust can contain silica, known to cause lung cancer. If you choose to apply granite dust, use respiratory protection during all handling and application activities.

Greensand is a naturally occurring sandstone found worldwide, often in association with other marine deposits of chalk and clay. It contains glauconite, a greenish-colored iron potassium silicate compound. Greensand typically contains about 6 percent K_2O and 1 to 2 percent P_2O_5. Both are very slowly available to plants, and greensand is best applied to an orchard during preplant preparation. Greensand is more effective on sandy soils than on heavier-textured soils.

Black mica, also known as biotite, refers to a class of micas, rather than to a specific chemical formula. These micas contain variable amounts of potassium, usually several percent, in a form that is quite available to plants in biologically active soils. If you need to add potassium and have ready access to inexpensive sources of biotite, it can be an effective soil amendment. Unfortunately, biotite fertilizers can be hard to find. Also, at least one European study found that one source of biotite included barium and strontium that could be taken up by horticultural crops in sufficient concentrations to possibly be toxic to humans.

More effective and cost-effective sources of potassium include the following: **Sulfate of potash-magnesia** (sul-po-mag or langbeinite) is a rich source of potassium that also supplies sulfur and magnesium. **Wood ashes** are rich in potassium and, when used cautiously (see page 92), can be beneficial, especially on acidic soils. Do not use wood ashes if the soil pH is 6.5 or above.

Potassium sulfate is a good source of rapidly available potassium, but ensure that your source meets organic certification standards. Some organic certifying organizations prohibit it altogether. One form of potassium sulfate fertilizer is created industrially using sulfuric acid and is not suitable for organic growers. Natural potassium sulfate is mined or extracted through evaporation of water from saline lakes.

Potassium chloride contains 60 percent potash that is rapidly available to plants. It also contains a high amount of chloride, so it's not a good idea to use it frequently or in large doses. The U.S. National Organic Program allows the use of potassium chloride only if it is "derived from a mined source and applied in a manner that minimizes chloride accumulation in the soil." Before applying potassium chloride, check with your organic certifying organization to ensure you can use this material.

Applying Potassium

Potassium is relatively immobile on heavy-textured soils and, in such cases, is best broadcast throughout the orchard before planting your trees. On sandy soils, potassium can be highly mobile and is best broadcast shortly before or at the time of planting. On loams and heavier soils, apply the potassium 1 or 2 years before planting fruit trees. Base your potassium applications on "available potash" or K_2O. As with phosphorus, many materials contain high concentrations of potassium, but not necessarily in forms that are available to plants.

Be cautious not to add too much potassium, as it can interfere with calcium uptake and metabolism in fruit trees. The problem is especially serious with apples, where excessive amounts of potassium may increase bitter pit problems in stored fruit.

Sulfur

Sulfur is seldom deficient in soils to the extent that it is necessary to apply sulfur amendments before planting. Deficiencies are most likely to occur on sandy soils that are low in organic matter, which allows the sulfur to leach from the root zone. If your soil test shows less than 10 ppm sulfur and your soil pH is 6.5 or below, add about 175 pounds of gypsum per acre any time before planting. Repeat annually, as necessary, until soil tests show 10 ppm or more sulfur. If your soil pH is above 7.0, add elemental sulfur a year before planting your trees. Table 4.4 shows the amount of elemental sulfur needed, depending on soil pH and soil type.

Calcium

Calcium is seldom deficient in soils, particularly in arid regions. It is most likely to be deficient in sandy, acidic soils. The amount of calcium available to crops, however, depends not only on the amount of calcium present in the soil, but also on the cation exchange capacity of the soil (CEC), which is reported in milliequivalents/100 grams of soil. CEC refers to the ability of negatively charged clay particles and humus in the soil to attract and hold positively charged cations, including calcium, which has two positive charges.

The higher the CEC, the greater the soil's ability to bind to cations. Likewise, soils with low CEC values have less binding ability. This is important because nearly all plant mineral nutrients are positively charged, so as the CEC increases, so does the soil's ability to bind to and store these nutrients. Nitrogen, being negatively charged, does not bind to the negatively charged soil particles. This is one of the reasons that nitrogen is easily lost from the soil and why it is the nutrient most likely to be deficient in soils.

Soils with low CEC values quickly saturate with cations and have high nutrient ratings, even when soil tests show low concentrations of those nutrients. This apparent contradiction arises because low-CEC soils have relatively few spaces for the nutrient cations to fill. On a low-CEC soil, all of the sites may be filled with calcium, leading to a high calcium saturation rating, even though the actual amount of calcium present in the soil is low. Soils with high CEC values require larger concentrations of calcium to bind to their available sites, and calcium ratings can be low, even when abundant calcium is present in the soil. Table 4.6 gives calcium ratings based on CEC and calcium soil test results.

If your soil's calcium rating is low and the pH is 6.5 or above, add gypsum as a soil amendment up to the time of planting (see the sulfur section above). Consult with your soil testing lab for recommended amounts of gypsum to add.

If the soil pH is less than 6.5 and soil magnesium levels are adequate, add limestone, preferably several months or more before planting your trees. If the pH is less than 6.5 and magnesium is also deficient, add dolomitic limestone. Remember that pH is an important factor in calcium management. Keep your soil between pH 6.0 and 7.0 and, ideally, around 6.5. Use the amounts of limestone or dolomite needed to adjust your soil pH to 6.5.

Magnesium

Magnesium deficiency is occasionally a problem in orchards and is known to cause leaf chlorosis in cherries and peaches. Deficiency problems are most likely to occur when soil pH is too low or soil calcium concentrations are too high. If your soil test shows less than 100 ppm magnesium, you should amend the soil with magnesium before planting your trees.

Determining Your Soil's Calcium Rating

Table 4.6

Soil calcium ratings as determined by soil calcium concentrations and cation exchange capacity (CEC). To determine your soil's calcium rating (low, medium, or high), first locate the soil CEC value in the top row, as shown in your soil test. Then drop down that column to find your soil calcium concentration, also shown on your soil test. Read off your calcium rating from the left side of the chart. For example, a soil with a CEC value of 15 meq and a calcium concentration of 4,500 ppm would have a calcium rating of "medium."

	SOIL CEC		
Calcium Rating	**< 10 meq**	**10–20 meq**	**> 20 meq**
Low	Ca < 1,000 ppm	Ca = < 3,000 ppm	Ca < 5,000 ppm
Medium	Ca = 1,000–1600 ppm	Ca = 3,000–5000 ppm	Ca = 5,000–8,000 ppm
High	Ca > 1,600 ppm	Ca > 5,000 ppm	Ca > 8,000

Adapted from *Soil Test Interpretations and Recommendation Guide: Commercial Fruits, Vegetables and Turf*, 1999, University of Missouri.

Epsom salts (magnesium sulfate) are an easy-to-use amendment. If sulfur and potassium are also low, sul-po-mag can be applied to add all three nutrients. Your soil test should give you suggested application rates for magnesium. Use the magnesium percentages in table 8.4 to determine how much of a particular fertilizer to add.

If both calcium and magnesium concentrations in the soil are low, dolomitic limestone can supply both nutrients while raising the soil pH. Use the amount of liming material recommended in your soil analyses. Do not apply dolomitic limestone if the pH is 6.5 or above, or when soil calcium concentrations are above 5,000 ppm.

Boron

Boron can be difficult for orchardists to work with because the trees need the nutrient only in very small quantities. Unlike most other nutrients, however, the adequate range of boron concentration within the tissues is very small, and boron becomes toxic at extremely low concentrations. If your soil test shows less than 0.5 ppm of boron, add 1 to 2 pounds of actual boron per acre.

Broadcast the boron amendment throughout the entire planting block. Never apply boron-containing materials to the soil in bands. According to the Organic Materials Research Institute, boric acid, hydrated forms of sodium tetraborate, sodium borate derivatives, disodium octaborate and its hydrated forms, and hydrated forms of colemanite may be used to supply boron but should be applied to the soil or plants only when you have a documented deficiency, as shown in soil or foliar analyses. The OMRI website lists recommended boron-containing amendments.

Organic Amendments

ORGANIC AMENDMENTS can improve water-holding capacity on droughty soils, improve tilth, and help you get a start on building concentrations of soil organic matter.

The best materials to add are those that have already been composted, for several reasons: Provided you have done the composting correctly, the process generates enough heat to kill many weed seeds, pathogens, and pests, while still leaving beneficial microorganisms. Raw organic materials can be sources of diseases, pests, and weeds. Properly composted materials are also more chemically stable and predictable than fresh materials. If you use animal manures in your compost, the process allows excess salts to leach out that might otherwise damage your crop.

And very importantly, good compost has a carbon-nitrogen balance that is suitable for plant growth — usually around 10 parts of carbon for each part of nitrogen. Green plant materials, such as green manure crops, are typically 12:1 to 17:1. Cattle and horse manures have ratios around 18:1. These materials can be added directly to soils with no fear of creating nitrogen deficiencies in your crops.

Materials that are rich in cellulose, such as straw (80:1) or sawdust (400:1), have high carbon-to-nitrogen ratios. When these materials are incorporated directly into the soil, microorganisms begin decomposing the carbon compounds. The microorganisms, however, require large amounts of nitrogen as they form structural and enzymatic proteins. Since the woody material has little nitrogen, the microorganisms take the nitrogen from the soil itself, which can create temporary deficits in available nitrogen. Once the microorganisms have completed decompos-

Off-Farm Amendments

In general, increasing soil organic matter concentrations by adding off-farm amendments is too expensive in terms of labor and transportation costs for any but small orchards. In some areas, community waste recycling programs supply mulch from tree trimmings and yard wastes. While the recycling is commendable and the materials are generally available in large quantities at low cost, beware! Such mulches have been found to be contaminated with pesticides from roadside and home herbicide and pesticide applications. Determining whether a municipal source of mulch or compost is pesticide-free may be impossible.

ing the organic material, they die and release the nitrogen back into the soil. Until that nitrogen is released, however, your crops can be deprived of adequate nitrogen and suffer stunting or even death.

If you choose to incorporate uncomposted woody materials into your soil, add nitrogen at the same time. For every cubic yard of bark, hardwood sawdust, or hardwood wood chips, add 4 to 5 pounds of actual nitrogen the first year and half that amount the following year. For straw and softwood sawdust or chips, apply 2 to 2.5 pounds of nitrogen the first year and half that amount the following year. In this case, if you were adding one cubic yard of hardwood sawdust and wanted to provide the nitrogen in the form of dry, composted cattle manure (approximately 0.5 percent nitrogen), you would need to add about 800 pounds of manure. Leaving uncomposted woody materials on the soil surface as mulches usually does not deplete soil nitrogen, and supplemental nitrogen often does not need to be applied.

Adding uncomposted woody materials as a soil amendment can require large quantities of the amendment and nitrogen fertilizer. Time is another factor. If you incorporate uncomposted woody materials into the soil, it is best to wait at least one full year (preferably two or more) before planting trees, even when you apply extra nitrogen. From these standpoints, amending soil with sawdust and similar materials is seldom feasible except for the smallest orchards.

Green Manure Crops

An effective way to add organic matter to the soils of orchards large and small is to grow one or more green manure crops and cultivate them into the soil before planting fruit trees. Although they require time to produce, green manure crops cost very little in terms of money and labor compared with purchasing, hauling, and incorporating sawdust or similar materials. Some green manure crops add nitrogen to the soil, and none require extra nitrogen in order to decompose.

Depending on where you live, many good green manure crops are suitable for preparing an orchard site. Look for crops that grow quickly, form dense stands that quickly cover the soil surface, produce large amounts of organic matter, out-compete most weeds, are easy to grow, have few pest and disease problems, and are easy to incorporate into the soil. The goals are to shade or crowd out weeds while producing large amounts of organic matter that will break down in the soil without causing nitrogen depletion.

Buckwheat, barley, clover, vetch, oats, beans, peas, and rapeseed are popular green manure crops. Avoid sod-forming grasses (hard to kill and incorporate) and row crops that leave bare areas of soil (poor weed competitors). In milder climates, you can often grow two or three green manure crops in a single season.

Green manure crops have other important advantages over other forms of organic amendments: Some crops serve as natural soil fumigants and can be valuable in reducing pest and disease problems. Particularly valuable are *Brassica* varieties (canola, mustard, or rapeseed) that have been selected for high concentrations of biofumigant compounds called glucosinolates. When glucosinolates come into contact with a group of plant enzymes called myrosinase during mowing or cultivating, the glucosinolates break down to form isothiocyanates, nitriles, epithionitriles, and thiocyanates. Isothiocyanates are the most common and abundant by-product and are considered highly toxic to nematodes and valuable as general biocides. The isothiocyanates are relatively volatile and remain in the soil for a few days to a few weeks.

Not all *Brassica* varieties are equally effective biofumigants. Canola varieties for human consumption have had most of the glucosinolates bred out of them. Two biofumigant varieties rich in glucosinolates, however, were developed at the University of Idaho: 'Humus' rapeseed and

'Ida Gold' mustard. If you are using open-pollinated seed, rather than named varieties, brown mustard appears to be a somewhat more effective biofumigant than white mustard or rapeseed.

Mustard, canola, and rapeseed crops produce seeds quickly and in enormous quantities. Be sure to till the green manure crops into the soil before they set mature seeds. Shredding the crop with a flail mower immediately before rototilling it into the soil maximizes its biofumigant value.

Instead of growing the *Brassica* plants, you can use cold-pressed meal from crushed seed. The biofumigant properties of the seed vary greatly, however, depending on the type and variety of the mustard, canola, or rapeseed used. While progress is being made in seed meal biofumigation, results often leave much to be desired, and it is best suited for small orchard applications.

Microorganisms

Mycorrhizae are soil fungi that form beneficial relationships with plants. The name literally means "fungus root," and these organisms are extremely valuable in helping plants take up moisture and nutrients by effectively extending the plants' root systems. These organisms can also, in some cases, help protect plants against pathogens and parasitic nematodes by competing with other microorganisms for space on plant roots and by forming a physical barrier to nematodes. There are many different mycorrhizal species. Some are specific to particular plant species or types, while others are generalists. Some mycorrhizae are adapted to acid soils, others to neutral or alkaline soils.

In a healthy soil, native mycorrhizae should already be present and do not need to be added as a soil amendment. When native mycorrhizae have been killed or damaged by fumigation, overuse of inorganic fertilizers, or other causes, it can be beneficial to inoculate an orchard soil at the time of planting. For example, you might want to place mycorrhizae in bands in your tree rows immediately before planting if you have grown biofumigant crops. Blends of several mycorrhizal species are generally more effective than applying a single species. You might also choose to dip the roots of bare root trees in a slurry of mycorrhizae at the time of planting. Be sure that the product you choose has been approved for use in organic production. Chapter 7 goes into detail on different mycorrhizal species and application methods for these products at the time of planting.

Adding algae and other microbes to the soil is seldom beneficial to organic fruit production. Despite producers' claims that their microbes fix nitrogen or somehow make nutrients more available to crops, the reality is that few of these products create significant benefits under the conditions that we want to maintain in our orchards. Nitrogen-fixing bacteria are very important in rice production, for example, but few of us would want to grow fruit trees in the wet, oxygen-poor environments needed by these microorganisms. An exception would be *Rhizobium radiobacter* strains that help protect roots against crown gall when used as root dips at the time of planting.

Other Soil Amendments

SOIL AMENDMENTS CAN BE BENEFICIAL when used for specific purposes, such as to correct a nutrient deficiency. They are not, however, a substitute for proper site selection, preparation, and maintenance. Far more important are correcting soil drainage problems, adjusting soil pH, developing and maintaining reasonable organic matter concentrations, utilizing green manure and cover crops, and creating moist, well-drained, biologically active soils. Until you have accomplished those goals, soil amendments are unlikely to be of great benefit.

Although some organic farmers and gardeners are convinced of the value of various "soil-building" amendments, the reality is that many of these products do little to improve an organic orchard. These products are typically very expensive for the small benefits that they provide.

Besides being ineffective, some products contain prohibited substances that can cost you your organic certification. Before applying any amendments, certified organic growers should be sure that those amendments meet their certifying agency's guidelines. For commercial products, purchase only those that you know, by brand name, are approved for organic production. If they are not specifically approved by OMRI or another recognized organic certification organization, avoid them.

There are a few exceptions, including the phosphorus and potassium amendments we discussed earlier. While rock dusts and humate products have little to offer orchardists, zeolites can be beneficial in cropping systems. According to the U.S. Geological Survey (USGS), "Zeolites are porous minerals with high cation-exchange capacity that can help control the release of plant nutrients in agricultural systems. Zeolites can free soluble plant nutrients already in soil, and may improve soil fertility and water retention. Because zeolites are common, these unique minerals could be useful on a large-scale in agriculture." Zeolites are most useful on soils with low CEC and act rather like sponges, holding and releasing plant mineral nutrients. In chapter 8, we will go into more detail on other plant nutrient sources and programs to use after you have planted your crops.

Weed and Pest Control

AT THIS POINT, WE HAVE CORRECTED SOIL DRAINAGE PROBLEMS, laid out planting blocks and other operating areas of the orchard, adjusted the soil pH, and amended the soil (according to soil tests!) to ensure adequate nutrition for our orchard crops. The next step is to address preplant weed control. Eliminating the most serious weeds or bringing the populations under control is a key step in establishing a healthy, productive orchard.

Until the late 1990s, we generally referred to managing vegetation on the orchard floor as weed control or weed management. Prior to the 1950s, trees were planted far apart in orchards to allow cross-cultivation and keep orchard floors bare of vegetation. As a result, the soil of these orchards was often heavily eroded and compacted, and it was difficult to access trees in wet weather. As preemergent and selective herbicides became available beginning in the 1950s, many orchardists began maintaining grass in alleyways and eliminating all grasses and broadleaf weeds in the tree rows. This remains the standard practice in conventional orchards today. Other benefits are reduced erosion and soil compaction, better access during wet weather, more soil organic matter, and a larger habitat for beneficial organisms.

As commercial organic orchards increased in size and number, it became more important to find cost-effective weed control methods. By studying the problem and testing different weed control strategies, we began to discover advantages in maintaining rather diverse vegetation on orchard floors. While it is relatively easy to manage alleyways in both organic and conventional systems, it is much more difficult and expensive for commercial organic growers to manage vegetation within tree rows.

Planting fruit trees into dense vegetation of any kind generally causes poor tree survival rates, and those trees that live are often stunted and unproductive. For that reason, it is important to eliminate, to the greatest degree possible, competitive plant species before planting your trees. Serious orchard weeds vary across the country, but they typically include aggressive perennial species such as quack grass, Bermuda grass, johnsongrass, nutsedge, bindweed, and Canada thistles.

Organic Herbicides and Thermal Controls

For many organic growers, bringing weeds under control before planting is the most difficult step in developing an orchard and requires varied cultural practices. Serious infestations of perennial weeds are hardest to control. Unfortunately, organic herbicides, including soaps, essential oils, and vinegar, are largely ineffective in controlling established perennial weeds. The same limitation applies to thermal weed control methods, such as flaming, infrared, and infrared plus steam. Only plant tissues directly contacted by the herbicides or heat are killed, and perennial weeds quickly resprout from underground organs. Thermal weeding and organic herbicides can be helpful in controlling annual weeds and young seedlings of perennial weeds.

Mechanical Cultivation

This method is more effective for annual weeds than perennial weeds. With aggressive perennial weeds, particularly those that spread by rhizomes, one or a few passes of a tiller or disk are more likely to increase the weed problem than to reduce it.

If your soil is deep enough and the weeds are confined to the top 4 inches of soil, try deep plowing with an inverting plow before adding the soil amendments we discussed earlier in this chapter. The goal is to bury the seeds and rhizomes so deeply that they cannot resprout. A typical moldboard plow may not bury the weeds and seeds deeply enough. California fruit specialists recommend using a Kverneland rollover plow for this operation. When farming on shallow soils, be careful not to turn your productive topsoil under and leave unproductive subsoil on the surface.

Repeatedly tilling or disking bare fallow ground for one or two growing seasons can also help destroy rhizomes, roots, and tubers from which perennial weeds resprout. Shallow disking or harrowing is less likely to create a hardpan or damage soil structure than is rototilling. Cultivate the field as early in spring as you can without causing the soil to compact. As soon as the field begins to green up, cultivate it again to kill the young sprouts and seedlings and gradually deplete the food reserves in the underground tissues. To reduce the adverse effects of leaving the soil bare for long periods of time, you can plant a biofumigant green manure crop in early spring and till it in before the seeds mature. Keep the soil bare using cultivation during the summer, and replant the biofumigant crop in fall as a late green manure crop. Repeat this process for 1 or 2 years.

If you choose to keep your orchard land bare using repeated cultivation to achieve weed control, follow up with 1 or 2 years of biofumigant and green manure crops. Use crops that form dense stands that shade out and out-compete weed seedlings. In warmer areas with prolonged growing seasons, you should be able to establish and till in at least two green manure crops in a single growing season. Be sure to till these crops under before they set seeds and become weeds themselves. Avoid using sod-forming crops, such as turf grasses.

Solarization

University of California orchard specialists recommend solarization as one means of reducing weed populations in the crop rows before planting trees. To solarize the soil, stretch one, or preferably two, layers of clear, 2 to 4 mil plastic film over bare, moist soil and cover the edges of the plastic with soil. Make sure that the soil is smooth, level, and free from clods and surface debris. The plastic should be as close to the soil as possible. Under ideal conditions, the top 2 inches of soil will be heated to around 140°F (60°C) and soil 18 inches deep will be as hot as 102°F (39°C). These temperatures are high enough to kill many seeds and pathogens.

Nematodes in the top several inches of soil may also be killed. The films may need to be left in place an entire summer.

Solarization works well in parts of California and other areas with hot summers and clear weather, but in cooler or cloudier areas, it has proven to be ineffective. In cooler climates, the plastic films serve as greenhouses, producing lush weed growth under the plastic. Solarization also creates a disposal problem for the plastic film, which is usually derived from a nonrenewable resource. Except for very small-scale growers in hot climates, solarization is not recommended.

Nonorganic Herbicides

At the risk of offending some readers, there is another strategy for bringing serious weed problems under control before establishing orchard trees. For sites heavily infested with hard-to-control perennial weeds, some orchardists elect to use synthetic, translocatable herbicides to kill the weeds during their transition to an organic orchard. While I do not promote this practice for organic fruit growers, it is an option for reclaiming a badly maintained site and converting it to a productive organic orchard.

Glyphosate, also known as Roundup and sold under a variety of brand names, is often the herbicide of choice because it moves from the leaves and stems of weeds into the roots and rhizomes, killing the weeds and preventing resprouting. Translocatable herbicides are often used as part of a season-long bare ground fallowing, followed by biofumigant and green manure crops, as we discussed above. The herbicides can be broadcast-sprayed throughout an entire field or spot-sprayed on weed patches.

Most organic certification programs require a wait of at least 3 years after the last application of a prohibited substance before that site can be certified organic. When used early in the site preparation process, herbicides provide a way to rapidly and economically control weeds, and organic certification is possible by the time the trees bear their first or second crops.

Pest Control

At the same time you begin your preplant vegetation management program, also begin controlling gophers, voles, and mice. We will discuss rodent control in more detail in chapter 11. For now, remove debris from the site, including piles of wood, boards, and anything these pests can hide under. Use repeated, shallow cultivation to destroy burrows and nesting sites. Keep vegetation around your planting site mowed very short to reduce cover. Especially for gophers, it is generally more effective to trap the animals than to poison them, and it poses little risk to pets and other off-target animals.

If you have birds of prey (raptors) that feed on small mammals in your area, install perches for them around the planting blocks. Owls and hawks are quite effective in hunting rodents.

Planting the biofumigant green manure crops we discussed earlier can help reduce nematode pest populations. Removing host plants from hedgerows and other non-crop areas on your site will help reduce pest and disease problems after planting your trees. Chapters 10 and 11 describe what plants serve as hosts of pests and diseases and what to remove from your property.

Congratulations! At this point, you are well on your way. By investing the time and effort to select a good location or evaluate your present site, create a great orchard design, and prepare or upgrade a planting site, you have given yourself excellent prospects for a healthy and productive organic orchard. Our next step will be what many growers consider one of the most enjoyable parts of fruit growing — selecting crops and varieties.

CHAPTER 5

SELECTING POME FRUIT CROPS AND VARIETIES

- Apple | 122
- Pear | 131
- Quince | 137
- Medlar | 140
- Mayhaws | 142
- Loquat or Japanese Medlar | 147
- Saskatoon | 150

Choosing orchard crops and varieties is one of the most enjoyable aspects of planning or expanding an orchard. Unfortunately, many people start with this step without considering all of the factors we have dealt with so far.

By this time, you should have chosen a great orchard site or thoroughly evaluated the site that you have. You have designed your new orchard or planned upgrades for an established orchard. Problems with soil drainage, pH, and nutrient concentrations have been identified and you know how to correct them. You have designed a soil-building program or your plans are already underway. With the groundwork behind you, it is time to start filling the orchard with trees.

Climate considerations. To reemphasize a critical point made in chapter 2, climate sets the limits within which all other factors operate. In order to give your organic orchard every chance of success, choose varieties that are well adapted to your climate and site. Use the Plant Hardiness Zone Map developed by the U.S. Department of Agriculture (see page 23) to identify the average minimum temperature for your area. For extra security, consider selecting varieties rated one or even two zones hardier than your site to account for unusually cold winters. We will use the USDA map in recommending fruit varieties.

Pest and disease considerations. In the ideal pome fruit climates of the western United States and parts of western Canada, even disease-prone varieties can be grown successfully by organic orchardists. In the more humid regions of North America, however, disease resistance is a critically important consideration for organic fruit growers. More than 260 species of bacteria and fungi cause economically destructive diseases on apples. Many of these or similar diseases also infect pear, quince, medlar, loquat, and mayhaw.

In North America, apple scab and fire blight are two of the most common and destructive diseases of apples. Pear scab infects pears, but it is seldom as devastating as apple scab is on apples. Pome fruits are also susceptible to powdery mildew; cedar-apple, cedar-quince, and cedar-hawthorn rust; viruses; and stem cankers. By selecting good varieties and employing effective orchard management practices, you can minimize these problems. Chapter 10 goes into detail on major diseases found in your region and recommendations on dealing with them.

Common insect pests on pome fruits include assorted mites, scales, aphids, mealybugs, leaf rollers, fruitworms, thrips, leafhoppers, leaf miners, lygus bugs, cutworms, and wood borers. Most of these pests will not be serious problems in a well-managed organic orchard. By far the most serious pests are plum curculio, apple maggot, and codling moth. In the not-distant past, these pests made pome fruit production of any kind

A Note on Propagation

When we plant a fruit tree, we generally want to know how tall the tree will grow; its susceptibility to pests and diseases; the bloom and ripening dates; the size, color, and flavor of the fruits; and so on. Most of us have our favorites – 'Bartlett' pear, 'Liberty' apple, or 'Madison' peach, for example. We know what to expect when we grow or buy these particular varieties. This predictability is achieved with orchard fruits by propagating our favorite plants vegetatively (exceptions sometimes include such crops as Nanking cherries, wild or semidomesticated bush cherries and plums, and Damson plum). (See page 47 for more information on the various propagation techniques.)

Other than fruit breeders and some rootstock nurseries, orchardists seldom grow fruit trees from seed. The reason is that every tree fruit seed produces a genetically unique individual that may, or may not, have desirable horticultural traits. Unlike certain vegetables, such as tomatoes and beans, most fruit crops do not breed true to variety (one exception is Damson plum).

Our most common domestic apple, *Malus domestica*, for instance, appears to be a complex hybrid involving several different species. A cross between two apple varieties usually produces thousands of widely differing offspring, few of which meet commercial standards. Obtaining one named apple variety from 1,000 to 10,000 individual offspring from a given cross is usually par for the course or better.

Many tree fruit varieties originate as sports or, more correctly, bud sport mutations. These mutations occur naturally in plants. With apples, for example, you may discover yellow fruit on a tree that normally bears red fruits. If the very difficult in some regions of North America. With the tools available today, however, these pests can be controlled in organic orchards throughout North America.

Plum curculio is primarily a pest in eastern North America, where it can cause severe damage to tree fruits when not controlled. It was long considered a leading obstacle to commercial fruit production within its range. The apple maggot's native range includes the Midwestern and eastern United States and eastern Canada, although the pest is now also

mutation is stable, wood from the affected branch can be propagated to produce the yellow fruit.

Plants that we have propagated vegetatively are clones. The term simply refers to the fact that an individual organism is genetically identical to one or more other individuals. Every 'Fuji' apple, for example, is theoretically identical to all other 'Fuji' apples of the same strain, regardless of where they are grown. The term "clone" does not, in any way, indicate that a tree is a genetically modified organism (GMO). As an organic grower, you do not need to shy away from clonal varieties, nor should you be locked into using old heirloom varieties. As a practical matter, we have been cloning fruit crops for thousands of years using the grafting, budding, and layering techniques described in chapter 3.

Where space is limited, such as in a home orchard, you can often produce trees bearing several different fruit varieties by budding or grafting those varieties onto established trees. You can produce a single tree with an assortment of apple varieties that successfully cross-pollinate one another. Sometimes, the varieties need to be the same crop. You cannot successfully bud or graft pears onto apples or vice versa. Quince rootstocks can be used for pear and some hawthorn species can be used for mayhaw.

You may wish to graft bud wood from an established tree onto a rootstock in order to create a new tree. For example, you may have a large, old tree that you love but you do not know what variety it is or that variety may no longer be available. You can purchase rootstock trees and bud or graft the old variety onto them, producing young trees in whatever size you wish and growing the fruits you've enjoyed for years.

"Variety" versus "Cultivar"

"Cultivar," coined from "cultivated variety," is a more correct term for fruit crops, and I would use it, not "variety," in a research journal article. "Cultivar" is specific in that it refers to a vegetatively propagated variety in which all individuals are genetically identical. "Variety" has several meanings in botany and taxonomy. The term "variety" is generally used by fruit tree nurseries and many orchardists and suits our purposes for this book. Variety (cultivar) names are shown with single quotation marks; for example, 'Red Delicious' apple.

found in California, Oregon, and Washington. Codling moth originated in Asia, but it is now found in all apple growing regions in the world. Chapter 11 goes into detail on pests and their management.

Financial considerations. For a home orchard where financial investments are small, feel free to push the limits and experiment. The worst that can happen is that you replace a few trees. For larger orchards where financial investments are greater, particularly for commercial orchards, it is extremely important to be conservative when selecting crops.

Ultimately, whether you have a large or a small orchard, you want varieties that:

- Will remain healthy and grow vigorously at your location
- Produce abundant, reliable yields of good-sized fruits that are appropriate for your needs
- Are resistant to serious diseases and suitable for organic production
- Are suitable for the training systems that you have chosen

Apple

Apples are the second most widely grown temperate fruit crop in the world and have been used for food for at least 8,000 years. Originating in Europe in the areas of Kazakhstan, the Black Sea, and the Caspian Sea, apples quickly spread and were well known in the Egyptian, Greek, and Roman empires.

Temperature concerns. Apples are one of the most adaptable temperate zone tree fruits. In North America, apples were traditionally grown in cooler areas from the central United States into Canada, due to the trees' chilling requirements. The dormant buds of most apple varieties must be exposed to temperatures between about 32 and 55°F (0 to 13°C) for 800 to 1,700 hours in the spring in order to begin growing normally. To extend apple production into southern states and other warm-climate regions, breeders developed low-chilling varieties. These varieties include 'Anna' and 'Ein Shemer' from Israel, 'Dorsett Golden' from the Bahamas, 'Tropic Sweet' from the Florida Agricultural Experiment Station, and 'Beverly Hills'. Some of these varieties can be grown successfully in areas that receive 100 to 200 chilling hours.

Cold hardiness is a concern for northern growers. Fortunately, we have many good cold-hardy apple varieties suitable even for the Prairie Provinces and parts of Alaska. Particularly cold-hardy varieties include 'Norson', 'Noran', and 'Norda'. With the introduction of extra-cold-hardy varieties for the north and very low-chilling varieties for the south, apples can now be grown reliably from Zones 3 to 9 and, with extra effort, in Zones 2 and 10.

Frost injury can be a serious problem for apples in many areas, although somewhat less so than for earlier-blooming stone fruits. Some apple varieties bloom up to 30 days later than others and are good for areas prone to late spring frosts. 'Rome Beauty' and 'Northern Spy' are popular late-blooming varieties. If you select one of these late-blooming varieties, make sure that you have a long enough growing season to mature the fruits. Also, in areas with severe fire blight problems, late-blooming varieties are more susceptible to the disease than earlier blooming varieties. In such areas, you might be better off planting early-blooming varieties and providing some form of frost protection, such as orchard heaters or wind machines.

Summer heat is critically important in developing and ripening your crop. Growing season length requirements vary tremendously among varieties. Some apple varieties mature in as little as 70 days from pollination, while others require 180 days or more. In areas with short growing seasons, varieties that ripen early are usually the best choice.

Long-season varieties 'Granny Smith' and 'Mutsu', for example, can be difficult to ripen in short-season locations.

Soil conditions. Apples are relatively adaptable to soil conditions, but they perform best on deep, well-drained soils with a pH between 6 and 7. Some rootstocks are better adapted than others to poorly drained soils, but such sites are best avoided or improved, as we discussed in chapter 4.

Apple Varieties

Cultivated apple varieties were described as early as 323 BCE by Theophrastus in Greece, and fruit specialists estimate that more than 10,000 varieties have been named. Until recently, however, very few apple varieties came from breeding programs; the vast majority arose as chance seedlings or bud sport mutations on existing trees. That began to change in the mid-1900s, and now university, government, and commercial breeding programs are developing many new apple varieties.

Without becoming too technical, the odd number of chromosomes in apples (× = 17) and the occurrence of four possible ploidy levels (the number of sets of chromosomes) greatly complicate breeding efforts by tending to preserve undesirable genes. Breeders call this process "enforced hybridity." While the genetic basis of the undesirable genes was unknown at the time, even ancient orchardists recognized the problem. Theophrastus noted that trees grown from seed almost always produce inferior-quality fruit. As early as 5000 BCE, fruit trees were being commercially grafted to preserve desirable traits.

Despite the long history of cultivation and the existence of approximately 25 apple species, advances in apple breeding have been remarkably limited. Although there are thousands of named apple varieties, they are derived from a rather narrow genetic base and one that has not drawn much on genes for pest and disease resistance. As a result, we have many beautiful, tasty apple varieties that are highly susceptible to pest and disease problems and necessitate enormous inputs of chemicals and labor to control. For susceptible varieties in humid areas, for exam-

Apples

AT A GLANCE

Chilling requirements: 32 to 55°F (0 to 13°C) for 100 to 1,700 hours from fall through late winter, depending on variety

Growing season: Length differs according to variety: anywhere from 70 days to 180 days.

Soil pH and type: Deep, well-drained soils with a pH between 6 and 7

Pollination: Most varieties require cross-pollination; place a pollinizer within 50 feet of tree.

Variety selection: Match your site's characteristics with the variety's chilling requirement, hardiness rating, and required growing season length. Emphasize disease resistance.

ple, a grower may need to apply 15 fungicide sprays per year to control apple scab.

In the 1990s, a group of some of the finest fruit scientists of the twentieth century stated in *Genetic Resources of Temperate Fruit and Nut Crops*: "In commercial apple orchards, spray chemicals plus their cost of application can often be as much as a quarter of the total cost of production. Unfortunately, sprays frequently do not give complete control of pests and therefore there is often economic loss despite spraying."

Alarmingly, the narrow genetic base is rapidly becoming narrower as growers focus on a few varieties with strong market appeal. Despite the fact that apples are grown on a large commercial scale in at least 63 countries and we have thousands of varieties to choose from, only a few dozen varieties are widely grown commercially worldwide, largely based on 'Delicious' and 'Golden Delicious' strains. While newer varieties are beginning to dominate the market, many of them also tend to be highly prone to diseases. Even worse, many heirloom varieties are being lost as germplasm collections are reduced due to budget shortages.

All is not gloom, however, and breeding programs are now producing high-quality, disease-resistant apples. One of the best programs for apples is the PRI Disease Resistant Apple Breeding Program, a cooperative among Purdue University, Rutgers, and the University of Illinois. This program has produced more than 1,500 selections and released

44 varieties, including 'Prima', 'Priscilla', 'Jonafree', 'Redfree', 'Dayton', 'William's Pride', 'Enterprise', 'GoldRush', and 'Pristine'. The Canadian Agricultural Research Station in Kentville, Nova Scotia, has released 'Novamac' and 'Nova Easygro.' Unfortunately, many of these fine apple varieties have yet to become accepted by large-scale apple growers.

Pollination. Most apple varieties require cross-pollination in order to achieve an abundant and reliable fruit set. Cross-pollination involves the female organ (stigma) on flowers on one plant being pollinated with pollen from a genetically different plant (different variety), usually of the same species. Because of this, apples are referred to as "self-unfruitful" or "self-incompatible," depending on the nature of the incompatibility. The possible exception is 'Yellow Transparent', a variety that has been reported to set crops reliably when self-pollinated. In all other cases, you will need to plant at least two compatible varieties within about 50 feet of each other or have two or more compatible varieties grafted or budded onto a single rootstock. Some apple varieties, including 'Jonathan', 'Rome Beauty', 'Oldenburg', 'Wealthy', 'Golden Delicious', 'Grimes Golden', 'York', and 'Granny Smith', are partially self-fruitful but still produce better and more reliable crops when cross-pollinated.

Not all combinations of apple varieties provide effective pollination. 'Cortland' and 'Early McIntosh' will not effectively pollinate each other, for example. Some varieties produce infertile pollen and cannot be used as pollinizers, as shown in the box at right. Very early-blooming and very late-blooming varieties do not effectively cross-pollinate each other because the flowering periods do not overlap sufficiently. Examples are shown in table 5.1.

Closely related varieties also tend to be poor cross-pollinizers. Strains developed from 'McIntosh', 'Red Delicious', and 'Golden Delicious' (including 'Jonagold') tend to be poor cross-pollinizers with their parents or other strains derived from

Apple Varieties Unsuitable as Pollinizers

- Bert's Special
- Bramley
- Gravenstein
- Jonagold
- Karmijn
- Mutsu (Crispin)
- Nured Stayman
- Nured Winesap
- Red Boskoop
- Rubinstar
- Shizuka

Suitability of Apple Varieties for Cross-Pollination

Table 5.1

Listed below are common early- and late-blooming apple varieties whose blooms do not overlap sufficiently to provide effective cross-pollination. Varieties in each group are suitable cross-pollinizers for other varieties in the same group. For example, 'Lodi' and 'Liberty' can pollinate each other, and 'Macoun' and 'Northern Spy' can pollinate each other. Varieties in each group are listed in order of ripening date, from earliest to latest.

EARLY BLOOMING		LATE BLOOMING
Lodi	Spartan	Macoun
Earliglo	Liberty	Red York
Jonamac	Mor-Spur McIntosh	Early Spur Rome
Nured McIntosh	Idared	Northern Spy
Improved McIntosh		Nured Rome

those parents. For example, 'Nured McIntosh', 'Jonamac', and 'Mor-Spur McIntosh' should not be used to cross-pollinate one another.

Other than the exceptions described above, most combinations of domestic apple varieties typically cross-pollinate well. Crab apples can serve as effective cross-pollinizers for domestic apples and are especially valuable in large, commercial orchards where varieties are planted in solid blocks to simplify cultural practices and harvest. Check with your nursery before making final choices to ensure that the trees you plant will effectively cross-pollinate each other.

In general, make sure you have a pollinizer within 50 feet of every tree. If your orchard has multiple tree rows, include pollinizers in every row and stagger their locations in every other row to form a repeating triangular pattern of pollinizers. When space is very limited, bud or graft a pollinizing variety onto your tree or trees.

Choosing a variety. It is extremely difficult to suggest apple varieties, due to the many thousands of varieties and the extensive geographic range over which apples are grown. Many new apple varieties have entered the market in recent years but have not been well tested in all growing regions. You will also find that certain varieties are popular in particular geographic areas, and nurseries focus on these relatively few varieties. The vast majority of apple varieties have been lost to time, and those available from nurseries are limited to a few hundred.

To simplify the selection process, use the USDA Plant Hardiness Zone Map and variety hardiness ratings from nurseries to narrow your list to varieties adapted to your climate. Be skeptical about exaggerated claims for plant hardiness, either north or south. It would be a rare fruit tree that can thrive in the warm, low-chilling Zone 9 as well as cold, short-season Zones 3 or 4.

Know the approximate chilling hours that your site receives and make sure they match your selections' dormancy requirements. Be cautious about growing low-chilling and medium-chilling varieties in high-chilling areas where the plants can break dormancy prematurely during winter and early spring warm spells. Select varieties that will ripen in your growing season. Apple varieties for cold climates are listed in table 5.3 (page 158). Medium-chilling varieties are shown in table 5.4 (page 161), and low-chilling apples for subtropical areas are listed in table 5.5 (page 163).

Throughout the process, emphasize disease resistance. Disease resistance is important to all apple growers, and especially so for organic growers. It is critically important that you consider resistance to apple scab and fire blight, particularly if your orchard is in a humid region. 'Liberty', 'Jonafree', 'Britegold', 'Macfree', and 'Florina' are scab-resistant or scab-immune varieties that have received favorable ratings for flavor and eating quality. 'Liberty' has proven especially tasty. Table 5.2 describes selected disease-resistant apple varieties.

Apple Rootstocks

The number of apple rootstocks is large and continuing to grow rapidly. During the late twentieth and early twenty-first centuries, many new apple rootstocks were developed that provide size control, disease resistance, hardiness, and adaptability to a wide range of growing conditions.

Unfortunately, a discourse on the history of apple rootstocks and the relative merits of the many available is beyond the scope and intent of this book. Instead, the following discussion provides rootstock recommendations for various tree sizes and training systems. Many rootstocks are designated by code letters followed by numbers. The letters refer to the place where the rootstock was developed. "CG" refers to selections from Cornell's Geneva Experiment Station. "EM" and "M" refer to East Malling rootstocks, and "MM" refers to Merton-Malling rootstocks, both developed in England in the early 1900s. "P" refers to rootstocks developed in Poland and "V" to Canada's Vineland series.

Very Small Trees (3 to 6 feet tall)

Some apple rootstocks provide extreme dwarfing, producing trees about 15 to 25 percent the size of standard seedling rootstocks. While not generally useful for commercial orchards, these very small trees have advantages for home orchards where the trees are grown in tubs or on trellises to form edible hedges. The **M27** rootstock has long been popular for very small trees. The **P22** rootstock produces trees about the same size as or a bit larger than M27 and is probably a better choice in colder areas and where soils are on the heavy side. The **Vineland 3** (**V3**) rootstock is especially suitable for cold climates (Zone 3). All these rootstocks require supports for the trees.

Small Trees (6 to 10 feet tall)

Trees that are 25 to 40 percent of standard size are very useful for modern, high-density plantings. The **M9** rootstock was long the industry standard for small trees and still works well for arid and semiarid regions where disease pressure is low. Problems with brittle roots and poor anchorage on M9 have been reported. A more cold-hardy and disease-resistant replacement is **Budagovsky 9 (Bud9)**, which also tolerates heavier soils better than other rootstocks in this size range. **Nic29** is a new M9-type rootstock from Belgium that, reportedly, produces a better root system and is stronger and more vigorous than M9 but still produces adequate dwarfing. Trees on Nic29 require support. Avoid the 'Mark' rootstock, which has proven susceptible to diseases and winter injury.

The **M26** rootstock has been widely popular, but its susceptibility to root rot and fire blight limit its value. Consider using **Geneva 11 (G11)** or **Bud9** instead of M26. For cold climates (Zone 3), consider **Vineland V2** and **V7** rootstocks, which are fire blight–resistant and produce trees toward the tall side for this range. The **Polish P2** rootstock is similar to M9 and reported to be very cold-hardy.

Ranetka crab seedling rootstocks are very cold-hardy and drought-resistant, and they grow well and produce good-sized fruit. These rootstocks reportedly survived a week of –40°F (–40°C) temperatures with no snow cover in Fairbanks, Alaska, but their compatibility with many apple varieties has not yet been reported.

The above rootstocks should support all but the most vigorous apple varieties. While early bearing is desirable, beware of overbearing, especially on very young trees. Plan on an aggressive fruit thinning program.

Medium-Sized Trees (10 to 16 feet tall)

Trees that are 40 to 65 percent of the standard size are valuable for freestanding or vertical axis systems. The **M7** rootstock and its virus-free version, **EMLA7**, have long been popular and remain fair choices, except when used with 'Red Delicious' strains and 'Idared'. They are quite winter-hardy and perform best on deep, well-drained soils. Trees on M7 rootstock can benefit from being staked for the first 4 or 5 years after planting.

Geneva 30 (G30) is more precocious and productive than M7, has good disease resistance, and is probably a better choice for organic growers, especially in eastern North America. Do not use G30 with 'Gala'. Avoid Ottawa 3 (O3), which has proven susceptible to fire blight, wooly apple aphids, and tomato ringspot virus and has been reported to provide poor rooting.

Large Trees (16 to 25 feet tall)

Large orchard trees are seldom recommended today, due to pest and disease problems, high labor requirements, poor labor efficiency, and poor yields per acre. The following rootstocks produce large trees, typically 16 to 25 feet tall in ideal conditions (25 feet being considered a full size or standard tree). Some organic growers continue to use these vigorous rootstocks in areas where cold, adverse climates cause poor tree growth on more dwarfing rootstocks. Some organic growers who maintain grass sod throughout their orchards and up to the tree trunks also prefer these vigorous rootstocks, which compete better against the sod than more dwarfing rootstocks.

The **MM106** rootstock produces trees 65 to 75 percent standard size and is resistant to wooly apple aphid. While the tree is vigorous and has good anchorage, its susceptibility to root rot and damage from fall frosts makes this a poor choice for organic growers.

The **MM111** rootstock produces trees 80 to 85 percent standard size, is well adapted to droughty soils, is moderately productive, and comes into bearing rather slowly. This rootstock is best used for adapting to droughty sites and with an interstem for size control. An M9 interstem on an MM111 rootstock produces a medium-sized, freestanding tree.

The **M4** rootstock provides slight to moderate size reduction, resists collar rot, and can be used for freestanding, spindle-trained trees. This rootstock has been popular in Denmark, Holland, Germany, and British Columbia.

Alnarp 2 (A2) rootstock is used mostly in Sweden, but it occasionally appears in North American nurseries. This rootstock provides good anchorage.

Pear

AS WITH APPLES, PEARS have been cultivated since prehistoric times. Although there are about 20 pear species worldwide, domestic pears derive primarily from three or four species and fall into two types: Asian pears and European pears. Worldwide, pear production statistics are not broken down into Asian or European types. In 2008, China was the overall leading pear producer with 11.5 million metric tons, followed distantly by Italy at 844,000 MT and the United States at 773,000 MT. Spain, Argentina, Germany, Japan, the Republic of Korea, and South Africa round out the top 10 pear-producing countries.

In general, pears have many of the same growing requirements and cultural practices as apples. Pears are generally more susceptible than apples to fire blight and can be challenging to grow in humid areas. Fortunately, blight-resistant varieties are available. Pears tolerate heavy and somewhat poorly drained soils better than most other tree fruits, but they still produce best on deep, well-drained soils.

Disease and pest resistance. With a few exceptions, Asian and European pears are equally susceptible to fire blight. In areas with fire blight, plan to treat your trees with copper and bacterial products during bloom, as we discuss in chapter 10. Bacterial canker (*Pseudomonas syringae*) is a serious problem on Asian pears in areas with cold springs. Collar rot (*Phytophthora* sp.) often occurs when the trees are planted too deeply.

Pear psylla damages tender, new plant tissues by feeding and can kill tissues by injecting toxic saliva that causes "psylla shock." Severe, repeated attacks over several years can weaken or kill trees. The honeydew produced by the psylla damages leaf and fruit tissues and supports the growth of sooty mold, which makes fruit unmarketable. More seriously, pear psylla transmits a mycoplasma disease called "pear decline" that has killed millions of California pear trees since the mid-1900s. The good news is that it can be managed in organic orchards. We'll discuss methods of controlling diseases in chapter 10.

Codling moth tends to be severe on Asian pears. Under some conditions, plant bugs, stinkbugs, two-spotted mites, and European red mites can create problems. All of these pests can be managed effectively using organic practices, as we discuss in chapter 11.

Pears are easier to grow under arid and semiarid conditions in western regions. In more humid locations, you will need to be especially vigilant in managing tree vigor, pests, and diseases.

Asian Pears

Asian pears have been grown for thousands of years in China. They were first introduced to North America as a garden curiosity around 1820, and seeds were later carried there by Chinese immigrants during the gold rushes of the mid-1800s. Asian varieties are also called Oriental pears, Chinese pears, Japanese pears, apple pears, sand pears, salad pears, Shalea, and Nashi. With the exception of the varieties 'Tsu Li' and 'Ya Li', Asian pear trees tend to be small, reaching heights and spreads of about 12 feet.

The fruit is called "apple pear" due to the fact that Asian varieties often resemble apples in shape and texture. They are generally rounder than typical European pears and have crisp, tart flesh, rather than the soft, buttery, melting quality of European varieties. European pears are usually harvested mature but not ripe and placed into storage to ripen. During storage, the fruits become softer and sweeter. Asian varieties, however, are picked ripe because they do not continue to ripen during storage.

Beginning in the 1950s, named Asian pear varieties were introduced to North America and planted in test blocks, but they proved difficult to produce and market, and most were removed. Around the mid-1980s, California fruit growers again introduced Asian pears to the United States and began

Asian Pears

AT A GLANCE

Chilling requirements: 32 to 55°F (0 to 13°C) or less for 450 hours

Region: USDA Zones 5 to 9. Humid areas can make Asian pear production challenging.

Soil pH and type: Deep, well-drained soils with a pH between 6 and 7, although pear trees tolerate heavy, poorly drained soils somewhat better than most tree fruits

Pollination: Partially self-fruitful but set larger and better crops when cross-pollinated

Other notes: Trees come into production slowly, bearing a few fruits in year 3 and 200 to 500 packed boxes per acre in years 5 to 7.

commercial plantings. Thanks to a greater understanding of the cultural practices required and greater care of the fruits, it is now feasible to produce Asian pears in North America. Consumers' recognition of, acceptance of, and demand for Asian pears are increasing in North America, but Asian varieties still represent a very small percentage of commercial and home pears. Although more than 3,000 named varieties exist in China, relatively few Asian pear varieties are commercially available in North America. The results of taste panels in the United States suggest that consumers prefer round, green- to yellow-skinned varieties best. Suggested Asian pear varieties appear in table 5.6.

Temperature concerns. A key concern with Asian pears is their limited cold hardiness: Asian varieties tend to be less cold-hardy than European pears. Most varieties are recommended for USDA Zones 5 to 9. Some nurseries rate 'Hosui', 'New Century', and 'Starking Hardy Giant' to Zone 4. Asian pear varieties typically require about 450 chilling hours at 32° to 55°F (0 to 13°C) or less to break dormancy. In Washington State University trials, the greatest challenges to growing Asian pears were susceptibility to disease and too short a growing season to mature the fruits of some varieties.

Susceptibility to frost can be troublesome on all pears, and particularly on early-blooming Chinese varieties 'Ya Li', 'Tsu Li', and 'Seuri', which bloom 10 to 14 days before 'Bartlett' in California. Later-blooming

Asian varieties bloom around the same time as European pears, which tends to be earlier than apples. Asian pears typically ripen from August through October, depending on variety and location. Be sure to select varieties that will ripen in your area.

Pollination. Generally, Asian pears are partially self-fruitful, but they set larger and better crops when cross-pollinated. The varieties '20th Century' and 'Shinseiki' can be planted alone in large blocks under ideal conditions (like those in California), but cross-pollination is recommended in areas with cool springs.

Of the commercially available varieties in North America, 'Korean Giant' (also known as 'Dan Bae') and 'Tsu Li' pollinate each other. 'Korean Giant' and 'Seuri' will pollinate each other, and all four varieties pollinate 'Yoinashi'. They do not, however, effectively pollinate other Asian varieties. 'Yoinashi' also tends to be a poor pollinizer. According to the University of California, early-blooming 'Ya Li', 'Tsu Li', and 'Seuri' are compatible. Among later-blooming varieties, 'Niitaka' does not produce viable pollen; 'Kikusui' cannot be used to pollinate '20th Century', and 'Seigyoku' and 'Ishiiwase' are poor pollinizers. The European pear 'Bartlett' is often recommended as a cross-pollinizer for many Asian varieties.

Asian pear yields are typically less than for 'Bartlett' or 'Bosc' European varieties because Asian pears must be thinned heavily to achieve market size. The trees are rather slow to come into production, bearing a few fruit in their third year and 200 to 500 packed boxes per acre yields in seasons five through seven. The trees usually mature between years 10 and 14 and produce 800 to 1,000 packed boxes per acre. Higher yields are possible, but the fruit may be too small to market.

European Pears

European pears are well known and well liked in the United States and Canada. Although fewer European pears are grown commercially in North America than apples and peaches, they are still very popular for home and small commercial orchards.

Pears were first introduced to North America around 1630, possibly by French and English colonists coming to Massachusetts. Commercial production, however, has become centered primarily in regions of Washington, Oregon, and California, where fire blight is fairly easily con-

European Pears

AT A GLANCE

Chilling requirements: 100 to 1,000 hours at 32 to 55°F (0 to 13°C)

Region: USDA Zones 6 to 8 for commercial growers, Zone 4 for the most cold-hardy varieties. Can be challenging to grow in humid areas.

Soil type: Deep, well-drained soils but tolerate heavier soils better than most other tree fruits

Pollination: Partially self-fruitful; sets larger and better crops when cross-pollinated

Variety selection: Select varieties resistant to fire blight and rust diseases

Other notes: Not all popular varieties are resistant to fire blight.

trolled. Eastern pear production declined significantly during the early 1900s, primarily due to pear psylla (an insect pest) and fire blight. Eastern pears are grown commercially mostly around the Great Lakes.

Temperature concerns. European varieties were developed from different species from Asian pears and are generally more cold-hardy than their Oriental cousins. The wood and buds of European pear trees are less winter-hardy than those of apples, and pear production is difficult in areas where winter temperatures fall below –20°F (–29°C). Hardiness ratings of varieties in some nursery catalogs range from Zone 4 in the north to Zone 9 in the south. Hardiness ratings of Zone 5 to Zone 8 are usually more accurate. A few low-chilling varieties have been developed that perform well in Zone 9. For commercial orchards, production in Zones 6 to 8 is generally most reliable.

Most European pear varieties require around 900 to 1,000 chilling hours to break dormancy, but breeders have developed varieties for southern growers that require only about 100 chilling hours. European varieties bloom somewhat later than Asian pears and stone fruits, and generally before most apple varieties. Consider providing frost protection if your site is susceptible to late spring frosts.

Ripening dates vary widely among European pear varieties. Be sure to select varieties that will ripen within the growing season at your location. 'Harrow Delight' and 'Moonglow' are cold-hardy, ripen early, and have some fire blight resistance or tolerance. 'Bartlett' and the somewhat blight-resistant varieties 'Anjou' and 'Seckel' can be difficult to ripen in short-season areas.

Pollination. Some varieties ('Bartlett', 'Comice', and 'Hardy') are reasonably self-fruitful under ideal growing conditions. Most other varieties are considered at least partially to completely self-unfruitful. For best results, plant with a pollinizer variety.

Varieties. Although nearly 3,000 European pear varieties have been named, worldwide only a handful are produced commercially and at home. 'Bartlett', known as 'Williams' in Europe and former British colonies, accounts for at least 80 percent of production in the United States and is very popular with overseas commercial growers. Other popular varieties in cool, high-chilling areas of North America include various strains of 'Anjou', 'Bosc', 'Clairgeau', 'Clapp', 'Comice', 'Duchess', 'Flemish Beauty', 'Kieffer', and 'Seckel'. Somewhat farther south, popular medium-chilling varieties include 'Ayers', 'Baldwin', 'Bartlett', 'Carnes', 'Garber', 'Kieffer', 'Orient', and 'Tennessee'. Low-chilling varieties 'Flordahome', 'Hood', and 'Pineapple' are popular in Florida and other areas that experience around 100 chilling hours or more. Not all popular pear varieties are blight resistant, however; a key consideration for organic orchardists.

'Kieffer' and 'Orient' are hybrids between European and Asian pear varieties. While their fruit is gritty and best used for canning and baking, the trees are highly blight resistant.

In Texas A&M evaluations of European pear varieties for southern growers, 'Warren' provided the best overall dessert-quality fruit and had the best blight resistance, followed by 'Ayers' and 'Magness'. Other blight-resistant European varieties recommended for southern growers include 'Carrick', 'Monterrey', 'Moonglow', and 'Morgan'. 'Honeysweet' was developed by Purdue University and is also resistant to fire blight. Unfortunately, some of these varieties are very hard to find in nurseries. Tables 5.7 and 5.8 list suggested European pear varieties.

Pear Rootstocks

Pear decline is a very serious disease that has killed many pear trees in North America. The disease is caused by a mycoplasma (bacteria-like) pathogen that is carried by the pear psylla insect pest. Some combinations of pear varieties and rootstocks have proven highly susceptible to the disease and other combinations quite resistant. Because pear decline involves rootstock/scion interactions, selecting the proper rootstock for your region is very important. Fire blight resistance and size control are other factors to consider when selecting rootstock.

For organic pear growers in North America, **'Old Home' × 'Farmingdale'** (OHxF) rootstocks are often the best choice. They are resistant to pear decline and fire blight and are cold-hardy, although they are moderately susceptible to pear root aphid and quite susceptible to nematode damage. OHxF clones 40, 69, 217, and 282 work well for most pear varieties. OHxF clones 277 and 97 are more vigorous and are recommended for Asian pears and for red European pear varieties.

California growers prefer **'Winter Nelis'** seedling rootstocks that have been developed using 'Bartlett' pollen for pear decline resistance. Unfortunately, these rootstocks are highly susceptible to fire blight and pear root aphid and are quite susceptible to nematode damage.

'Angers Quince' and **'Provence Quince'** provide dwarfing and resistance to pear decline, pear root aphid, and nematodes, but they are quite susceptible to fire blight and are highly susceptible to cold injury.

For very cold climates, seedling *Pyrus ussuriensis* rootstocks probably offer the greatest chance of success, but they are highly susceptible to pear decline, as are *Pyrus serotina* rootstocks. *Pyrus calleryana* rootstocks are very susceptible to cold injury. 'Old Home' rootstocks are susceptible to root aphids.

Quince

QUINCE IS A PEAR-LIKE FRUIT native to Armenia, Azerbaijan, Georgia, Iran, and Pakistan. It is the only species in the genus *Cydonia* (*C. oblonga*). The ornamental Japanese quince, Chinese quince, and common flowering quince (*Chenomales* sp.) produce very small, hard, nearly inedible fruits and are seldom used for fruit production.

Quince was well known in ancient times and may have been cultivated before the apple; references to its cultivation date to perhaps 2000 BCE or earlier. From its origins in the Middle East, quince spread throughout the known world and was important in both Greek and Roman civilizations. Quince was, and remains, culturally important in many parts of the world.

Although quince has long been grown in North America and remains a popular garden fruit, commercial demand and cultivation have been very limited here. What little cultivation there was declined sharply following the introduction of fire blight to the United States in the 1830s. Argentina is a major quince producer and exporter, and Chile and Uruguay also have notable amounts of quince production. The fruits are popular in Mexico and other Latin American countries where they are known as "membrillo."

Only a few varieties of quince are sweet enough to eat fresh; most are too hard and astringent to eat right off the tree and must be cooked before eating. Some varieties are allowed to soften by frost and mild decomposition (bletting) before harvest. Quince fruits tend to have strong aromas and flavors and are often used in small quantities to flavor apple and pear preserves and pastries. Immature quince fruits are covered with dense, grey to brown fuzz (pubescence). As the fruits ripen, they turn yellow, and the fuzz often rubs off. The trees generally begin bearing 3 or 4 years after planting and reach maturity after 10 years. At full production, you might expect around 45 to 50 pounds of fruit from a healthy tree.

Quince trees are naturally small, growing 8 to 15 feet tall with a 15-foot spread. Rather than growing on short, long-lived spurs like apples and pears, quince fruits develop at the tips of current-season branches. Not only does this habit greatly limit yields, but it also produces rather gnarly-shaped trees. The trees can be trained similarly to peaches, or they can be trained to be bush-like trees. Some of the newer apple training systems described in chapter 12

Quinces

AT A GLANCE

Region: USDA Zones 7 to 9 for commercial growers
Soil type: Deep, moist, well-drained soils; tolerate wet soils and drought better than apples
Pollination: Entirely self-fruitful, partially self-fruitful, or not self-fruitful, depending on variety
Other notes: The greatest challenges to growing quince involve pests and diseases. Thoroughly evaluate your market before moving into quince.

might be adaptable to quince production. Relatively little pruning is required to maintain the trees. Trees are typically spaced 15 feet apart and rows 20 to 25 feet apart.

Climate and habitat concerns. Quince trees are not noted for their cold hardiness; most varieties are rated to Zones 5 to 6 in colder areas and Zone 9 in the south. For commercial orchards, production in Zones 7 to 9 will be most reliable. In general, if you can grow peaches successfully in your area, you should also be able to grow quinces reasonably well. Some Russian and Asian varieties are reported to be cold-hardy to Zone 4.

It is important to select the right site for growing quinces. While they tolerate wet soils and drought somewhat better than apples, quinces grow best on deep, moist, well-drained soils. As with all other tree fruits, full sun is highly recommended. Protected locations are also best because the trees do not tolerate rapid changes in temperature and exposure. Quinces bloom rather late, providing them some frost protection.

Pollination. Some quince varieties are reported to be entirely self-fruitful, others partially self-fruitful, and others not self-fruitful. Recommendations also vary from one authority to another. If you are unsure as to whether the variety you select is self-fruitful, plant another variety within 50 feet to ensure fruit set.

Pests and diseases. The greatest challenges in quince culture involve pests and diseases. Cornell University reports that quinces were once widely grown in New York, but flower bud injury due to winter temperatures, fire blight, borers, codling moth, plum curculio, scale, and tent caterpillars led to cutbacks in production.

Quinces can be grown organically, but they are more challenging than some other tree fruits. It is very important to manage tree vigor by maintaining low to moderate growth. Too much nitrogen or pruning can stimulate lush, fire blight–susceptible foliage. Because quinces bloom rather late in spring, weather at that time tends to be warm and humid, favoring fire blight. Plan to use an aggressive pest and disease control program that may include applications of antibiotics and beneficial bacteria during bloom.

Consumer demand. Although quinces can be rather expensive in food markets, remember that consumer demand for these fruits is also low in most parts of Canada and the United States. Washington State University reports some commercial producers are having success marketing to niche ethnic markets that favor Mediterranean cuisine. Markets in and near Mexico might also provide opportunities, as quince (membrillo) is popular there.

Be sure you have thoroughly evaluated your market before moving heavily into commercial quince culture. Customer education will be an important part of your marketing, as relatively few people are familiar with the fruits and how to use them. Table 5.9 lists suggested quince varieties.

Medlar

MEDLARS ARE QUINCE- OR PEAR-LIKE POME FRUITS that are actually more closely related to hawthorn (*Crataegus*) and serviceberry (*Amelanchier*) than to pears and quince.

Common medlar, *Mespilus germanica*, is native to Europe and Asia Minor, although authorities differ on where it actually originated. This fruit has been cultivated for thousands of years and was well known to the ancient Greeks and Romans.

Stern's medlar, *Mespilus canescens*, was first described as a species in 1990 and consists of about 20 trees and large shrubs growing in Konecny Grove, Arkansas. The genetics of Stern's medlar was debated for years. In 2009, the Arkansas Natural Heritage Commission reported on research that suggests Stern's medlar is actually a hybrid between common medlar and native blueberry hawthorn, *Crataegus brachy-*

Medlars

AT A GLANCE

Region: USDA Zones 5 to 9, with some production in Zone 4
Soil: Well-drained soil
Pollination: Most varieties are self-fruitful.
Variety selection: 'Breda Giant' is noted for being particularly flavorful.

acantha. At this point, Stern's medlar remains a curiosity rather than a source of fruit.

Japanese medlar or **loquat**, *Eriobotrya japonica*, is also a pome fruit related to apples, pears, quince, and common medlar. We will discuss loquat later in this chapter.

Common medlar forms small trees or large multistemmed shrubs that can grow to about 25 feet tall. Cultivated varieties are more likely to be 8 to 20 feet tall and produce both showy blossoms in spring and attractive red foliage in fall. The trees are normally grown freestanding with single trunks and are spaced 7 to 20 feet apart within rows that are 15 to 20 feet apart.

The fruit strongly resembles rose hips in shape and color, grows from 1 to about 3 inches in diameter, and contains large, hard seeds. The blossom ends flare open, creating a distinct shape, and the skin is usually russetted.

Medlars are relatively hardy, surviving well in USDA Zones 5 to 9, with some production in Zone 4. Choose a sunny location on well-drained soil. The fruit is not ready for harvest until late in the year, typically September through November.

The fruit, like many quince varieties, is too hard and astringent to eat fresh, at least in temperate climates. In Mediterranean areas, the fruit is said to be edible directly from the tree. In temperate zones, medlar fruits are typically not harvested until after one or more hard frosts and are then allowed to "blett" indoors in a cool, dry location for 2 to 4 weeks. Bletting (see page 487) reduces the fruit's acidity and produces a mushy, apple sauce–like texture. While some people find the appearance of the bletted medlar flesh unappetizing, the flavor is generally likened to spicy, old-fashioned apple sauce. The fruit can be eaten fresh, sometimes with cheese, or processed into jam, syrup, or wine. Medlars can also be roasted or baked into tarts. One recipe calls for combining the fruit pulp with eggs and butter to make a dish similar to lemon curd. Each fruit produces relatively little edible pulp.

Medlars are relatively pest free, although fire blight will be a problem and little information is available on blight resistance. Few varieties are available in North America, the most common being 'Breda Giant', 'Dutch' (a.k.a. 'Giant' or 'Monstrous'), 'Hollandia', 'Large Russian', 'Nottingham', 'Royal', and 'Russian'. 'Breda Giant' is noted for being particularly flavorful. Most medlar varieties are self-fruitful, usually allowing you to produce crops with a single variety.

Medlars are normally grafted onto quince or hawthorn rootstocks. Quince A and Quince BA29 rootstocks provide some dwarfing and provide good tree stability, with freestanding trees reaching 13 to 20 feet in height and spread. Susceptibility to fire blight is a concern with quince rootstocks, as is cold hardiness. For Zones 4 and 5, hawthorn rootstocks are recommended.

If you grow medlars commercially, you will face several challenges, primarily because the fruit has not been widely popular since Victorian times. To market the fruits, you will need to educate customers in what medlars are and how to prepare and use the sauce-like fruits. It may be that selling medlar products will be easier than selling the fruit itself.

Mayhaw

MAYHAWS (*Crataegus aestivalis* and *Crataegus opaca*) are hawthorns, of which hundreds of species have been described in North America. While other North American species of hawthorn produce fruits and are often used as landscape ornamentals, the abundance of high-quality, large-fruited pome crops makes hawthorn fruit culture, outside of mayhaws, generally unrewarding.

Mayhaws are native to the southern United States and produce small trees or tree-like shrubs that are 20 to 30 feet tall. The rounded trees produce showy displays of white flowers and yellow, orange, or red fruits. They are suitable for use in edible landscapes, although some varieties are thorny. Mayhaws are grown as freestanding trees.

Mayhaw fruits generally resemble crab apples with prominent blossom (calyx) ends and sometimes vertical ridges down the sides. The fruits range from about ¼ to 1 inch in diameter and usually have red or yellowish-red skins.

Mayhaws have long been popular fruits collected from the wild in the southern United States and mayhaw jelly is Louisiana's official jelly. Since the 1980s, interest in domesticating mayhaws has grown, and the fruits are now produced commercially in orchards. While the small, tart fruits are not desirable for fresh use, popular products include flavorings, butters, jams, jellies, marmalades, sauces, condiments, pie filling, coffee cakes, ice cream, syrups, wine, and desserts. The Louisiana Mayhaw Association provides many delightful recipes on their website (see Resources), as well as information on mayhaw production.

According to the Association:

"Historically, mayhaws have been harvested in backwoods sloughs, swamps, and river bottoms. Boats are sometimes utilized in the harvest of wild mayhaws. Limbs are shaken over the boat and nets are used to scoop them out of the water. Accessibility has dwindled over the years as developers have cleared the woodlands. Commercial and home orchards are now being created with grafted mayhaws."

Climate and habitat concerns. Mayhaw trees have very short chilling requirements and typically bloom sometime between late January and early March. The fruits usually ripen in late April-May, which explains the name mayhaw. Later-blooming and later-ripening varieties are being selected for colder areas. The early bloom and fruit set make this crop susceptible to frost damage. While grafted trees bear fruits within 1 or 2 years of planting, commercial harvests typically start 4 to 5 years after planting.

Unlike many other tree fruits, mayhaws are well adapted to low-lying and rather wet soils. In their native range, the trees are found growing in the acidic soils alongside streams, ponds, and swamps from Texas and Arkansas east to North Carolina and south to Florida. The trees are susceptible to frost during bloom, but they are relatively hardy after bloom. Some varieties are surprisingly hardy in the winter. Trees that were exposed to −13°F (−25°C) have been observed blooming, and 2-year-old trees are reported to have survived −25°F (−32°C).

Mayhaws suitable for fruit production appear best adapted to USDA Zones 8 and 9, and it is possible to produce select varieties in Zones 6 and 7. For some varieties, the lack of chilling can be a problem in Zone 9. While some nurseries market the trees as adaptable to Zone 3, frost and freeze damage are likely to be severe. In areas colder than Zone 8, test a few mayhaw trees for at least 5 years before investing heavily in planting stock and establishing an orchard.

Mayhaw trees tolerate full sun or partial shade, but they produce best on well-drained upland soils in full sun. Soils that are too moist and fertile produce vigorous trees, but fruit develops late, and there can be smaller yields. Such sites are also likely to have increased pest and disease problems. Young trunks and scaffold limbs are susceptible to sunscald in locations with intense, direct sunlight. In nonorganic orchards, interior white latex paint thinned with water is used to paint the trunks and reflect the heat and light. Although effective in reducing sunscald, this practice is not allowed in certified organic orchards. Instead, organic growers may wish to screen the lower trunks and scaffold limbs from direct sun during the first 3 to 4 years from planting. PVC pipes split lengthwise, thin boards tied to the trunks, or commercially available tree wraps serve this purpose.

Propagation. Mayhaws are easily propagated. Unlike most other fruits, trees grown from seed often produce desirable fruit that resemble those of the mother tree.

Softwood stem cuttings, hardwood stem cuttings, and root cuttings from mayhaw can be rooted using root-promoting hormones such as indole-butyric acid (IBA), although IBA and similar rooting compounds are not considered organic. Around one-third of softwood cuttings root. Although you can produce rooting compounds organically using willow water, such compounds are not likely to be potent enough to propagate the rather hard-to-root mayhaws. According to the USDA National

Mayhaws

AT A GLANCE

Chilling requirements: Very short chilling requirements; typically blooms during late January to early March and fruits ripen during late April to May

Region: USDA Zones 8 and 9. Young trunks and scaffold limbs are susceptible to sunscald.

Soil: Best grown on well-drained upland soils but grows wild on low-lying, rather wet soils

Pollination: Varieties range from self-fruitful to self-unfruitful. Cross-pollination is recommended.

Variety selection: Because it's a new crop, many named varieties consist of vegetatively propagated clones of trees originally collected from the wild.

Other notes: Trees are easily propagated, and trees grown from seed often produce desirable fruit. Grafted trees bear fruits within 1 or 2 years of planting, but commercial harvests typically start 4 to 5 years from planting.

Organic Program, "Nonorganically produced planting stock to be used to produce a perennial crop may be sold, labeled, or represented as organically produced only after the planting stock has been maintained under a system of organic management for a period of no less than 1 year." If you want to grow mayhaws on their own root systems without rootstocks, you can root your own cuttings with IBA or buy cuttings rooted with IBA. Just be sure to document that you could not find organically produced rooted cuttings, and keep the cuttings in an organic nursery bed for at least 1 year before planting in the orchard.

Rather than growing mayhaws on their own roots, perhaps the best choice is to graft a desirable variety to a rootstock. These trees are easily grafted and are generally compatible with any hawthorn species. While many different combinations have been tried, there is no data on long-term performance on non-mayhaw rootstocks. At this time, mayhaw seedling rootstocks are probably the best choice.

Planting, training, and harvesting. Because mayhaws are grown as free-standing trees and can develop a spread of 30 feet or more, allow plenty of space for the trees. Although early recommendations were to plant the trees 15 to 18 feet apart, experienced growers now more typically recommend planting the trees 25 feet apart in rows that are 25 to 30 feet apart. Rather than having the trees directly opposite each other, you may want to stagger the trees in adjacent rows to maximize the distance between trees and enhance light penetration and air movement. The distance between tree rows will depend on the type of harvesting equipment you use. For mechanically harvested trees, it's best to space the trees 25 feet apart in rows that are also 25 feet apart.

The development of tree management systems is in its infancy, but the central leader and modified central leader training systems that we discussed for apples in chapter 3 work well. Prune annually to keep the tree canopies open.

Because the fruits are many and small, hand harvesting is not economical on a large commercial scale. For commercial orchards, train your trees with scaffold limbs positioned to allow mechanical harvesting using a shaker and ensure the trunk and limbs are strong enough to withstand the shaking. Train to a single trunk, and develop the first permanent scaffold limbs about 48 inches above the ground. Train the main scaffold branches upward to at least a 45-degree angle from the central leader.

Pests and diseases. Pests are common on mayhaw, plum curculio being the most serious. In most areas of the southeastern, eastern, and south-central United States, you need aggressive control measures to manage this pest. Wooly apple and other aphids, flat-headed apple borers, tent caterpillar, whiteflies, and various foliage-feeding insects are occasional problems. Deer and rabbits can also cause severe damage in some locations.

Rust diseases caused by fungi attack mayhaws. Cedar-apple rust, cedar-hawthorn rust, and cedar-quince rust are closely related and produce similar symptoms on mayhaw, apple, pear, quince, and other pome fruits. Fire blight can be a problem (although usually not severe) on mayhaw, particularly when susceptible pear and apple trees are growing nearby. Brown blight also attacks mayhaw. We will discuss disease control in chapter 10.

Varieties. Because mayhaws are a new crop, many named varieties consist of vegetatively propagated clones of trees originally collected from the wild. Not all varieties grow and produce equally well, so choose varieties suitable to your location. If you live in the more southerly and warmer areas of the mayhaw range, choose varieties with very low chilling requirements. For those living in more northerly and cooler areas, varieties that bloom late and are cold-hardy will reduce spring frost and winter freeze damage.

While 70 or more varieties have been named, most are being discontinued as superior selections consistently outperform them. 'Super Spur' and 'Texas Star' have been the standard industry varieties in Louisiana. 'Red Majesty' and 'Maxine' are promising new varieties now being tested.

Georgia fruit specialists recommend that growers in central and north Georgia "plant only the highest-chilling, latest-blooming mayhaw varieties available, such as 'Saline', 'Crimson', and 'Texas Star', to reduce the chance of spring frost damage. According to Texas A&M University, 'Super Spur' and 'Super Berry' are best grown in east central Texas and southeastern Texas. 'Super Spur' has chilling problems during mild winters in the Beaumont area. 'Big Red', 'Winnie Yellow', 'Highway Red', 'Highway Yellow', 'T.O. Warren Superberry', 'Angelina', 'Harrison', 'Big Mama', and the '#1 Big' varieties usually bloom later and are better adapted to northeastern Texas.

Table 5.10 lists some popular mayhaw varieties. At this time, few nurseries carry mayhaw, and some varieties can be very difficult to locate. The names are also commonly confused, due to several variations on the name "Superberry." Expect dramatic changes in cultivar recommendations with improvements in fruit size, fruit quality, and tree performance over the next couple of decades. Disease resistance is one of the key criteria in selecting new varieties.

Loquat or Japanese Medlar

LOQUAT OR JAPANESE MEDLAR, *Eriobotrya japonica,* is native to China, and has long been cultivated in Japan. Japan is the world's leading producer, followed by Israel and Brazil. California is the leading North American producer; the crop is also grown in Florida, the southern portions of the Gulf states, and Hawaii.

Loquats are evergreen and grow as large shrubs or small trees that can reach 30 feet tall. In cultivation, we normally keep the trees around 10 feet tall. Loquat seedlings are the most common rootstocks, although quince is sometimes used to reduce tree size. Because the trees are spreading, rather like a magnolia, allow plenty of space in your planting:

plant trees 20 to 25 feet apart in rows that are 25 to 35 feet apart.

Loquat fruits grow about 1 to 2 inches long, are rounded to pear-shaped, and are usually yellowish-orange, often with a red blush. The ripe flesh is very soft and juicy. Depending on variety, it can be white, yellow, or orange; sweet or tart; and contain from one to many seeds. Most cultivated varieties have one to five seeds. Loquats can be eaten fresh, but they bruise easily and can be hard to find in the marketplace. You can also process the fruits into jellies, pastries, sauces, wines, and other culinary items.

Harvesting. Ripen loquats on the trees because these fruits do not continue to ripen after they are harvested. Unfortunately, the ripe fruits are very soft and fragile. Rather than picking them by hand, as you would an apple or pear, clip the fruits individually from the clusters. More information on harvesting, handling, and storing is provided in chapter 13.

Commercially, loquats are very labor intensive. Given a suitable climate, however, local niche production can be profitable. The fact that loquats ripen in what is an off-season for many fruits provides a market advantage. In most areas of North America, customer education will be a big part of your marketing effort.

Climate and habitat concerns. Loquats extend pome fruit production into warm areas where apple and pear production can be challenging due to chilling requirements. They are strictly grown in subtropical to mild temperate climates, USDA Zones 8b to 12, where temperatures remain above about 12°F (−11°C). Flower buds can be damaged or killed at 18 to 28°F (−8 to −2°C) and fruit can be damaged at 25°F (−4°C). Outside of Florida, Hawaii, parts of California, and coastal Georgia, loquat trees may survive but produce few fruits. Even in Florida, trees grown north of Jacksonville may survive but fail to bear fruit.

Sites with full sun are best, but the trees tolerate partial shade. Loquats tolerate sandy to clay soils at various pH values, but the University of Georgia recommends growing this crop on well-drained loamy soils with a pH of 6–7. Soil drainage is important; loquats do not tolerate standing water.

Loquats or Japanese Medlars

AT A GLANCE

Region: USDA Zones 8b to 12, where temperatures remain above about 12°F (−11°C). Flower buds can be damaged or killed at 18 to 28°F (−8 to −2°C), and fruit can be damaged at 25°F (−4°C).
Summer heat: Suffer in extreme heat, and leaf scorch is common in areas with hot, dry winds
Soil pH and type: Grow best on well-drained loamy soils with a pH of 6–7 but will tolerate sandy to clay soils and various pH values. Loquat do not tolerate standing water.
Pollination: Self-fruitful to self-unfruitful, depending on variety; cross-pollination improves production and helps ensure large fruits
Other notes: It can be difficult to find plants in North America.

Although adapted to subtropical climates, loquats suffer in extreme heat, and leaf scorch is common in areas with hot, dry winds. High heat and intense sunlight during the winter can sunburn the fruit. White-fleshed varieties are best grown in cool, coastal areas.

Pests and diseases. Various species of fruit flies trouble loquat and can be serious in some areas. Codling moth is a serious pest in California, and scale insects can create trouble wherever loquats are grown in North America. Fire blight is a serious problem and limits production in humid regions. Phytophthora crown rot and Pseudomonas cankers are occasional problems.

Pollination and varieties. The number of seeds in a fruit largely determine the fruit's size, and the quantity of seeds often relates to how well a flower is pollinated. Some loquat varieties are self-fruitful, but cross-pollination improves production and helps ensure large fruits. Plant two varieties within about 50 feet of each other to provide cross-pollination, and try to provide good bee pollination during bloom.

While more than 800 varieties have been reported in Asia, very few are available in North America. Finding loquats in North American nurseries can be difficult. Table 5.11 lists suggested loquat varieties.

Saskatoon

SASKATOON FRUITS RESEMBLE BLUEBERRIES IN SIZE, shape, and color, but they are actually pomes, not berries. The flavor is rather like blueberries, but it has been likened to apples. The plants grow as 6- to 30-foot-tall shrubs, depending on the species and growing conditions.

We recognize about 20 species of saskatoons (a.k.a. serviceberries or Juneberries) in the genus *Amelanchier*. Most cultivated varieties come from *A. alnifolia*, which grows wild from the Desert Southwest to Alaska and east to Manitoba and the Dakotas. Saskatoons were important to many Native American and First Peoples groups in North America. They are grown commercially, mostly in Canada's Prairie Provinces, and are becoming popular ornamental plants with showy white blossoms in early spring.

Climate and habitat concerns. Saskatoons are remarkably adaptive and hardy plants. Some varieties tolerate winter temperatures around −60°F (−51°C). Orchardists in Fairbanks, Alaska (65° North latitude, Zone 2a), grow saskatoons, with the varieties 'Martin' and 'Smokey' being popular. Saskatoons are among the earliest fruit crops to bloom, and frost damage can be a problem, so locate saskatoons on your least frost-prone site.

The plants tolerate a wide range of soil types and pH but generally perform best on loams and sandy loams. I have grown them on silt loams, but good drainage is necessary. If your soil is heavy or otherwise poorly drained, grow saskatoons on raised beds 10 to 12 inches high and 3 feet wide. Choose a full-sun location.

Planting. Saskatoon bushes resemble blueberries in appearance and how they are managed. The canes are generally upright and form vase-shaped bushes. In cultivation, most varieties are kept about 6 to 10 feet tall with rows that are 10 to 12 feet apart. If you plan to harvest mechanically, you may want to space the rows a bit wider, depending on the harvester you are using.

Saskatoons are sometimes spaced as closely as 1 foot apart, so that orchards can be established quickly and fruits can be harvested rapidly using over-the-row blueberry harvesters. At this density, crowding soon develops and

Saskatoons

AT A GLANCE

Region: Very hardy and adaptive; some varieties survive and are productive in Fairbanks, Zone 2 (−40 to −50°F [−40 to −46°C]).

Soil pH and type: Performs best on loam and sandy soils but tolerates a wide range of soils. Needs good drainage.

Pollination: Generally self-fertile. Cross-pollination can help improve fruit set and quality in some locations.

Variety selection: Only about 26 varieties have been named.

pest and disease problems increase. For an organic orchard, particularly one that will be handpicked, space the plants 4 to 5 feet apart in rows.

Pests and diseases. Despite their hardiness and adaptability to a wide range of climates, saskatoons are susceptible to a number of diseases and pests. Compounding the problem is that wild saskatoons, which are abundant across much of North America, serve as reservoirs of these diseases and pests, which affect not only cultivated saskatoons but other cultivated fruit crops as well. Eliminate wild saskatoons from in and around your orchard. As with other pome fruits, removing junipers and eastern red cedar from your property can help reduce rust disease problems.

Saskatoon diseases include saskatoon-juniper rust, Entomosporium leaf and berry spot, and powdery mildew. Fire blight, brown rot, and Cytospora canker can become serious in some locations and during some years. Typical orchard pests are wooly elm aphid, saskatoon bud moth, apple curculio, sawfly, mites, leaf rollers, pear slug, and shoot borer.

Varieties. Although saskatoons have been harvested from the wild for centuries and probably millennia, they have only been cultivated as an orchard crop since the early- to mid-1900s. Only about 26 varieties have been named, and they developed as chance seedlings in the wild or on farms. On the positive side, that leaves saskatoon breeders with much room for improvement. In addition to culinary saskatoons, several varieties have been selected for ornamental use.

Pollination. With the exception of the ornamental variety 'Altaglow', saskatoons are generally self-fertile and can be planted in solid blocks. Cross-pollination can help improve fruit set and quality in some locations.

Disease-Resistant Apples

Table 5.2

	Blooms	*Ripens*	Apple Scab	Cedar-Apple Rust	Fire Blight	Powdery Mildew
Bramley's Seedling Zones 5–8	*n/a*	*very late*	R	S	R	R
Britegold Zones 5–8	*n/a*	*late*	VR	S	R	R
Dayton Zones 3–8	*early to mid-season*	*early to mid-season*	VR	MS–R	MR–R	MR–R
Enterprise Zones 4–7	*mid- to late season*	*late to very late season*	VR	R–VR	R–VR	MR–R
Fireside/Connell Zones 3–8	*mid- to late season*	*late to very late season*	I	I	S	n/a
Freedom Zones 4–7	*early to mid-season*	*mid- to late season*	VR	S–R	MR–R	MR–R
Goldrush Zones 5–8	*late season*	*late season*	VR	VS–S	MR–VR	S–R
Honeycrisp Zones 3–6	*mid-season*	*mid- to late season*	R	n/a	R	S

VS = very susceptible, S = susceptible, MS = moderately susceptible, MR = moderately resistant
R = resistant, VR = very resistant, I = immune, n/a = data not available

Notes

The large, green, red-streaked fruits have firm, coarse flesh. Keeps well and is very popular for baking and cider. Ripens very late. Resistant to black rot/bitter rot and moderately resistant to white rot.

The medium-sized, green, red-blushed fruits are sweet with little acidity. Considered by some to be a replacement for 'Golden Delicious'.

The large, glossy red fruits are firm and have a mild flavor and good dessert quality for a summer apple. Hangs well without overripening. Consistent producer. Somewhat susceptible to bitter pit and black rot/bitter rot. Can be hard to find in nurseries. Perhaps best suited for home and market orchards.

The large, bright red fruits are suitable for fresh use and cooking. Stores well. Among the most disease-resistant apples. Resistant to black rot/bitter rot. Suitable for commercial orchards. Pollinate with other mid- to late-blooming varieties.

'Fireside' has red-striped skin. 'Connell Red' is a red-color sport of 'Fireside'; otherwise, the two varieties are the same. The large fruits are aromatic with a distinctive flavor. Suitable for fresh use and processing. Very popular with Minnesota commercial apple growers. Will not pollinate one another.

The large, red fruits are good for cooking. Among the more disease-resistant apples but susceptible to black rot/bitter rot and white rot (soft rots). Good pollinizer for 'Liberty'.

A large, yellow 'Golden Delicious' hybrid. A winter apple with excellent storage properties. Suitable for fresh use and cooking. Susceptible to biennial bearing. Requires careful and diligent thinning, beginning in early bearing years. Pollinate with 'Enterprise', 'Gala', and 'Golden Delicious'.

The medium-sized to large, green, orange-red mottled fruits can develop solid red in full sun. Exceptionally crisp and juicy. A Minnesota introduction and ranked as one of the highest-quality apples. Excellent storage characteristics.

Disease-Resistant Apples, *continued*

	Blooms	*Ripens*	Apple Scab	Cedar-Apple Rust	Fire Blight	Powdery Mildew
Hudson's Golden Gem Zones 5–8	*n/a*	*late to very late season*	R	S	R	R
Irish Peach (Early Crofton) Zones 5–8	*n/a*	*mid-season*	R	n/a	MR–R	R
Jonafree Zones 5–8	*mid-season*	*mid-season*	VR	VS–MR	S–MR	MR–R
Liberty Zones 4–9	*mid-season*	*mid- to late season*	VR	R–VR	MR–R	S–R
Nova Easygro Zones 3–6	*early to mid-season*	*mid- to late season*	VR	VR	MR–R	S–R
Novamac Zones 4–7	*mid- to late season*	*mid-season*	VR	VR	R	MS–MR
Orleans Reinette Zones 5–8	*n/a*	*late to very late season*	MR	R	MR	MR

VS = very susceptible, S = susceptible, MS = moderately susceptible, MR = moderately resistant
R = resistant, VR = very resistant, I = immune, n/a = data not available

Notes

The medium-sized to large fruits are conical and somewhat ribbed with a golden, heavily russetted skin. The flesh is firm, coarse, sweet, juicy, and reminiscent of pears in aroma and taste. For fresh use. Resistant to black rot/bitter rot and moderately resistant to white rot. The russetted skin should reduce problems with sooty blotch and flyspeck, but the rough appearance may make fresh marketing challenging.

The small, green fruits have red stripes with aromatic, rather tart flesh. Good for baking and dessert use. An old Irish variety more popular in the UK than North America. The trees are slender and upright. Fruits tend to be borne at the branch tips. Moderately susceptible to white rot.

The medium-sized, red-blushed fruits are suitable for fresh use, cooking, and cider. Resembles 'Jonathan', but not susceptible to Jonathan spot or bitter pit. Moderately resistant to white rot and resistant to black rot/bitter rot. Pollinate with 'Goldrush' or 'Enterprise'.

The medium-sized, red-striped fruits resemble 'McIntosh' but have better flavor. Suitable for fresh use, cooking, canning, and desserts. Does not store well but is suitable for commercial direct-sales. One of the best disease-resistant varieties. Reported to be moderately to very susceptible to soft rots and may be challenging to grow in the southern United States. Pollinate with other mid- and late-blooming varieties, such as 'Dayton', 'Prima', and 'William's Pride'.

The large, greenish-yellow, red-striped fruits are suitable for fresh use and cooking. Quality improves during storage. Among the most disease-resistant apples. Moderately resistant to black rot/bitter rot.

A medium-sized, medium red fruit similar to its 'McIntosh' parent but stores well. Susceptible to preharvest drop. Highly susceptible to quince rust and moderately resistant to black rot/bitter rot and white rot.

The small, greenish-red, lightly russetted fruits have firm, fine-textured, very juicy flesh and a flavor reminiscent of sweet oranges. The trees are moderately vigorous and may not produce crops reliably. Moderately susceptible to white rot. Best grown in warm regions.

Disease-Resistant Apples, *continued*

	Blooms	*Ripens*	Apple Scab	Cedar-Apple Rust	Fire Blight	Powdery Mildew
Prima Zones 5–7	*mid-season*	*early to mid-season*	VR	VS–S	MR	R
Priscilla Zones 3–7	*early to mid-season*	*early to mid-season*	VR	R–VR*	MR–R	MR–R*
Pristine Zones 5–8	*early season*	*early to mid-season*	VR	S–R	MR	S–VR
Redfree Zones 5–8	*mid-season*	*early season*	VR	VR	S–R	S–MR
Trent Zones 4–7	*mid- to late season*	*late season*	VR	R	MR	R
William's Pride Zones 4–9	*early season*	*mid-season*	VR	S–VR	R	MS–R

VS = very susceptible, S = susceptible, MS = moderately susceptible, MR = moderately resistant
R = resistant, VR = very resistant, I = immune, n/a = data not available

Notes

The medium-sized to large, yellow, dark red–blushed fruits are mild-flavored and do not store well. Susceptible to bitter pit, winter injury, quince rust, and white rot. May be susceptible to powdery mildew in some regions.

The medium-sized, bright, red-blushed fruits are similar to 'Jonathan'. The flesh is sweet, aromatic, juicy, and crisp. Stores extremely well but susceptible to cracking when overripe. Low yields. Highly susceptible to quince rust and susceptible to white rot. *Has been reported to be susceptible to cedar-apple rust and powdery mildew in some regions. Resistant to black rot/bitter rot.

The medium-sized to large yellow, orange-blushed fruits have good quality, shelf life, and storability for a summer apple. Suitable for fresh use or cooking. Ripens early, with 'Lodi', but is less susceptible to bruising. Requires two to three pickings. Early bearing. Pollinate with 'Pristine', 'William's Pride', 'Redfree', 'Jonafree', or 'Liberty'.

The medium-sized, bright red fruits have good flavor and are suitable for fresh use and cooking. Fair storage. Ripens in early mid-season. Susceptible to small fruits and biennial bearing and requires aggressive thinning. Some limbs produce bare wood. Susceptible to black rot/bitter rot and white rot. Pollinate with other mid- and late-blooming varieties.

The dark red fruits store well. Susceptible to bitter pit and highly susceptible to quince rust.

The medium-sized to large, dark red, slightly striped fruits are suitable for fresh use and cooking. Stores well. Susceptible to water core and requires two or three pickings. Resistant to black rot/bitter rot and moderately susceptible to white rot. Not recommended on MM111 rootstock due to bitter pit. Pollinate with other mid- and late-blooming varieties.

Cold-Hardy Apples for Northern Regions

Table 5.3

Battleford Zones 2–6	The medium-sized to large fruits have yellow, red-blushed skins. Very tart and best used for cooking. Developed in Manitoba. Ripens in mid-season to late season (mid-September in Manitoba).
Beacon Zones 3–6	The medium-sized to large, dark red fruits tend to be soft but are suitable for fresh use and cooking. Ripens in early to mid-season.
Boughens' Delight Zones 2–6	The medium-sized, yellow, red-streaked fruits have a mild flavor and are considered excellent for fresh use and fair for cooking. Does not store well. Susceptible to fire blight.
Carlos Queen Zones 2b–6	A large, green, red-blushed apple.
Carroll Zones 3–6	The medium-sized fruits are mottled red over green and are excellent for fresh use but only fair for cooking. Ripens in late season.
Collet Zones 2–6	The medium-sized to large, red fruits are suitable for fresh use or cooking and store well. Ripens late.
Dexter Jackson Zones 3–6	A medium-sized to large, red apple. Suitable for fresh use and cooking and stores well. Reportedly good resistance to fire blight and apple scab.
Fall Red Zones 2–6	A red-blushed fruit suitable for fresh use. Stores well. Ripens in mid-season to late season.
Gemini Zones 2–6	The medium-sized to large, light red fruits are suitable for fresh use. Stores well. Ripens in early to mid-season.
Goodland Zones 3–6	The medium-sized to large fruits are green with a red blush. Suitable for fresh use and cooking.

Haralson / Haralred Zones 3–6	The medium-sized, bright red fruits are suitable for fresh use, cooking, and baking. Stores well. 'Haralred' is a red-color sport of 'Haralson'; otherwise the two varieties are the same. Blooms late and ripens late. Popular in Minnesota.
Harcourt Zones 3–6	A greenish-yellow, red-blushed fruit with crisp, sweet, juicy flesh of fair quality and suitable for fresh use, cooking, and juice. Developed in Alberta. Somewhat resistant to fire blight.
Heyer #12 Zones 2–6	A small, yellowish-green apple suitable for pies and sauce. Not for fresh use and does not store well. Developed in Saskatchewan and suitable for cold arid conditions. Reportedly resistant to apple scab.
Keepsake Zones 3b–7	The small to medium-sized, mostly red fruits are irregular in shape and have firm, fine-textured, crisp, juicy flesh suitable for fresh use. Stores well. Ripens late. Very cold-hardy. Resistant to fire blight and cedar-apple rust. Unattractive appearance limits commercial value.
Mantet Zones 3b–5	A small to medium-sized, red-blushed apple suitable for fresh use and cooking. Ripens early.
Norda Zones 2–6	A medium-sized, greenish-yellow apple. Popular in the Prairie Provinces and southern Alaska.
Norkent Zones 2–6	The large, yellow, red-blushed fruits are suitable for fresh use. Ripen in mid-season to late season (August in Manitoba).
Parkland Zones 3–6	The medium-sized, light red over green fruits are suitable for fresh use and cooking. Does not store well. Ripens early. Considered a replacement for 'Norland'. Tends to bear biennially. Reported to be resistant to fire blight.

Prairie Magic **Zones 3–6**	The large, yellow, red-blushed fruits are suitable for fresh use or cooking. Ripens late (mid-September in Manitoba). Reportedly disease resistant and considered by some to be one of the best cold-climate apples.
Prairie Sensation **Zones 3–6**	The large, greenish, red-blushed fruits are highly rated for flavor and store well. Produced crops following severe winters in Saskatchewan.
Prairie Spy **Zones 3–8**	The large, red-blushed fruits are suitable for fresh use and baking. The flavor improves with cold storage. Ripen very late.
Prairie Sun **Zones 3–6**	The medium-sized to large, yellowish, pink-blushed fruits have moderately sweet flesh that resists browning. Suitable for fresh use and processing. Ripens in mid-season. The trees are semispur, semidwarf, very hardy, and easily grown but require heavy thinning.
Red Baron **Zones 3b–6**	A medium-sized red and yellow apple suitable for fresh use, baking, and cider. Does not store well. Ripens in early to mid-season. Resistant to fire blight.
Red Sparkle **Zones 3–6**	The medium-sized to large, dark red fruits are suitable for fresh use and considered to be excellent dessert apples. Ripens in mid-season to late season.
September Ruby **Zones 2–6**	The small to medium-sized, dark red fruits are suitable for fresh use or cooking and store well. Ripens in mid-season to late season.
Zestar **Zones 3–7**	The medium-sized to large, mostly red with some green fruits are suitable for fresh use and cooking. Stores well. Ripens early. Developed at the University of Minnesota.

Medium-Chilling Apple Varieties

Table 5.4

These varieties require 400 to 600 hours of chilling and are suitable from the mid-southern United States into parts of Canada. None is highly rated for disease resistance.

✿ = Blooms, ● = Ripens; E = *Early,* M = *Mid-season,* L = *Late,* VL = *Very Late*

Arkansas Black Zones 4–8 ✿ M ● VL	The medium-sized, dark red fruits are very firm and tart and reach peak flavor after 30 days in storage. Excellent keeping qualities. A chance seedling of 'Winesap'.
Fuji Zones 6–9 ✿ M ● L–VL	The medium-sized to large, green, red-striped fruits have crisp, sweet flesh. Suitable for fresh use and stores well. Very popular among western apple growers.
Gala Zones 5–8 ✿ E–M ● L	The small to medium-sized, yellow, red-striped fruits have firm, crisp flesh suitable for fresh use, sauces, baking, and freezing. Very popular with growers in Washington State. Requires heavy thinning to avoid biennial bearing.
Ginger Gold Zones 4–7 ✿ M ● E–M	This greenish-yellow, 'Golden Delicious' offspring has excellent quality for an early-ripening apple. Mildly tart and best used fresh. Stores well for a summer apple. A popular early-season commercial variety in Michigan.
Golden Delicious Zones 5–7 ✿ M–L ● M–L	The medium-sized to large, yellowish-green to yellow fruits are suitable for fresh use and cooking. One of the most popular varieties worldwide.
Granny Smith Zones 5–8 ✿ M ● L–VL	The medium-sized to large, green fruits have tart flesh that is excellent for fresh use. Stores well. Resistant to cedar-apple rust.
Jonagold Zones 5–8 ✿ M ● M–L	The medium-sized to large, yellow, light red–striped fruits are suitable for fresh use and cooking. Stores moderately well. Not suitable as a pollinizer.

Jonathan Zones 5–8 ❀ E–M ● L	The medium-sized, green, red-blushed fruits are suitable for fresh use.
Lodi Zones 4–8 ❀ E ● E	The medium-sized, light yellow apples have a firm, tart flesh and are best used for cooking. This is a summer apple and does not store well. The trees tend to overcrop and develop biennial bearing if not thinned aggressively.
McIntosh Zones 4–8 ❀ M ● M	The medium-sized, deep red fruits are suitable for fresh use, drying, and cider. Does not store well.
Mollies Delicious Zones 5–8 ❀ E ● E–M	The large, light yellow, red-blushed fruits have excellent flavor and are suitable for fresh use. Can be stored for 8–10 weeks. Excellent pollinizer.
Mutsu (Crispin) Zones 4–8 ❀ M–L ● L	The large to very large, greenish-yellow fruits have excellent flavor. Suitable for fresh use, cooking, cider, and sauce. Not suitable as a pollinizer.
Red Delicious Zones 5–8 ❀ E–M ● M	The small to medium-sized, deep red fruits have sweet, mild-flavored flesh suitable for fresh use. Many strains are available. Widely planted worldwide.
Rome Zones 4–8 ❀ M–L ● L	The medium-sized to large, deep to bright red, sometimes striped fruits have a tangy flavor and dense, starchy flesh. Suitable for fresh use (after storage), drying, baking, cooking, and cider. Very popular for baking and sauces.
Royal Gala Zones 5–8 ❀ E–M ● E–M	The medium-sized, bright red fruits are rather tart. Requires heavy thinning.
Stayman Zones 5–8 ❀ M ● L	The medium-sized to large, red-blushed fruits are suitable for fresh use, baking, and cider. Popular in the southern United States.

Winesap Zones 3–9 ✿ L 🍎 L	The medium-sized, dark red fruits have crisp, juicy flesh with a spicy flavor and aroma. Suitable for fresh use, cooking, and cider. Stores very well. An old favorite in the southern United States.
York Zones 5–8 ✿ L 🍎 L	The deep red, green-streaked fruits are used primarily for commercial processing but can be used fresh. Stores well.

Low-Chilling Apple Varieties

Table 5.5

These varieties require 300 chilling hours or fewer and are suitable for coastal southern California, parts of the Desert Southwest, Florida, and along the Gulf of Mexico. None is highly rated for disease resistance.

Anna Zones 8–9	The large, yellow, red-blushed fruits have a sweet, mild flavor. Imported from Israel with 'Ein Shemer' in 1967. Fruits well in central Florida, ripens in Gainesville in late June to early July. Pollinate with 'Dorsett Golden'.
Beverly Hills Zones 8–9	The medium-sized, greenish-yellow, orange-red–striped fruits russet at maturity. The fruits are tart and suitable for fresh use and cooking. Ripens about mid-August. Quickly loses flavor and texture if left on the tree too long. Popular in southern coastal California. Sometimes reported to be self-fruitful but is only partially so. Pollinate with another early to mid-season, low-chilling apple.
Dorsett Golden Zones 8–9	The small, yellow, light pink–blushed fruits have firm, crisp, sweet flesh. Stores in refrigeration for 2 weeks. Ripens in late June in Gainesville, Florida. Introduced from the Bahamas. Very popular in southern California and Florida. Pollinate with 'Anna'.

Ein Shemer Zones 8–9	The small to medium-sized, greenish-yellow fruits have a mildly sweet flavor that resembles 'Golden Delicious'. Imported from Israel. Needs no chilling. Very popular in southern California. Only conditionally recommended by University of Florida Extension because it blooms late and is less resistant to insects and diseases than are 'Anna', 'Dorsett Golden', and 'TropicSweet'.
TropicSweet Zones 8–9	Resembles 'Anna' (see above) but is firmer and sweeter, with low acidity. Ripens 5–7 days earlier than 'Anna' in Gainesville. An offspring of 'Anna' released by the Florida Agricultural Experiment Station and recommended for Florida. Blooms at the same time as 'Anna'.

Suggested Asian Pear Varieties

Table 5.6

Atago Zones 5–8	The brownish, oblong fruits are sweet and juicy. Better adapted to cool climates than most Asian pears. Requires cross-pollination.
Chojuro Zones 5–8	The round, yellowish-orange, somewhat russetted fruits have crisp, rather coarse flesh with a sweet, mild flavor. For fresh use. Ripens in late mid-season. Pollinate with 'Bartlett' or another Asian pear.
Hamese Zones 5–8	The medium-sized, yellow fruits have good flavor and ripen early to mid-season. Requires cross-pollination.
Hosui Zones 5–9	The round, greenish-yellow fruits are tangy-sweet and juicy. Best for fresh use. Ripens mid-season to late season. Resistant to fire blight. Has proven popular as a commercial variety. Pollinate with another Asian pear.

Ichiban Zones 5–8	The brownish-golden fruits have fine-textured flesh and ripen early. Has some resistance to *Pseudomonas* blight. Has proven reliable in the Pacific Northwest. Requires cross-pollination.
Kikisui Zones 5–8	The large, yellow fruits have a good, sweet flavor and crisp flesh. Ripens in late mid-season to late season. Bears early. Resistant to fire blight. A particularly good choice for organic growers with sufficiently long seasons to ripen the fruits. Requires cross-pollination.
Korean Giant (Large Korean) (Olympic) (Dan Bae) Zones 5–8	The large to very large, greenish-orange fruits have crisp, sweet, white flesh. Ripens very late and stores well. Resistant to fire blight. A particularly good choice for organic growers in long-season locations. Requires cross-pollination.
New Century Zones 5–9	The small to medium-sized, yellow fruits have sweet, fine-grained flesh. Ripens in mid-season. Requires cross-pollination.
Seuri Zones 5–8	The large, orange fruits ripen very late. Resistant to fire blight. Requires cross-pollination.
Shinko Zones 5–8	The medium-sized to large, brownish-green fruits ripen in mid-season to late mid-season and store well. Ripens very late. Requires around 450 chilling hours. Highly resistant to fire blight. An excellent choice for organic growers in long-season areas.
Shinseiki Zones 5–8	The medium-sized to large, yellow fruits are mild-flavored, crisp, and very juicy. Ripens in mid-season. Requires cross-pollination.

Medium- and High-Chilling European Pear Varieties

Table 5.7

These varieties are suggested for moderate and cold climates.

Ayers Zones 5–7	The medium-sized, yellow, red-blushed fruits are very sweet and suitable for fresh use or drying. Blooms early and ripens in late mid-season to late season. Fire blight–resistant. Medium-chilling and adaptable from the mid-southern United States to the Pacific Northwest and southwestern Canada.
Bartlett (Williams) Zones 5–8	The large, yellow, pink-blushed fruits have smooth, firm flesh suitable for fresh use and canning. Ripens in late mid-season. Makes up 80 percent of commercial pear production in North America and is very popular in Europe, where it is known as 'Williams'. Do not pollinate with 'Seckel'.
Bella di Guigno Zones 5–8	The medium-sized to large, greenish-yellow, red-blushed fruits ripen very early. Requires cross-pollination.
Blake's Pride Zones 5–8	The medium-sized, yellow, russetted fruits have a sweet, rich taste and aroma and ripen late, three weeks after 'Bartlett'. Produces high yields. Highly resistant to fire blight. Best for long-season areas. Pollinate with 'Bartlett'.
Bosc Zones 5–8	The medium-sized, long-necked, brown, russetted fruits are firm and crisp. The flavor improves during 4–8 weeks of storage. Although an old favorite and often recommended, 'Bosc' is susceptible to fire blight and not the best choice for most organic growers. Best suited to arid and semiarid locations. Several strains of 'Bosc' are available. Requires cross-pollination.
Comice Zones 5–8	The large, greenish-yellow fruits have excellent flavor and texture, which reach their peak after

	4 weeks of storage. Best for fresh use. Ripens very late and can be challenging in short-season locations. Has a medium-chilling requirement. Resistant to fire blight. Requires cross-pollination.
D'Anjou Zones 5–8	The medium-sized, yellow fruits are firm, juicy, and develop excellent flavor during 4–8 weeks of storage. Stores well. A leading commercial variety. Red-skinned strains are available.
Harrow Delight Zones 4–8	The medium-sized, green, red-blushed fruits have smooth flesh with good flavor. Among the most cold-hardy pears and ripens late. Resistant to fire blight and pear scab.
Honeysweet (Honey Sweet) Zones 5–8	The small to medium-sized, brownish-golden, russetted fruits have smooth flesh. Resembles 'Seckel' but the fruit is larger. Resistant to fire blight and leaf spotting diseases. Produces best when cross-pollinated.
Moonglow Zones 5–8	The large, green, red-blushed to nearly red fruits have soft, sweet, mild-flavored flesh. Ripens in midseason. Resistant to fire blight. A good pollinizing variety and requires cross-pollination.
Red Sensation Zones 5–8	A red-skinned strain of 'Bartlett'. Particularly recommended for roadside stands and farmers' markets.
Seckel Zones 5–8	The small, round, brownish-golden, heavily russetted fruits have excellent, very sweet flavor. Ripens late to very late. Sets heavy crops. Resistant to fire blight. Requires cross-pollination. Do not pollinate with 'Bartlett'.
Spalding Zones 5–9	The medium-sized, green fruits are crisp and ripen late. Resistant to fire blight. A southern United States introduction that has proven adaptable in the Pacific Northwest. Self-fruitful.

Warren Zones 5–9	The medium-sized to large, green fruits have good flavor and keep well. In Texas A&M evaluations, 'Warren' provided the best overall dessert quality and blight resistance for southern growers. A Mississippi introduction, well suited to eastern North America and tolerates temperatures to −20°F (−29°C). Highly resistant to fire blight.

Low- and Medium-Chilling European Pear Varieties

Table 5.8

These varieties are suggested for warm climates.

Baldwin Zones 7–9	The large, yellow, red-blushed fruits are suitable for fresh use or baking. A medium-chilling variety and well adapted to the southern United States from south Texas to southern Florida.
Flordahome Zones 6–9	The medium-sized to large, green fruits are almost seedless. A low-chilling variety (100–150 hours) that is well adapted to the southern United States. Pollinate with 'Hood' or 'Pineapple'.
Hood Zones 7–9	The large, yellowish-brown fruits have smooth, white, sweet flesh with a crisp texture. An excellent low-chilling variety (100–150 hours) for the southern United States. Resistant to fire blight. Pollinate with 'Flordahome' or 'Pineapple'.
Kieffer Zones 5–8	The medium-sized to large, yellowish-green fruits have gritty flesh best suited to baking or canning. A medium-chilling European-Asian pear hybrid adapted to the southern United States and more northerly climates. A reliable producer with excellent resistance to fire blight.

Orient Zones 6–8	The large to very large, greenish-yellow fruits are best used for baking and canning. A good medium-chilling European-Asian pear hybrid for growers in the southern United States but requires careful training and crop management. The trees are large and produce heavy crops. Excellent resistance to fire blight.
Pineapple Zones 8–9	The medium-sized brown-skinned fruits have an elongated gourd shape. The flesh remains very hard, even when ripe. Best used for preserves, as it disintegrates during canning. A low-chilling variety commonly grown in the southern United States. Pollinate with 'Flordahome' or 'Hood'.

Suggested Quince Varieties

Table 5.9

Aromatnaya Zones 5–8	The medium-sized, round, yellow fruits have a pineapple-like flavor and can be used fresh or cooked. The fruits ripen very late. A Russian introduction. Reported to be very disease-resistant.
Ekmek Zones 5–8	The medium-sized, yellow fruits are best used for cooking. Ripens late. Originated in Turkey.
Havran Zones 5–8	The fruits are large, pear-shaped, and quite sweet. Ripens late to very late. A Turkish variety.
Karp's Sweet Zones 6–9	The flesh is sweet and tender when the fruits are grown in warm climates. The flavor is less sweet in cooler climates.
Kaunching Zones 4–8	The medium-sized fruits are sweet and tender enough to eat fresh or can be cooked. This variety originated in central Asia and is among the most cold-hardy quinces available in North America. Reported to be self-fertile.

Orange Zones 5–8	The large bright yellow fruits have orange-yellow flesh. Suitable for cooking. Ripens late to very late. Adaptable to a wide range of growing regions. Reported to be self-fertile.
Pineapple Zones 5–8	The large fruits have tart, pineapple-flavored flesh suitable for cooking. This low- to medium-chilling variety (300 hours) is surprisingly cold-hardy. Produces heavy crops. Reported to be self-fertile.
Smyrna Zones 5–9	The large, yellow fruits are mild-flavored and suitable for cooking. An old, Turkish variety that is widely recommended. Reported to be self-fertile.
Portugal Zones 5–8	The large, strongly pineapple-flavored fruits are suitable for cooking and turn deep red during cooking.

Mayhaw Varieties

Table 5.10

Varieties are listed in order of ripening.

T.O. Superberry	The dark red fruits average about 0.7 inch in diameter and have pink flesh. Blooms in late February and ripens in late April. Early bloom is a problem.
Mason's Superberry (Texas Superberry)	The dark red fruits average about 0.7 inch in diameter and have pink flesh. Blooms in mid-February and ripens in late April. Early bloom is a problem. Highly rated for yield and desirable tree form. The fruits hold tightly on the trees. Thorny. Self-fertile.

Superspur	The light yellowish-red fruits average 0.7 inch in diameter and have soft, yellow flesh. Blooms in late February and early March and ripens in late April and early May. Highly rated for yield and tree form. Early bloom is a problem, as is inadequate chilling in warmer areas and during mild winters. Fruits fall prematurely (shatter) when ripe. Rust-susceptible.
Saline	The red fruits average 0.6 to 0.75 inch in diameter and have firm, pink flesh. Blooms in early to mid-March and ripens in late April through early May. Productive, and fruit retention is good.
Big Red	The red fruits average about 0.75 inch in diameter and have pink flesh. Blooms in early March and ripens in late April through early May. The trees are upright and spreading. Needs a pollinizer.
Crimson	The mostly red fruits average about 0.7 inch in diameter and have light pink flesh. Blooms in mid-March and ripens in late April through early May.
Big V	The light red fruits average about 0.7 inch in diameter and have soft, pinkish-yellow flesh. Blooms in late February and early March and ripens in late April through early May. The fruits are susceptible to shattering when ripe.
Betsy	The large, red fruits have red flesh. Blooms in late February and early March and ripens in late April through early May. Reported to be a heavy bearer. Needs a pollinizer.
Turnage 57	The red fruits average about 0.5 inch in diameter and have soft, yellow flesh. Blooms in early to mid-March and ripens in early to mid-May. Susceptible to shattering when ripe.

Texas Star	The red fruits average 0.75–0.9 inch in diameter and have yellow flesh. Blooms in early to mid-March and ripens in early to mid-May. A promising new variety but not yet extensively tested.
Turnage 88	The red fruits average 0.65 inch in diameter and have rather soft, yellow flesh. Blooms in early to mid-March and ripens in mid-May.
Georgia Gem	The light red fruits average about 0.75 inch in diameter and have red flesh. Blooms in mid- to late March and ripens in mid- to late May. Requires a pollinizer.
Reliable	The red fruits average about 0.75 inch in diameter and are reported to have excellent flavor. Blooms in mid- to late March and ripens in mid- to late May. Requires a pollinizer.
Red Majesty	A new, as yet not widely tested, variety developed for commercial fruit production by the Louisiana Mayhaw Association. Reported to be late-blooming and easy to harvest. Extends the mayhaw harvest season.

Adapted, in part, from J. A. Payne and G. W. Krewer. 1990. Mayhaw: A new fruit crop for the south. p. 317–321. In: J. Janick and J. E. Simon (eds.), *Advances in new crops*. Timber Press, Portland, OR.

Loquat Varieties

Table 5.11

Advance	The medium-sized to large, rounded to pear-shaped, yellow fruits have white pulp and are sweet with excellent flavor and good commercial quality. The skin is tough, an advantage when shipping the fruits. Ripens in mid-season to late season. A naturally small tree growing to about 5 feet high. One of the best and most popular loquats. Highly resistant to pear blight. Requires cross-pollination ('Thales' works well) and is a good pollinizer for other varieties.
Big Jim	The large to very large, orange fruits are pear-shaped and have sweet, orange-yellow flesh. Ripens in mid-season. The trees are vigorous, erect, and highly productive. Has performed well in southern Louisiana. Requires cross-pollination.
Champagne	The medium-sized to large, pear-shaped, yellow fruits have white flesh and a tough, astringent skin. The fruits are sweet and have excellent flavor and quality. Ripens in mid-season to late season. One of the best loquats. Requires cross-pollination.
Early Red	The medium-sized to large, orange fruits have orange flesh with tough, astringent skins. The flesh is sweet, and flavor ranges from fair to excellent. Ripens very early, making it desirable for southern California, where later-ripening varieties suffer sunburn. Requires cross-pollination.
MacBeth	The large to very large, yellow fruits have white flesh. Has performed well in southern Louisiana. Requires cross-pollination.

Oliver	The medium-sized to large, orange fruits have sweet, pale orange flesh with good flavor. Ripens in mid-season to late season. Popular in southern Florida but has proven susceptible to fire blight in southern Louisiana. Requires cross-pollination.
Tanaka	The small, dark yellow to orange fruits have yellowish-orange flesh with excellent flavor. Ripens in mid-season to late season; too late for southern California. Trees are of medium height. Among the more cold-hardy loquats. Popular in Florida and recommended for trial plantings in Louisiana. Partially self-fruitful but produces better with cross-pollination.
Thales (Gold Nugget and Placentia)	The small to medium-sized, dark yellow to orange fruits have sweet, yellowish-orange flesh with a good flavor that has been likened to apricot. Stores and ships well. Similar to and may be the same as or descended from 'Tanaka'. Ripens late. Bears few fruits per cluster, reducing the need for thinning. Trees are vigorous and erect. Recommended for trial plantings in southern Louisiana. Self-fruitful or partially self-fruitful.
Vista White	The small to medium-sized, round, yellow fruits have sweet, white flesh. Ripens late to very late. Requires cross-pollination.
Wolfe	The large, yellow fruits have fairly thick skin and white to pale orange flesh with excellent flavor. Particularly well suited for cooking. Reliably productive. Requires cross-pollination.

Suggested Saskatoon Varieties

Table 5.12

CULINARY VARIETIES

Honeywood	Produces medium-sized to large, blue fruit and abundant crops. Grows about 15 feet tall with a 12-foot spread and produces few suckers. Blooms as much as a week later than most other varieties and is a good choice for frosty sites. Appears to have some resistance to powdery mildew.
Martin	Produces large, blue fruit. A seedling of 'Thiessen' and generally resembles that variety. Ripens uniformly. An excellent choice for extremely cold sites.
Northline	The medium-sized to large, blue fruits are sweet and flavorful and resist cracking. Produces abundant crops. Grows to 12 feet high with an 18-foot spread and suckers heavily. Very popular. Wooly elm aphid can be a problem.
Pembina	The blue fruits have excellent flavor. The fruits are susceptible to cracking if the plants receive too much water during ripening. The plants are productive, grow about 15 feet high with a 15-foot spread, and produce few suckers. Entomosporium leaf spot and wooly elm aphid can be problems.
Smokey	The small, blue fruits are flavorful, sweet, and seedy. Can ripen unevenly. Grows about 14 feet high with an 18-foot spread and suckers heavily. Once popular commercially, it is losing ground to varieties with better fruit qualities. Entomosporium leaf spot and Cytospora canker can be problems. Used as a pollinizer for 'Martin' in Alaska.

Thiessen	The large fruits have excellent flavor and can ripen unevenly. Grows about 15 feet tall with an 18-foot spread and suckers moderately. Blooms earlier than most other varieties, making it more susceptible to frost damage. Somewhat resistant to powdery mildew, but Cytospora canker can be a problem. Very popular for U-pick orchards.
ORNAMENTAL VARIETIES	
Altaglow	The large, white fruits are sweet but bland. Grows about 20 feet tall with a 10-foot spread and forms a columnar shape. Produces brilliant fall foliage. Best used in large landscapes. Requires cross-pollination in order to set fruit.
Paleface	The large, white fruits are mild-favored. Grows about 6 feet tall and produces few suckers. Suitable for smaller landscapes.
Regent	The small, dark blue fruits have a sweet, mild flavor. Grows about 6 feet tall with a 6-foot spread and suckers moderately. May have some resistance to Entomosporium leaf spot and saskatoon-juniper rust. Produces attractive fall foliage.

SELECTING STONE FRUIT CROPS AND VARIETIES

- Apricot | 178
- Cherry | 181
- Peach | 188
- Nectarine | 191
- Plum | 192

We have already discussed the importance of matching crops to your climate and location. This step is especially important for stone fruits. With a few exceptions, stone fruits tend to be less cold-hardy than apples and pears, are more susceptible to spring frosts, and are less tolerant of poor drainage. Stone fruits are also susceptible to many pests and diseases. Notwithstanding, they remain popular choices for organic orchardists in many areas of North America.

As we discussed in chapter 5, the vast majority of fruit trees are propagated vegetatively by budding or grafting desirable varieties onto suitable rootstocks (see page 120).

Apricot

ORIGINATING IN NORTHEASTERN CHINA, apricots are among the more cold-hardy stone fruits. They were cultivated in China about 3,000 years ago. By around 50 BCE, apricots were grown in the Roman Empire. They thrive in areas with long, hot, dry summers and cool, wet winters.

Today, apricots are widely grown throughout the world. In terms of quantity produced, Turkey is, by far, the world leader, followed by Iran, Pakistan, Uzbekistan, Italy, Algeria, Japan, and Morocco. As of 2008, the United States ranked number 16.

Apricots were, reportedly, introduced to North America by English settlers and/or Spanish explorers. Some of the first orchards were in Virginia. Apricots do not perform particularly well in humid climates, however, and did not come into their own in North America until they arrived in the arid American West. About 94 percent of the commercial apricots in North America are grown in California, with nearly all commercial apricot production being in the San Joaquin Valley. Washington produces about 6 percent of U.S. commercial apricots, and Utah

Apricots

AT A GLANCE

Region: Hardiest varieties tolerate −20 to −30°F (−29 to −34°C) and grow best in USDA Zones 5 to 9; a few varieties can be grown in Zones 3 and 4. Warm, dry summers and few spring frosts are highly desirable.

Soil pH and type: Neutral to slightly acidic, deep, well-drained, light- to medium-textured soil

Pollination: Mostly self-fruitful, but some experts recommend planting two varieties close together to ensure good fruit set

Other notes: Open canopies, good light penetration, and good air movement are critically important. Sulfur is toxic to apricots; apply fixed copper fungicides instead.

less than 1 percent of the total crop. Canada imports more apricots than it exports, and it buys approximately 50 percent of U.S. apricot exports. The United States is also a major importer of apricots. Most commercial apricot orchards in North America are family-owned and about 50 to 60 acres in size, although there is a trend toward fewer and larger apricot orchards.

Due to declining consumption and increased imports of inexpensive apricots, some U.S. producers have moved away from apricots. According to the University of California, 75 to 90 percent of U.S. apricots are used in the processing market, but the consumption of canned apricots has declined sharply. Consumption of fresh apricots in the United States has remained relatively stable at about 0.2 pound per person per year, with consumption of dried apricots increasing. Total apricot consumption in the United States varies from about 0.9 to 1.6 pounds per person per year. The best opportunities for new growers are local, direct sales.

Climate concerns. The hardiest apricot varieties available in North America tolerate temperatures between –20 and –30°F (–29 to –34°C) and are well suited to USDA Zones 5 to 9, with a few varieties rated to Zones 3 and 4. While apricots are usually somewhat more cold-hardy than the closely related peaches, they bloom earlier in the spring, making them highly susceptible to frost injury. Over the past 20 years or so, U.S. commercial production has fluctuated between about 55,000 and 155,000 tons annually, due to frost injury and the tendency in some areas for the trees to bear fruit biennially or triennially. Apricots are also susceptible to winter injury in areas where winter temperatures fluctuate widely. Areas with warm, dry summers and few spring frosts are best.

Planting, cultivating and harvesting. Apricots do not tolerate wet feet, so plant on deep, well-drained, light- to medium-textured soil that is neutral to slightly acidic. Place the trees on slopes away from frost pockets.

Wild apricot trees in their native range can grow 30 to 45 feet tall, but cultivated apricots are generally kept to around 12 feet in height. The trees tend to spread and are typically planted about 18 to 24 feet apart within and between rows.

The fruits are 1.5 to 2.5 inches in diameter and have skins that are yellow to orange, often with a red blush. The skins range from fuzzy to nearly smooth. Most varieties have yellow flesh, but a few white-fleshed varieties are available. The trees should bear their first marketable crops 3 to 5 years after planting. The fruits are usually borne on rather short-lived spurs but can be produced on vigorous 1-year-old shoots. A key purpose in pruning and training is to keep the spurs well exposed to sunlight.

Fresh market fruit is hand harvested when the fruits are mature but firm. Plan on picking each tree two to three times as the fruits mature. Apricots destined for drying are allowed to fully ripen before harvest. Mechanical shakers are sometimes used for harvesting apricots destined for processing, although the trees are susceptible to damage from the shakers. Apricots do not store well and should be refrigerated after harvest and used within 1 to 2 weeks.

Because of susceptibility to fungal and bacterial pathogens, open canopies, good light penetration, and good air movement are critically important in organic orchards. NOTE: Sulfur is toxic to apricots and must not be applied to the trees for disease control. Apply fixed copper fungicides instead.

In general, apricots are managed similarly to peaches, although they require less pruning. The trees form heavier branches than peaches and can be trained to either open-centered vase shapes (like peaches) or modified central leaders (like apples). Open center training is the most common. Pest and disease problems are similar to those for peaches.

Rootstocks. Apricot trees are most commonly budded onto apricot or peach seedling rootstocks using seed from varieties adapted to a particular growing area. Seedling peach rootstocks such as 'Lovell' are sometimes employed on sandy soils in mild climates. In Michigan, 'Manchurian' and 'South Haven No. 6' apricot rootstocks have performed well. For sandy soils where nematodes are a problem, apricot or nematode-resistant peach rootstocks are preferred. Peach rootstocks are not well suited to cold climates. On heavier soils, myrobalan plum rootstocks can be used, but they produce weak unions, gradual tree decline, and slow production.

Pollination. Most apricot varieties are self-fruitful and can be grown commercially in solid blocks, but some experts recommend planting two varieties close together to ensure good fruit set. The varieties 'Goldrich', 'Perfection', and 'Riland' require cross-pollination.

Unlike apples, relatively few apricot varieties are available in North America. 'Blenheim' (also known as 'Royal') is the dominant commercial variety grown in western North America. Despite the limited number of varieties, apricots are grown successfully from the deep southern United States into the Canadian Prairie Provinces and southern Alaska. For southern locations with limited chilling temperatures, consider 'Zard', 'Katy', 'Goldkist', 'Newcastle', and 'Golden Amber'.

Cherry

EVIDENCE SHOWS THAT CHERRIES HAVE BEEN CONSUMED by humans for at least 6,000 to 7,000 years and have been popular orchard and roadside fruit trees since the days of the Roman Empire. Cherries were introduced to North America during the mid-1600s and became popular garden crops, following settlers westward. Commercial cultivation of cherries in North America lagged behind many other fruits, largely beginning in the

1920s and 1930s. Because of pest and disease problems, as well as chilling requirements, cherries in North America are primarily grown north of the Mason-Dixon Line (approximately 39 degrees north latitude).

Turkey is the world leader in cherry production, followed by the United States, Iran, Italy, Ukraine, Romania, Russian Federation, Spain, and Uzbekistan. In North America, Washington is the leading commercial sweet cherry producer, followed by Oregon, California, and Michigan. Michigan is the leading tart cherry producer, followed distantly by California and New York. Canada grows about 5 percent of the commercial cherries produced in North America, primarily in British Columbia.

Although there are many cherry species, only a few have been used to produce most modern varieties. Sweet cherries are diploid (two sets of chromosomes) and primarily derived from *Prunus avium*. Tart cherries are tetraploid (four sets of chromosomes) and are classified as *Prunus cerasus*, although they probably originated as complex natural crosses between sweet cherry and European ground cherry species. (Duke cherries are hybrids between sweet and tart cherries and have largely fallen out of favor as better-quality cherries have been developed.) Several species of bush cherries are used for fruit production, including the cold-hardy Nanking bush cherry, *Prunus tomentosa*, which is used as an ornamental and sometimes for fruit. Other bush cherries and hybrids have been developed for specific growing conditions, such as the Canadian Prairie Provinces.

Although cherries have been grown for millennia, there have been some recent improvements that are largely confined to Europe, including systematic breeding and the use of genes for pest and disease resistance and climatic adaptation. There are relatively few cherry varieties available in North America, and fewer still that are worthy of cultivation.

Climate concerns. Despite their long popularity, cherries are among the more challenging tree fruits to grow. Cherries have rather limited cold hardiness; wood, bark, and bud hardiness limit winter survival and crop production in Canada and the northern United States. Depending on variety, cherries typically need 800 to 1,300 hours of chilling and are difficult to produce in southern areas with limited chilling. Cherries bloom very

Cherries

AT A GLANCE

Chilling requirements: 800 to 1,300 hours of chilling at 32 to 55°F (0 to 13°C)

Region: USDA Zones 5 to 9, with a few varieties rated to Zone 4

Soil pH and type: Deep, well-drained loam or sandy soils

Pollination: Most sweet cherries are self-unfruitful and require cross-pollination; plant varieties from different pollination groups together to ensure cross-pollination. Tart cherries tend to be self-fruitful.

Variety selection: It is very important to carefully select the right scion and rootstock combination for your particular area and training style.

Other notes: Tart cherries are more cold-hardy than sweet cherries, bloom somewhat later, and are better suited to cold climates than their sweeter cousins.

early and are highly susceptible to spring frost damage everywhere they are grown.

Site selection is particularly critical with cherries. Avoid planting in a frost pocket and plant on deep, well-drained soil. Benches and gentle slopes above temperature inversion layers are often the best sites for cherries. Loam and sandy loam soils provide good drainage and adequate water-holding capacity. Planting in areas near large bodies of water helps reduce spring frost damage.

Pests and diseases. Pseudomonas blight, brown rot, viruses, and other fungal, bacterial, and viral pathogens cause serious disease problems, particularly in humid climates and even more particularly for organic growers. Given careful planning and maintenance, however, organic orchardists in many parts of North America can enjoy these fruits.

Rootstocks. Cherry varieties, rootstocks, training systems, and climate interact in complicated ways, and tree performance is strongly dependent on these interactions. It is very important to carefully select the right scion and rootstock combination for your particular area and training style. For organic cherry production, particularly in humid regions, select combinations and training styles that produce relatively small trees with open canopies to reduce disease problems and facilitate harvesting.

Two rootstocks have long dominated cherry production in North America: Mazzard and Mahaleb seedlings. Mazzard seedlings have been the most common, producing large, vigorous, productive, and long-lived trees resistant to many pests and diseases. Mahaleb seedlings are hardier, are better adapted to droughty areas with extreme temperatures, and produce trees that are about 90 percent of the size of those on Mazzard rootstock. A third choice, Stockton Morello clonal rootstock, is used to a much lesser degree than the other two and provides some dwarfing for sweet cherries.

The picture, however, has changed markedly. While Mazzard and Mahaleb rootstocks still have their uses, many new dwarfing, highly productive rootstocks have become available and are being tested commercially. Relative heights of selected cherry rootstocks are shown in figure 3.5.

Some of the more popular new rootstocks include various Gisela, Krymsk, and Weiroot selections, as well as Colt and Maxima 14. Some of the new rootstocks produce smaller, more manageable trees that come into bearing very early. The main downside has been overbearing trees that produce fruit too small to be marketable. Gisela 5 (50 to 60 percent of standard size) and Gisela 6 (85 to 90 percent of standard size) received great attention when they were first released and have been tested across North America. Initial results that showed a reduction of tree size and early bearing were impressive, but problems maintaining fruit size have limited widespread adoption of these rootstocks for large, commercial orchards. Gisela 12 (80 to 90 percent of standard size) comes from the same rootstock program. Newer pruning techniques, described in chapter 12, have been developed to help maintain fruit size on these highly productive rootstocks.

Krymsk 6 (65 to 70 percent of standard size) and Krymsk 5 (85 to 90 percent of standard size) from a Russian breeding program are seen, by some, as replacements for the Gisela selections. According to the research station where these selections were developed, the Krymsk rootstocks are compatible with sweet and tart cherries, encourage wide branch angles, provide good adaptation to cold climates and wet soils, and produce early bearing and good fruit size. There needs to be much more testing on these and other dwarfing cherry rootstocks, however, before we have a clear picture of which rootstocks are best for particular cherry varieties, training systems, and growing regions.

For small commercial, market, and home orchards, your best results will probably come from the use of Gisela 5, Gisela 12, or Gisela 6 (in order of increasing tree size). Of the three, Gisela 6 appears to be somewhat more commercially accepted. Trees on Gisela 5 should be supported, and supports are also often recommended for Gisela 12 and Gisela 6. Krymsk 5 and Krymsk 6 appear worth testing. Weiroot 72 rootstock produces very small cherry trees about 6 to 9 feet tall, but it requires excellent soils and growing conditions, as well as support for the trees. This rootstock is not likely to be used for commercial orchards, but it could be valuable for home orchards where space is limited.

Adara rootstock provides no dwarfing but is adaptable to heavy, poorly drained, and calcareous soils. Likewise, Colt and Maxima 14 produce standard-sized trees. Maxima 14 is precocious and can shorten the time from establishment to harvest.

Pollination. With a few exceptions, sweet cherries are self-unfruitful and require cross-pollination. Unfortunately, many sweet cherry varieties do not effectively cross-pollinate each other. Efforts have been made to cluster incompatible varieties into "pollination groups," of which there are at least nine. To ensure effective cross-pollination, you should include in your orchard a pollinizing variety from another pollination group. Many newer cherry varieties, however, have not yet been listed in appropriate pollination groups. Use table 6.2 as a guide in selecting pollinizing varieties.

You should purchase your trees from a reputable fruit tree nursery and ask the nursery about suitable pollinizing varieties for the trees you purchase. Tart cherries are mostly self-fruitful and usually do not require cross-pollination.

Sweet Cherry

Despite at least 1,100 sweet cherry varieties having been named (certainly an underestimate), only about 65 are commonly recommended in North America, and perhaps a dozen make up the vast majority of sweet cherries produced commercially in North America. Table 6.2 lists some of the more popular sweet cherry varieties for North America.

Sweet cherry varieties are typically rated as suitable for USDA hardiness Zones 5 to 9, with a few varieties rated to Zone 4. Production is easiest and most reliable in Zones 6 to 8, in arid and semiarid regions with mild climates and abundant irrigation water. Sweet cherry trees can grow very large unless grown on dwarfing rootstocks and the size managed with pruning and training.

Sweet cherries can be divided into two groups: Heart and Bigarreau types. Heart cherries tend to be heart-shaped, although not always, and tend to be rather soft. The color of the fruit ranges from dark with red juice to light-colored with colorless juice. With the exception of 'Black Tartarian', Heart cherry varieties are normally limited to home production. Bigarreau cherries tend to be rounder and firmer than Heart cherries and make up most of the economically important sweet cherry varieties. 'Bing' and 'Lambert' are older, commercially important dark Bigarreau cherries, while 'Napoleon', 'Rainier', and 'Gold' varieties are yellow to red blush-colored members of the group.

Newer sweet cherry varieties have been developed, but they have not been widely tested. 'Stardust', 'Sumele', 'Sumleta', 'Sumnue', and 'Sumste', for example, are recommended for commercial production in British Columbia but do not yet have much of a following in the United States.

Until recently, most sweet cherries were destined for fresh use. Today, they are increasingly used for processing. 'Anderson', 'Black Gold', 'Black York', 'Blushing Gold', 'Corum', 'Emperor Francis', 'Gold', 'Nugent', 'Sam', 'Ulster', and 'WhiteGold' are popular commercial processing varieties.

Tart Cherry

Tart cherries (also known as sour cherries and pie cherries) are more cold-hardy than sweet cherries, bloom somewhat later, and are better suited to cold climates than their sweeter cousins. Most tart cherries are self-fruitful and usually produce smaller trees than sweet cherry varieties.

'North Star' and 'Meteor' are small trees that are very cold-hardy and well suited to home and market orchard use in northern states and southern Canada. Tart cherries were traditionally used only for processing, but some newer varieties are sweet enough to also eat out of hand. Table 6.3 lists suggested tart cherry varieties.

Tart cherries are most commonly grown on Mahaleb rootstock (see page 184), but Ontario trials show promise for other rootstocks. In those trials, Mahaleb produced the largest cumulative yields, followed closely by Wieroot 10, Wieroot 13, Gisela 6, with Wieroot 158 producing lower yields. Wieroot 10 and Wieroot 13 tended to develop suckers at the bases of the trees. Small fruit size is a concern with Gisela rootstocks.

Bush Cherry

In addition to sweet and tart cherries, several species of cherries native to North America are used for fruit production. These bush cherries are mostly used for home production or small-scale production of niche products. Most varieties are self-unfruitful. Table 6.4 lists suggested varieties of bush cherries.

Western sandcherry (*Prunus besseyi*) is native to the Central Plains states, Upper Midwest, Great Lakes states, Arkansas, Oregon, Saskatchewan, Manitoba, and Ontario. The plants are spreading, multistemmed shrubs about 3 to 8 feet tall. The tart fruits are yellow, purple, or black cherries about ½ inch in diameter. The plants are hardy to −30°F (−34°C) and perform best on well-drained soils in areas with 12 to 24 inches of annual precipitation. They are susceptible to root rot on heavier soils, as well as fungal and bacterial diseases. Some attempts have been made to use western sandcherries as dwarfing plum rootstocks, although the results have generally been poor. Several named varieties are available and described in table 6.4.

Cherry-plum hybrids are not to be confused with cherry plum (*Prunus cerasifera*), which is also called myrobalan plum and is discussed in the plum section. Cherry-plum hybrids are hybrids between western sandcherry and other species. Many of these interspecific hybrids have been introduced in South Dakota and are very cold-hardy. Suggested varieties are listed in table 6.4.

Nanking cherry (*Prunus tomentosa*) is native to northwestern China and has long been cultivated for its flowers and fruits. It was first introduced to North America in 1892. The plants are erect to somewhat spreading, multistemmed shrubs that typically grow 6 to 10 feet tall. Most often used as an ornamental in North America, Nanking cherry bears an abundance of white flowers and tart, red fruits ¼ to ½ inch in diameter. The plants are very cold-hardy (rated to Zone 2) and perform best on well-drained loam soils. Because Nanking cherries are typically sold as seedlings, there is substantial variability between plants. Some plants are extremely susceptible to brown blight and other diseases, and the crop is best grown organically in areas with warm, dry summers.

Dwarf sour cherry hybrids are crosses between domestic tart cherries (*Prunus cerasus*) and European dwarf cherries (*Prunus fruticosa*). European dwarf cherries are native to northern Europe and Mongolia. Other names for European dwarf cherry are Mongolian cherry, Steppe cherry, and European ground cherry. Available crosses come from the University of Saskatchewan and are very cold-hardy. Suggested varieties are listed in table 6.4.

Peach

PEACHES (*Prunus persica*) appear to have originated in China, where they have been cultivated since around 2000 BCE. The fruit spread widely and is now cultivated worldwide in temperate regions on both sides of the equator, between about 24 and 45 degrees north and south latitude, due to chilling requirements and cold hardiness. Oceanic factors, as well as other factors that affect climate, can extend production a bit beyond these latitudes.

Italy and the United States are the leading peach producers, followed by China, Greece, Spain, and many other countries. In North America, California is the leading peach producer. Other states with significant commercial peach production include South Carolina, Georgia, New Jersey, Pennsylvania, Michigan, Washington, North Carolina, Texas, and Virginia. In Canada, peaches are grown commercially around the Great Lakes and are well suited to portions of the Okanogan and Fraser Valleys of British Columbia.

Climate concerns. In general, peaches are best adapted to warm and moderate climates. They are among the least cold-hardy of the temperate-zone fruits and are difficult to grow where winter temperatures fall below –15°F (–26°C). In colder areas, even the hardiest varieties tend to be short-lived. They bloom early in spring and are best planted well above frost pockets where air drainage is good.

As with the other stone fruits, peaches are susceptible to root diseases and disorders caused by poorly drained soils; they grow best on well-drained loams, sandy loams, and loamy sands. Soil pH should be slightly acidic. On alkaline soils, peaches and nectarines are particularly susceptible to iron chlorosis.

Peaches

AT A GLANCE

Region: Depending on variety, peaches are rated from Zones 4 to 9 but are best adapted to warm and moderate climates (Zones 6–8 commercially). They tend to be difficult to grow where temperatures are below −15°F (−26°C). Challenging to grow in humid climates.

Soil pH and type: Slightly acidic, well-drained loams, sandy loams, and loamy sands

Pollination: Most varieties are self-fruitful, but a few varieties are self-unfruitful.

Variety selection: Varieties cover a wide range of growing regions and ripening times. Consult with local, state, and provincial fruit specialists to select varieties.

Pests and diseases. In humid regions of the southeastern United States, peaches and nectarines are considered to be high to very high management crops because of disease and pest problems. Organic peach production in humid climates can be particularly challenging due to the peach's susceptibility to a host of bacterial and fungal diseases. Two species of peach tree borers, plum curculio and Oriental fruit moth, represent major pests. From an organic perspective, peaches and nectarines are much easier to produce in the arid western parts of North America than in eastern regions. Select disease-resistant varieties, and use aggressive pest and disease programs, particularly in humid climates.

Training. Of all temperate zone tree fruits, peaches are probably the most labor intensive. Unlike most other tree fruits, peaches do not form fruits on spurs. Instead, the fruits form on 1-year-old shoots, and the trees require extensive annual pruning to keep them productive. To maintain large size, the fruit must be thinned, and particular care must be taken during harvest not to damage the delicate fruit.

On the positive side, peaches and nectarines mature quickly and begin bearing marketable crops 3 to 4 years after planting. The trees are naturally small and can be grown with or without trellising.

Varieties. Of all the fruits described in this book, peaches are, by far, the hardest to make recommendations for. Many peach varieties are available, covering a wide range of growing regions and ripening times, from spring to fall. In 1995, *Modern Fruit Science* listed 102 peach and 25 nectarine varieties recommended for commercial and home production. As the authors predicted, the number of commercially available, high-quality varieties is far greater today. There are several reasons for the dramatic increase in peach varieties:

- The trees are relatively short-lived, and there is constant pressure from growers, processors, and marketers to develop new varieties.
- Peaches are probably the easiest tree fruit to breed as, unlike apples or pears, a high percentage of the seedlings strongly resemble their parents and meet selection standards.
- Peaches flower at a young age, allowing for early evaluations and selections.

In short, there are far more good peach varieties available than we can cover in this chapter. In areas with large-scale peach production, such as California or New Jersey, commercial peach growers are advised to consult with Cooperative Extension fruit specialists in selecting varieties. In addition, a fairly recently released series of 'Flamin' Fury' peach and nectarine varieties merits testing. They are not listed in table 6.5 because they have not yet been widely tested and are not recommended by many fruit specialists outside of New Jersey.

Arkansas and North Carolina have taken the lead in developing peach varieties resistant to bacterial blight. Varieties that show good overall quality as well as high resistance to bacterial blight include 'All Gold', 'Candor', 'Clayton', 'Derby', 'Dixiered', 'Emery', 'GoldJim', 'Norman', 'Pekin', 'Roygold', and 'Surecrop'.

Low-chilling varieties (100 to 400 hours) recommended by fruit specialists in California, Florida, Georgia, Louisiana, and/or Texas include 'Babcock', 'Delta', 'Earligrande', 'Flordacrest', 'Flordaglo', 'FlordaGrande', 'Florida Prince' ('FlordaPrince'), 'Gulf Crest', 'Gulf King', 'Gulf Prince', 'Sam Houston', 'Tropical Beauty' ('TropicBeauty'), 'Tropic Snow', and

'ValleGrande'. Many of these low-chilling varieties are hardy only in USDA Zones 8 to 9.

Flat or "doughnut" peach varieties have become popular in some markets. Varieties include 'Galaxy' and 'Saturn', as well as selections from breeding programs in New Jersey. While the fruits are an interesting novelty and can provide niche market advantages, the trees have only produced low to variable yields, and their disease resistance is only moderate. Be conservative with these varieties, and evaluate them thoroughly before planting substantial numbers of trees. Table 6.5 provides a balance of old favorites and promising, newer varieties of peaches. Most peach varieties are self-fruitful and do not require cross-pollination.

Rootstocks. Peaches and nectarines are usually grown on peach seedling rootstocks, most often Halford, Lovell, and Bailey. Bailey is reported to be, possibly, the most cold-hardy of these rootstocks. Particularly in colder climates, some fruit tree nurseries use rootstocks from seedlings of locally adapted varieties. In medium- to high-chilling climates where nematodes are a problem, the clonal rootstocks Nemaguard and Nemared are sometimes used, although Nemaguard is suitable for only very well-drained soils. In low-chilling climates, Flordaguard is a good nematode-resistant rootstock. Dwarfing rootstocks are not commercially available for peaches. Given the naturally small tree size, dwarfing rootstocks are not really required.

Nectarine

A COMMON MISCONCEPTION is that nectarines are hybrids between plums and peaches. In reality, nectarines and peaches are the same species; nectarines simply lack fuzzy skins. Nectarines can develop from peach seeds or bud sport mutations on peach trees, and vice versa. In general, nectarine fruits are somewhat smaller than peaches and ripen earlier in the season. The lack of fuzz on the skins makes them more attractive for fresh use, but it also increases damage from insects such as plum curculio, thrips, and green peach aphid. The fuzzless skins may also make the nectarines more susceptible to bacterial and fungal fruit diseases. Nectarines tend to be more susceptible to brown rot than peaches and are harder to produce organically in humid regions.

California produces about 98 percent of the nectarines grown in North America, mostly in the southern San Joaquin Valley for the fresh market. Culturally, nectarines are treated the same as are peaches. In terms of varieties, there are far fewer nectarines than peaches. As with peaches, most nectarine varieties are self-fruitful and do not require cross-pollination. Table 6.6 lists some suggested varieties.

Plum

PLUMS ARE AMONG THE MOST, IF NOT THE MOST, adaptable tree fruit crops in the world, with commercial production in such diverse climates as those found in Israel, Egypt, China, Norway, Tunisia, and Hungary. This is due to their widely diverse genetic base. Around 15 species are used either for cultivation or breeding, and there are more than 2,000 named plum varieties.

World leaders in commercial plum production are Spain, Chile, the United States, South Africa, Italy, the Netherlands, Poland, and Serbia. Fewer prunes (plums that develop enough sugar to allow the fruit to dry without rotting around the pit) are produced than plums. In North

Plums

AT A GLANCE

Region: USDA Zones 8 or 9 in southern United States to Zone 3 in northern United States and Canada, depending on variety

Soil pH and type: Slightly acidic, well-drained loams, sandy loams, and loamy sands

Pollination: Self-fruitful to partially self-fruitful to self-unfruitful, depending on type and variety

Variety selection: Types include European, Damson, Japanese, and North American plums.

Other notes: European plums benefit from having beehives in the orchard; Japanese varieties tend to overset and often perform well in commercial settings with only wild bees for pollination.

America, most commercial plums and prunes are grown in the western United States, with California being the leading producer, followed by Idaho, Michigan, Oregon, and Washington. California produces more prunes than plums. Texas and areas adjacent to the Great Lakes produce plums for local markets. Commercial production in eastern North America can be challenging due to insect and disease pressure.

Cold hardiness ranges from USDA hardiness Zones 8 or 9 in the southern United States to Zone 3 in the northern United States and Canada. Native soils vary from droughty beach sands to deep, heavy river bottoms. Although somewhat more tolerant of heavy and poorly drained soils than other stone fruits, most plums are still highly susceptible to root diseases and disorders and are best planted on slightly acidic, well-drained loams, sandy loams, and loamy sands.

Plums bloom very early in spring and are susceptible to frost damage. They are also susceptible to brown blight and other fungal and bacterial diseases. To avoid frost and reduce disease pressure, plant plums on sites with good to excellent air drainage.

Types of Plums

Because of the diversity of species and varieties, it is hard to make generalizations about plums. Plants range from small sprawling shrubs to 20-foot-tall trees. Plum varieties vary greatly in their suitability for fresh use, freezing, canning, and drying. Pay close attention to variety descriptions to ensure your selections will meet your planned needs.

European Plums

European plums (*Prunus domestica*) are, by far, the most important commercial plums and have been cultivated in Europe for 2,000 years or more. In North America, they are the most adaptable and most commercially important plums. This crop likely developed from a hybrid between other species, as no wild form has been discovered. European plums are typically divided into four or five groups: prune, greengage, yellow egg, and Imperatrice and Lombard.

Prunes. 'Stanley', 'Sugar', and 'Italian' are well-known varieties. Prune-type plums can be used fresh or for drying or canning.

Greengage. These plums are round, sweet, tender, juicy fruits with greenish-yellow or golden skins. Popular varieties include 'Reine Claude', 'Imperial Gage', 'Jefferson', and 'Washington'. These varieties are used primarily fresh and for canning.

Yellow egg. Not widely grown, these plums are used for canning. The best-known variety is 'Yellow Egg'.

Imperatrice- and Lombard-type. These plums are sometimes lumped together under the Lombard group. The two are similar in most respects, except that the Imperatrice plums have blue skins with heavy, waxy blooms on the skin and the Lombard plums are red. These varieties are mostly used fresh and include 'Bradshaw', 'Diamond', 'Grand Duke', 'Lombard', and 'President'.

European plums require similar site conditions as other stone fruits, although European varieties tend to be more cold-hardy than sweet cherries, peaches, and nectarines. Particularly in cooler climates, such as the Northeastern United States, European plums are longer-lived and provide more consistent cropping than Japanese varieties, partly because their later blooming allows them to better escape frost damage. Trees often perform poorly in southern, warm-climate areas. Pest and disease problems are similar to other stone fruits.

European plums are naturally rather small trees, and rootstocks that provide some dwarfing are available. For home orchards, plant the trees at least 18 to 20 feet apart. For high-density commercial orchards, trees can be spaced as closely as 6 feet apart in rows that are 12 to 15 feet apart, although spacings of 10 to 20 feet in rows that are 16 to 20 feet apart are more common. Training methods include modified central leader for upright varieties to open vase shapes. Plums are normally not grown on supports.

Many European plum varieties are available, some of which have been cultivated for centuries. There is a relatively small number of commercially acceptable European plum varieties and an even smaller number for organic growers. It is critically important that you select disease-resistant varieties.

In the fruit-growing regions of British Columbia, 'Demaris' and 'Greata' are sometimes recommended. California varieties include

'Empress', 'Express', 'French Prune', 'Imperial', and 'Sugar'. 'Edwards' and 'Empress' are recommended in New Mexico.

For cold (Zone 4) locations, consider growing 'Bavay's Gage', 'Geneva Mirabelle', 'Golden Transparent', 'Imperial Epineuse', 'Kuban Comet', 'Mirabelle de Metz', 'Mount Royal', 'President', 'Reine de Mirabelle', 'Seneca', 'Victory', or 'Brooks'. Suggested European plum varieties are listed in table 6.7.

Damson Plum

Damson plum (*Prunus insititia*) resembles *P. domestica,* and some authorities consider it a subspecies of European plum. The crop was well known in ancient Rome and the name "Damson" supposedly relates to the city of Damascus. The fruits are small and tart and have limited uses for eating out of hand. They are used to make preserves suitable for home use and niche markets.

The trees resemble their European cousins and are cold-hardy and resistant to diseases. Although named varieties exist, the trees come nearly true to seed. The most important variety, 'St. Julien', is used to produce seedlings for use as plum rootstocks. For culinary use, 'Damson', 'French Damson', and 'Shropshire' are often recommended. These trees are small and ripen in the middle to later part of the season. They are considered self-fruitful.

Japanese Plums

Japanese plums (*Prunus salicina*) probably originated in China. This group produces high-quality fruits that are primarily used fresh. The crop was brought to North America around 1870 and was a favorite of Luther Burbank, who introduced many named varieties. Popular Japanese plums include 'Santa Rosa', 'Queen Ann', and 'Shiro'.

Most Japanese plums tend to be less cold-hardy than European varieties, bloom earlier, do not live as long, are less reliably productive, and are more susceptible to damage from fluctuating winter temperatures. Japanese varieties are best grown in the same climates as peaches. While you will see nurseries rate Japanese plums suitable for USDA Zones 4 to 9, most varieties are far better adapted to Zones 6 to 8 and grow best on sites free of spring frosts. Recently, varieties have been developed in Florida and Georgia for low-chilling (150 to 400 hours, Zone 9) and medium-chilling (400 to 700 hours) regions. Some of the newer selections have also proven to be quite disease resistant and well adapted to humid eastern North America. These varieties offer great opportunities for organic plum growers.

Japanese plums have been hybridized with various plum species native to North America to produce very cold-hardy selections for the northern United States and southern Canada, as well as low-chilling, disease-resistant varieties for the southeastern United Sates. Suggested Japanese plum and hybrid varieties are listed in tables 6.8 and 6.9.

North American Plums

Various North American plum species are cultivated on small, local scales.

American plum (*Prunus americana*) is a cold-hardy species native from the east coast of the United States to the Rocky Mountains and from Florida to Montana. The crop is cultivated as far north as south-central Canada and proved hardy in my northern Idaho trials. The trees are rather slow to come into production and are typically self-unfruitful. American plum fruits have tough yellow to orange skins and yellow flesh. Named varieties include 'De Soto', 'Hawkeye', 'Wyant', 'Weaver', and 'Terry'. Plant two varieties together for cross-pollination.

Canada or black plum (*Prunus nigra*) is a cold-hardy species native to Canada from New Brunswick to southeastern Manitoba. 'Cheney' is the best-known variety of this crop and is cultivated from the central Mississippi Valley into southern Canada.

Hortulan or wild plum (*Prunus hortulana*) is native to the central United States from Texas to Nebraska and eastward to Maryland and Virginia, excluding the southeastern states. While it lacks the cold hardiness and flavor of American plum, it is more resistant to brown blight. This crop is used mostly for processing; varieties include 'Wayland' and 'Golden Beauty'. Wild plums have been valuable as parents in developing low- and medium-chilling, disease-resistant Japanese hybrids.

Munson's or wild goose plum (*Prunus munsoniana*) is cultivated in the southern Mississippi Valley, where it has proven to be resistant to frost and brown blight. 'Wild Goose' is a named variety for this crop.

Beach plum (*Prunus maritima*) is native along the Atlantic Seaboard from Virginia to New Brunswick and is considered hardy in USDA Zones 3 to 7, depending on seed source. The dark blue to purple fruits are quite tart, are generally used for processing, and have been the subject of repeated efforts at cultivation. In 1948, for example, the Cape Cod Beach Plum Growers Association was formed in Massachusetts but proved short-lived. Beach plums today are mostly used as edible ornamentals in landscapes. Fruits are harvested from the wild and sold commercially in products targeting the tourist industry. The fruits range in color from red to blue to purple to black and are about the size of a cherry. Flavors are tart and resemble plums or grapes. Four named varieties of beach plum exist, 'Autumn' probably being the most reliable producer. The bushes flower each year but often bear only every 2 to 3 years. Beach plums are quite susceptible to the same pests and diseases that affect peaches and plums.

Oregon, Sierra, or **Klamath plum** (*Prunus subcordata*) is native along the Pacific coast from southern California to northern Washington. The small, tart fruits are used for sauces and preserves similarly to cranberries. Susceptibility to brown blight has made cultivation difficult, but the species is being tested as a dwarfing rootstock for peaches and plums.

Pollination

Plums range from self-fruitful to partially self-fruitful to self-unfruitful. Unfortunately, self-fruitfulness in some varieties seems to vary from region to region and can be affected by weather. Unless you know that your selections are self-fruitful in your area, plant at least two compatible varieties together to ensure cross-pollination. If space is limited, you can bud a pollinizing variety into your trees.

Within European plums, the variety 'Esperen' cannot be used as a pollinizer. 'Coe's Golden Drop' and 'Allgrove's Superb' cannot pollinize one another, and 'Cambridge Gage', 'Late Orange', and 'President' cannot pollinize one another. Note that some early-blooming varieties cannot effectively pollinate some late-blooming varieties, and vice versa. European plums are not consistently effective pollinizers for Japanese plums, partly due to compatibility problems and partly due to the fact that Japanese varieties bloom as much as 3 to 4 weeks earlier than most European varieties.

Japanese plums tend to be self-unfruitful and either require or benefit from cross-pollination. Some varieties are unsuitable as pollinizers. For the varieties listed in table 6.8, 'Shiro' is not suitable as a pollinizer.

'Burbank Elephant Heart', 'Redheart', and 'Santa Rosa' are considered particularly good pollinizers.

For commercial plantings, European plums benefit from having beehives in the orchard. Japanese varieties tend to overset and often perform well commercially with only wild bees for pollination.

Rootstocks

Plums are normally budded onto plum seedling rootstocks, peach seedling rootstocks, or a relatively small number of clonal rootstocks. A few of the more common rootstocks are listed here. When selecting a rootstock, consider your soil drainage, winter temperatures, nematodes in your soil, diseases in your area, and the training system you plan to use.

Myrobalan plum (cherry plum, *Prunus cerasifera*) **seedlings** produce large, vigorous trees and are cold-hardy, long-lived, and well adapted to most soils, including wet winter soils. They are somewhat susceptible to root rot organisms.

Myrobalan 29C rootstock produces trees that are somewhat less vigorous than those on myrobalan seedlings. This rootstock is immune to root knot nematode but susceptible to root lesion nematode, oak root fungus, and root rot. Trees on Myrobalan 29C can be poorly anchored and blow over in windy areas.

Marianna 2624 rootstock is considered by some fruit specialists to be the best overall choice for plum rootstocks. It is resistant to oak root fungus, root rots, root knot nematodes, and crown gall, but it is susceptible to bacterial canker and root lesion nematode. This rootstock produces a small, rather shallow-rooted tree and is, perhaps, the best plum rootstock for heavy soils and rather poorly drained soils. It tends to produce suckers.

St. Julien is a Damson plum variety (*Prunus insititia*) whose seedlings are used for rootstocks. These rootstocks tend to be hardy and relatively disease resistant, while providing some dwarfing.

Peach seedlings are compatible with most plum and prune varieties and produce moderately large trees that bear early and set consistent crops. 'Lovell' peach seedlings are often used as rootstocks for peaches and plums. Peach rootstocks are somewhat resistant to bacterial canker but cannot tolerate heavy or otherwise poorly drained soils. They are

Western Sandcherry Rootstocks

Western sandcherry (*Prunus besseyi*) seedlings serve as semidwarfing rootstocks. Unfortunately, these rootstocks sucker badly, produce inferior-quality fruit, and are partially incompatible with plum varieties. They are not recommended.

generally susceptible to oak root fungus and only recommended for well-drained soils. Some selections are resistant to nematodes. Plums are not used as rootstocks for peaches.

Citation is a peach-plum hybrid rootstock that produces full-sized trees and is quite tolerant of heavy or otherwise poorly drained soils.

American plum (*Prunus americana*) can be a good rootstock for plums in general, but it may sucker heavily. Use American plum seedlings as rootstocks for Japanese-American plum hybrid varieties.

Suggested Apricot Varieties

Table 6.1

This list is based, in part, on recommendations by Bob Purvis, chair of the North American Fruit Explorers Apricot Interest Group.

Alfred Zones 4–8	The small to medium-sized fruits have good flavor. The trees are reportedly very hardy and dependably productive.
Autumn Royal Zones 5–9	The medium-sized to large fruits are suitable for fresh use, canning, and drying. Ripens late.
Blenheim (Royal) Zones 5–9	The medium-sized to large fruits are considered by some to be the best-flavored and most popular apricot in the world. Very popular in California for commercial production. Self-fruitful.
Earlicot Zones 5–9	This large, highly-colored, orange fruit has firm, freestone flesh with good flavor. Ripens early and is popular in California for commercial use.
Goldcot Zones 4–8	The medium-sized, yellowish-orange fruits are firm and suitable for fresh use and canning. Does not have the best fruit quality. Cold-hardy and dependable.
Golden Amber Zones 5–9	The large, uniform, yellow fruits have firm, yellow, slightly acid flesh. Suitable for fresh use, canning, and drying. Requires about 500 hours of chilling and is suitable for areas with mild climates.
Goldkist Zones 7–9	This large, yellowish-orange, freestone variety requires only about 300 hours of chilling and is well suited for home production in the southern United States.

Goldrich Zones 5–9	The large fruits have firm, fine-textured, deep orange flesh. Vigorous, hardy, and productive but blooms early and is sensitive to apricot ring pox disease. Resists splitting after rain. Requires cross-pollination and will not pollinate 'Perfection'.
Harcot Zones 4–8	The medium-sized to large, slightly red-blushed fruits have sweet, smooth, fine-grained flesh with good flavor. Ripens early. Suitable for local direct sales but does not ship well. Reportedly has moderate to good resistance to perennial canker, bacterial spot, and brown rot. Disease resistance makes this variety a good choice for organic growers.
Harglow (Harglo) Zones 5–9	The medium-sized, bright orange fruits have firm, freestone flesh. Suitable for fresh use, U-pick, and home canning. Trees are compact. Blooms late. Moderately resistant to perennial canker, brown rot, and bacterial spot. Disease resistance makes this a good choice for organic growers. Yields best with cross-pollination.
Hargrand Zones 4–7	The large fruits have smooth, orange, freestone flesh. Best for fresh use but can be processed. Somewhat tolerant of brown rot, bacterial spot, and perennial canker. Disease resistance makes this a good choice for organic growers.
Harlayne Zones 4–7	The medium-sized, red-blushed fruits have firm, orange, freestone flesh with good flavor. Suitable for fresh use and home processing. Requires careful thinning. Reportedly resistant to perennial canker, moderate to good resistance to brown rot and bacterial spot. Disease resistance makes this a good choice for organic growers.

Harogem Zones 4–7	The small to medium-sized, red-blushed fruits have firm, orange, freestone flesh with good flavor. Developed for the fresh market and stores well. Reportedly resistant to perennial canker and brown rot. Disease resistance makes this a good choice for organic growers.
Harojoy (formerly HW 446) Zones 4–7	The medium-sized to large fruits have firm, freestone flesh. Excellent for fresh use. Ripens early. More cold-hardy than 'Veecot' (see below). Reported to be tolerant of bacterial spot, brown rot, canker, and cracking. Disease resistance makes this a good choice for organic growers.
Harostar (formerly HW 436) Zones 4–7	The medium-sized to large fruits have firm, freestone flesh. Ripens uniformly. Reportedly resistant to bacterial spot, brown rot, and cracking. Disease resistance makes this a good choice for organic growers.
Katy Zones 5–9	The large fruits have mild-flavored, freestone flesh. Requires only about 300 hours of chilling and is popular in southern locations.
Large Early Montgamet (Chinese Golden) (Mormon) Zones 4–9	The fruits are medium-sized and ripen over an extended period. Late bloom helps reduce spring frost damage. Self-fruitful.
Moongold Zones 4–8	The medium-sized fruits have yellowish-orange, freestone flesh with a sweet, slightly acid flavor. Suitable for fresh use and processing. May ripen unevenly, and fruit cracking and premature fruit drop have been reported. Trees are vigorous. Disease resistance makes this a good choice for organic growers. Needs 'Sungold' as a pollinizer.

Newcastle Zones 5–9	The medium-sized, yellow, red-blushed fruits are sweet and suitable for fresh use. A low-chilling variety that requires 300–400 hours of chilling. Can be grown in the southern United States and has been recommended for testing on high-elevation sites in Hawaii. Reportedly produces prolific crops.
Perfection Zones 5–9	The large, yellowish to deep orange fruits have yellowish to deep orange flesh with good fruit quality and flavor. Very hardy and productive. Will not cross-pollinate 'Goldrich'.
Riland Zones 5–9	The fruits color up before fully ripening, have good quality, and ship well. One of the earliest-ripening commercial varieties. Very popular for commercial production in Washington State. Requires cross-pollination and is a good pollinizer for 'Perfection'.
Sungold Zones 4–8	The medium-sized fruits have orange, freestone flesh with a sweet, mild flavor. Suitable for fresh use and processing. Heavy bearer. Pollinate with 'Moongold'.
Tilton Zones 4–9	The medium-sized, yellowish-orange fruits have firm, sweet-tart flesh suitable for fresh use, canning, drying, and freezing. A leading variety for commercial freezing, drying, and canning. The trees are vigorous and productive. Blooms very early.
Veecot Zones 4–8	The deep orange fruits are attractive and hang well after ripening. Suitable for fresh use and canning. Bacterial spot has been a problem in Canada during some years, but tree health tends to be excellent. The traditional standard apricot for colder climates.
Wenatchee (Wenatchee Moorpark) Zones 5–9	The large, yellow fruits have yellow flesh with good flavor. Excellent for home fresh, dried, or canned uses. Compatible with 'Perfection' and 'Tilton' for cross-pollination.

Suggested Sweet Cherry Varieties

Table 6.2

Bing Zones 5–8	The large, dark, mahogany-red fruits are very sweet and excellent for fresh use. Most suitable to the western United States from Arizona and New Mexico to Washington. Not recommended for much of the eastern United States, particularly for fresh markets. Not considered to be of commercial quality in British Columbia. California growers pollinate with 'Early Burlat'.
Blackgold Zones 5–7	The dark red fruits have red flesh and are suitable for fresh use and processing. An introduction from Cornell University from a cross between 'Stark Gold' and 'Stella'. Late-blooming and reported to be disease resistant. Popular in Michigan, Pennsylvania, and Iowa. Use as a pollinizer for late-blooming varieties. Self-fruitful.
Black Tartarian Zones 5–8	The medium-sized, purplish-black fruits have dark red flesh. Produces vigorous growth. Widely adaptable and recommended by fruit specialists in California, Illinois, New Mexico, and Virginia.
Chelan Zones 5–8	The large, dark fruits have red flesh. Resembles 'Bing' (see above) but ripens 10–14 days earlier. Developed in Washington State and popular in the Pacific Northwest and Michigan.
Emperor Francis Zones 5–7	The medium-sized to large, yellow, red-blushed fruits are suitable for fresh use and processing. Reported to be more resistant to cracking than 'Napoleon'. Widely adaptable and popular in the eastern United States from Virginia to Delaware and New York, westward to Michigan and Indiana. Do not use as a pollinizer for 'Napoleon', 'Bing', or 'Lambert'.

Gold Zones 5–8	The small, golden yellow fruits have firm skins and soft flesh and are suitable for fresh use and processing. Produces good yields and can be mechanically harvested. Among the most cold-hardy sweet cherries. Widely adaptable and popular from Virginia to the northeastern United States and Upper Midwest. Not the best fresh-market cherry for the eastern United States due to rain cracking problems. Serves as a pollinizer for most sweet cherry varieties.
Hardy Giant Zones 5–8	The large, dark red fruits have dark red flesh. Best used for processing. An older variety popular in the Mid-Atlantic states. Pollinate with 'Stella'.
Hartland Zones 5–8	The medium-sized to large, dark red fruits have moderately firm flesh that is less sweet than other varieties. Best used for processing. Ripens early and resists cracking. Recommended from Virginia to Pennsylvania and New York. Susceptible to rain cracking problems in rainy areas. A good pollinizer for other sweet cherries.
Hedelfingen Zones 5–8	The small, black fruits are soft and suitable for processing. Resists cracking. An old variety that produces reliable crops. The trees are easy to train. Widely adaptable and popular from the Mid-Atlantic states to New Hampshire and westward to Indiana and New Mexico.
Hudson Zones 5–7	The black fruits are sweet and more resistant to cracking than many other sweet cherries. Moderately heavy crops ripen late in the season. Popular in the Mid-Atlantic states, the northeastern United States, and the Upper Midwest.
Kristin Zones 4–8	The large, purplish-black to black fruits are sweet. Particularly suitable for roadside stands and U-pick farms. Among the most cold-hardy sweet cherries.

Lambert **Zones 6–8**	The large, black fruits are suitable for fresh use or freezing. An older, late-season variety that remains popular in the western United States. Not generally recommended for eastern North America. Pollinate with 'Hardy Giant'.
Lapins **Zones 5–8**	The fruits are large and mahogany red. A newer variety from British Columbia that is becoming popular in the northwestern United States from Montana to Oregon and Washington. Also recommended for Indiana, Virginia, and British Columbia. Not the best fresh-market cherry for the eastern United States due to soft fruit and rain cracking. Self-fruitful and reported to be an excellent pollinizer.
Napoleon (Royal Ann) **Zones 6–8**	The small, yellow, red-blushed fruits have soft, light-colored flesh. Excellent for canning. Widely planted and recommended from Virginia to California and from New Mexico to Oregon. Susceptible to rain cracking. Not recommended by Pennsylvania fruit specialists. Do not use to pollinate 'Bing', 'Lambert', or 'Emperor Francis'.
Rainier **Zones 6–8**	The large, yellow, red-blushed fruits are sweet and flavorful. A Washington variety now grown from Georgia and Virginia westward to Arizona and New Mexico and northward from the Central Plains to the Upper Midwest and Pacific Northwest. Not recommended by Pennsylvania fruit specialists.
Sam **Zones 5–8**	The large, dark fruits ripen in mid-season. An older, cold-hardy variety still recommended in the northeastern United States, Central Plains, Upper Midwest, Intermountain states, and New Mexico. No longer considered commercially suitable in British Columbia. Pollinate with 'Van'.
Stark Gold **Zones 5–7**	The large, yellow fruits ripen in early mid-season. Popular in the Central Plains and New Mexico.

Stella Zones 6–7	The large, medium to dark red fruits are firm and sweet and resist cracking. Widely popular across the United States from Virginia to California and from Arizona to Utah. Grown in British Columbia. Rather cold-tender. Self-fruitful.
Sweetheart Zones 5–7	The medium-sized, bright red fruits have good flavor. Ripens very late. Produces vigorous growth and is prone to over-cropping and small fruit size. Requires careful management to maintain training and acceptable fruit size. Popular in Virginia, Michigan, and the Pacific Northwest. Self-fruitful.
Ulster Zones 5–8	The large, dark fruits have sweet flesh and are suitable for fresh use and processing. Resists cracking. Popular in the Eastern and Midwestern United States.
Van Zones 4–8	The medium-sized, black fruits have firm flesh and resemble 'Bing' (see above) but are smaller. An older, cold-hardy, vigorous variety that is often planted with 'Sam' (see above). Bears heavily. Most popular in the Northern and Central Plains westward and is also grown in California and Arizona. No longer considered commercially suitable in British Columbia. Can be difficult to grow for the fresh market in the eastern United States due to rain cracking. A good pollinizer for many sweet cherries, but do not use to pollinate 'Regina'.
Whitegold Zones 5–7	The medium-sized, yellowish-red fruits ripen in mid-season. A new variety gaining popularity. Reported to be disease- and crack-resistant. Self-fruitful.

Suggested Tart Cherry Varieties

Table 6.3

Balaton Zones 5–8	The medium-red fruits are larger and firmer than 'Montmorency' (see below) and are sweet enough to be eaten fresh. Ripens over an extended period and is well suited to farmers' markets. Popular in the Midwest and Upper Midwest, Northeastern United States, and the Mid-Atlantic states. Self-fruitful but produces larger and better crops with cross-pollination from any other tart cherry.
Danube Zones 5–8	The dark red fruits have dark red flesh and juice. Sugar concentrations are high, and the fruit can be eaten fresh. Less cold-hardy than 'Montmorency' (see below). A new introduction from Hungary. Recommended by fruit specialists in Michigan, Pennsylvania, and Virginia. Self-fruitful.
Early Richmond Zones 4–8	The small, red fruits are tart and suitable for cooking. Ripens early. Cold-hardy, but also recommended by fruit specialists in California and Georgia. Self-fruitful.
English Morello Zones 4–8	The dark red to nearly black fruits have dark red juice. Used primarily for cooking but can sometimes be eaten fresh when fully ripe. Suitable for canning, drying, freezing, and liqueurs. Self-fruitful.
Jubileum Zones 5–8	The dark red fruits have dark red flesh and juice and are sweet enough for fresh use. A new introduction from Hungary. Recommended by fruit specialists in Michigan and Pennsylvania.
Mesabi Zones 4–8	The fruits resemble 'Montmorency' (see below) but are sweeter and have smaller pits. Sugar concentrations are intermediate between typical sweet and tart cherries. Self-fruitful.

Meteor Zones 4–8	The medium-sized to large, dark fruits have yellowish flesh. The trees are small and very cold-hardy. Widely popular for home and market orchards and smaller U-pick operations. Self-fruitful.
Montmorency Zones 4–7	The dark fruits have yellow flesh and colorless juice. A very old variety that remains the standard for tart cherries and is still widely popular. Self-fruitful.
North Star Zones 4–8	The dark red fruits have dark red flesh and juice. Ripens in mid-season. The trees are 6–8 feet tall, productive, and very cold-hardy. Widely popular for home and market orchards and U-pick operations. Self-fruitful.
Surefire Zones 4–8	The red fruits have red flesh and juice, are larger than for 'Montmorency' (above), and ripen about 2 weeks later. Yields are lower than for 'Montmorency', and the trees are more open and easier to harvest. Blooms very late, providing some frost protection. Self-fruitful.

Suggested Bush Cherries

Table 6.4

CHERRY-PLUM HYBRIDS

Compass Zones 3–8	The small, dark purple fruits have yellowish, tart, juicy flesh. Used for canning. Grows 3–8 feet tall. Pollinate with another cherry-plum hybrid.
Opata Zones 2–8	The green-blushed fruits are suitable for processing. Grows 15 feet tall with a 10-foot spread. Pollinate with another cherry-plum hybrid.
Red Diamond Zones 3–8	The 1-inch-diameter, dark red fruits have sweet, smooth-textured flesh that is suitable for processing. Grows 6–10 feet tall. Pollinate with another cherry-plum hybrid.

Sapa Zones 2–8	The small, dark purple fruits are sweet and suitable for processing. Grows to 15 feet tall with a 10-foot spread. Pollinate with another cherry-plum hybrid.
Sapalta Zones 3–8	The medium-sized, purple fruits have sweet red flesh of excellent quality. Used for processing. Grows to 15 feet tall with a 10-foot spread. Pollinate with another cherry-plum hybrid.
DWARF SOUR-CHERRY HYBRIDS (*P. cerasus* × *P. fruticosa*)	
Carmen Jewel (SK Carmen Jewel) Zones 2–8	The dark red fruits are suitable for processing. Grows 6–8 feet tall. Train as an open-centered bush. A University of Saskatchewan introduction. Related varieties in the Saskatchewan releases that are suitable for testing include 'Crimson Passion', 'Juliet', and 'Romeo'. Self-fruitful.
Evans Zones 3–8	Slightly larger fruits than for 'Carmen Jewel' (above). The bright red fruits have clear juice and are suitable for processing. Grows 12–14 feet tall. Train to a single trunk. A University of Saskatchewan introduction. Self-fruitful.
NANKING CHERRY (*P. tomentosa*)	
Nanking Zones 2–7	The ¼- to ½-inch diameter fruits are bright red with yellow flesh. Sweet to tart, juicy, and have a pleasant flavor. For fresh use or processing. The bushes grow 6–10 feet tall with up to a 15-foot spread. Grown from seed. Can be highly susceptible to brown blight and is challenging to grow in humid climates. Requires cross-pollination from other Nanking cherries.

WESTERN SAND CHERRIES (*Prunus besseyi*)	
Black Beauty Zones 3–6	The fruits are ¾ inch in diameter, black, and thin-skinned. They are sweet and have fair to good quality for fresh use and processing. Grows to 8 feet tall. Considered self-fruitful.
Golden Boy Zones 3–6	The fruits are yellow and sweet. Suitable for fresh use or processing. Grows 4–5 feet tall.
Hansen's Zones 3–6	The small, dark purple fruits are suitable for fresh use or processing. Grows 3–6 feet tall. Requires cross-pollination from another western sand cherry.

Suggested Peach Varieties

Table 6.5

Candor Zones 5–8	The medium-sized fruits have yellow, semifreestone, medium-firm flesh. Resistant to bacterial blight and bacterial spot and popular in the southeastern United States. Ripens early.
Belle of Georgia (Belle) (Georgia Belle) Zones 5–8	The medium-sized to large fruits have white, freestone flesh with mild flavor and excellent dessert quality. Ripens in late mid-season. Widely recommended because of its cold hardiness. Resistant to bacterial spot but highly susceptible to bacterial blight. Not a good choice for organic growers in humid areas.
Biscoe Zones 5–8	The medium-sized to large fruits have deep yellow to orange, freestone flesh that resists browning and has fine texture and excellent quality. Excellent for canning. Ripens in late mid-season. Good winter hardiness. Highly resistant to bacterial spot. Widely popular. Requires thinning to maintain fruit size.

Contender Zones 4–8	The large fruits have firm, yellow, freestone flesh with good flavor and few split pits. Very productive and ripens in mid-season. Cold-hardy and widely recommended but moderately susceptible to bacterial blight.
Cresthaven Zones 5–8	The fruits are medium-sized to large and firm with yellow, freestone, browning-resistant flesh of excellent quality. Excellent for fresh use, canning, or freezing. Best for home use and local markets, as the fruit lacks sufficient color to compete against other cultivars. Ripens in early mid-season. The trees are vigorous and early bearing. Widely recommended but moderately susceptible to bacterial blight.
Derby Zones 5–8	The medium-sized to large fruits are medium-firm to firm with yellow, semiclingstone flesh. Noted to be very resistant to bacterial blight in North Carolina. Can be hard to find in nurseries.
Dixiered Zones 5–8	The medium-sized, round, clingstone fruits have red skins and yellow flesh with good quality. Ripens in mid-season. Noted as being resistant to bacterial blight and bacterial spot. Popular in the southeastern United States.
EarliGrande Zones 8–9	The small to medium-sized fruits have yellow, semifreestone flesh with excellent flavor. Low-chilling (200 hours) and ripens very early.
Flameprince Zones 5–8	The large, firm fruits have yellow, freestone flesh. Ripens late to very late. Widely popular in the eastern and southeastern United States.
Flavorcrest Zones 5–8	The large, firm fruits have yellow, freestone flesh. Ripens in mid-season. Widely recommended in the eastern and southcentral United States.

Florida Prince (FlordaPrince) Zones 8–9	The small, clingstone fruits ripen very early. Low-chilling (100 hours). Recommended in California, Florida, Louisiana, and Texas.
Gala Zones 5–8	The large, firm fruits have yellow, semifreestone flesh. Ripens in early mid-season to mid-season. Widely recommended in the eastern and south-central United States.
Garnet Beauty Zones 5–8	Fruit size ranges from small to large, depending on growing region. The yellow-fleshed, semifreestone fruits have good dessert quality. Resistant to bacterial spot. Ripens very early to early. Widely recommended.
Glohaven Zones 5–8	The large, firm fruits have yellow, freestone flesh that resists browning and has excellent dessert quality. Cans and freezes well. Ripens in late mid-season, and the mature fruits hang well on the trees. Popular in the Midwestern and western United States.
Harbelle Zones 5–8	The medium-sized to large fruits have yellow, semifreestone to freestone flesh of fair quality. Produces some split pits. Ripens in early season to early mid-season. The trees are small, cold-hardy, moderately vigorous, reliable, and consistently productive. Resistant to bacterial spot. Widely recommended.
Harbinger Zones 5–8	The medium-sized fruits have yellow, freestone flesh with good dessert quality. Resistant to bacterial spot. Ripens very early to early. Requires careful thinning.
Harken Zones 5–8	The medium-sized fruits have yellow, semifreestone to freestone, browning-resistant flesh of good quality. Produces some split pits. Ripens in mid-season. Productive, moderately vigorous, and cold-hardy. Resistant to bacterial spot. Widely recommended.

Harrow Beauty Zones 5–7	The medium-sized fruits have firm, freestone flesh. Ripens in mid-season. A Canadian introduction popular in the northeastern United States.
Harrow Diamond Zones 5–7	The medium-sized, yellow, red-blushed fruits have yellow, nearly freestone flesh. Ripens early. A Canadian introduction popular in the northeastern United States. Reportedly has good resistance to bacterial spot, brown rot, and perennial canker.
Harvester Zones 5–8	The medium-sized to large, bright red fruits have moderately firm to firm, yellow, semifreestone to freestone flesh of excellent quality. Popular in the southeastern and southcentral United States.
Junegold Zones 7–9	The large, moderately firm fruits have yellow, fine-textured, clingstone flesh. Ripens in early mid-season. Recommended in the southeastern and southcentral United States. Medium-chilling (650 hours).
Klondike Zones 5–7	The large to very large yellow, red-blushed fruits have firm, freestone, subacid, white flesh. Ripens in mid-season. Popular in the northeastern United States. Requires 700–800 chilling hours.
La Feliciana Zones 7–8	The medium-sized to large fruit has firm, freestone flesh with excellent flavor. Ripens in late mid-season. Medium-chilling (550 hours). Popular in California, Louisiana, and Texas.
Loring Zones 6–8	The large to very large fruits have yellow, freestone flesh with red around the pit. Not as attractive to consumers as red-skinned varieties. Suitable for fresh use, freezing, and canning, but the flesh can be soft. Ripens in late mid-season to late season. Resistant to bacterial spot. A reliable producer and widely recommended.

Madison Zones 5–8	The medium-sized, firm fruits have yellow, freestone flesh with fine texture and excellent dessert quality. Ripens in late mid-season. Has excellent winter hardiness and good frost tolerance. Resistant to bacterial spot. An old standard that remains widely recommended.
Majestic Zones 7–8	The medium-sized, firm fruits have yellow, freestone flesh that resists browning. Ripens in late mid-season. Medium- to high-chilling (800 hours). The trees produce heavily. Recommended in the southeastern and southcentral United States.
O'Henry Zones 7–8	The large fruits have very firm, yellow, freestone flesh of high quality. Ripens late to very late. Best for warmer climates.
Pekin Zones 5–9	The fruits are medium-sized and semiclingstone. Ripens in early mid-season to mid-season. Very resistant to bacterial blight. A North Carolina introduction.
Redglobe Zones 5–8	The medium-sized to large fruits have firm, yellow, freestone flesh of good quality. Ripens late to very late. Medium- to high-chilling (850 hours). Recommended in the southeastern and southcentral United States.
Redhaven Zones 5–8	The medium-sized fruits have firm, yellow, semi-freestone flesh that resists browning and has good quality. An industry standard and excellent for canning, freezing, and fresh use. Pits can be hard to remove and pit splitting can be excessive. Ripens in early mid-season. Susceptible to brown rot blossom and fruit rot but resistant to bacterial spot. Requires heavy thinning and pruning and is susceptible to frost. Once the most widely planted freestone peach and still widely recommended. Being replaced by newer varieties. Not the best choice for organic growers. Best grown in arid regions.

Redskin Zones 5–8	The medium-sized to large fruits have firm, yellow, freestone flesh with red around the pit. Excellent dessert quality and suitable for fresh use, canning, and freezing. Medium- to high-chilling (750 hours) and one of the most widely recommended varieties in North America. Resistant to bacterial spot.
Reliance Zones 4–8	The medium-sized fruits have rather soft, yellow, freestone flesh of fair to good quality. Blooms late but ripens early for some frost resistance. A New Hampshire introduction that is often regarded to be the most cold-hardy peach. Requires heavy thinning. Widely recommended in cooler climates where cold hardiness is critically important. Has been reported to bear fruit after exposure to −25°F (−32°C). Resistant to bacterial blight.
Sun Haven Zones 5–8	The large fruits have yellow, freestone flesh that resists browning and has excellent dessert quality. Ripens very early. The trees are large, vigorous, productive and require little thinning. Resistant to bacterial spot.
Surecrop Zones 5–8	The medium-sized fruits have yellow, semiclingstone, medium-firm flesh. Ripens in mid-season. Resistant to bacterial blight and very popular in eastern North America. One of the better peaches for organic growers.
Tropical Beauty (TropicBeauty) Zones 8–9	The medium-sized, semifreestone fruits ripen very early. Low-chilling (150 hours). Recommended in Florida and Texas.
White Lady Zones 5–9	The large fruits have white flesh and excellent flavor. Ripens in mid-season.

Suggested Nectarine Varieties

Table 6.6

Arctic Glo Zones 5–8	The small, red fruits have red-mottled, white flesh and excellent flavor. Ripens early.
Armking Zones 6–9	The medium-sized fruits have yellow flesh. Ripens early. A low- to medium-chilling variety suited to the southern United States.
Fantasia Zones 6–8	The large, brightly colored fruits have firm, yellow, freestone flesh with excellent flavor and very high quality. Ripens late. A consistent producer and the most widely recommended nectarine in North America.
Flavortop Zones 5–8	The large fruits have yellow, freestone flesh with excellent flavor. Ripens in mid-season.
Hardired Zones 5–8	The medium-sized fruits have good dessert quality. Ripens in mid-season. Consistently productive in the cool, maritime climate of the northeastern United States and southeastern Canada. Some resistance to bacterial spot and brown rot but highly susceptible to leaf curl. Develops some skin russet.
Mericrest Zones 5–8	The medium-sized fruits have yellow, freestone flesh with excellent dessert quality. Some resistance to bacterial spot and bacterial blight. A New Hampshire release considered to be among the most cold-hardy nectarines. Widely recommended.
Red Gold (RedGold) (Redgold) Zones 5–8	The large fruits have yellow, freestone flesh with excellent flavor and dessert quality. Fruit set is moderate, and the trees are vigorous and time-consuming to prune. Second only to 'Fantasia' nectarine in recommendations by fruit specialists.
Summer Beaut (Summer Beauty) Zones 5–8	The medium-sized fruits have yellow, freestone flesh with excellent dessert quality. Ripens in mid-season. Widely recommended.

Sunglo Zones 5–8	The large fruits have yellow, freestone flesh. Ripens in early mid-season. The trees are hardy and vigorous. Widely recommended.
Sunhome Zones 8–9	The fruits have yellow, semifreestone flesh. A low- to medium-chilling variety (250 hours) recommended for central Florida and coastal Texas and Louisiana.
Sunraycer Zones 8–9	The reddish-orange fruits have yellow, semifreestone flesh with good flavor. A low- to medium-chilling variety recommended for southern and central Florida and along the Gulf Coast.
Sunred Zones 6–9	The medium-sized, yellow-blushed red fruits have firm, yellow flesh with good flavor. An older low- to medium-chilling variety (200–300 hours) adapted to the southern United States.

Suggested European Plum Varieties

Table 6.7

Bavay Gage Zones 4–9	The medium-sized, gage-type, yellowish-green fruits have yellow flesh. Ripens late and hangs well on the tree. The trees are small. Best for home orchards. Generally self-fruitful.
Bluefre Zones 5–8	The large, blue-skinned fruits have firm greenish-yellow, freestone flesh. Comes into production quickly. The trees are vigorous, spreading, and moderately productive but have been reported to be inconsistent producers in New Jersey. Widely recommended. Self-fruitful but produces better crops with another European plum as a pollinizer.
Brooks Zones 4–8	The large, blue fruits have firm, greenish-yellow, freestone flesh with tart but good flavor. Ripens in late part of mid-season to early part of late season. The trees are vigorous and upright to somewhat spreading. Self-fruitful.

Earliblue Zones 5–8	The blue, prune-type fruits have soft, yellowish-green, semifreestone flesh and good flavor. Ripens very early. The trees are upright, vigorous, and moderately productive. Requires cross-pollination. Widely recommended.
Early Italian Zones 5–9	The medium-sized to large, blue, prune-type fruits have yellowish-green, freestone flesh. Ripens in mid-season. The trees are vigorous and upright to somewhat spreading. Pollinate with another European plum.
Geneva Mirabelle Plum (Mirabelle 858) Zones 4–9	The small, yellow fruits have yellow flesh. Can be eaten fresh but are primarily used for preserves, pastries, and canning. Requires cross-pollination.
Golden Transparent Zones 4–9	The yellow, gage-type fruits have yellow flesh. Ripens late to very late. Self-fruitful.
Greengage (Reine Claude) Zones 6–9	The small, greenish, gage-type fruits have yellow, semiclingstone flesh with excellent flavor. Suitable for fresh use or processing. The trees are moderately vigorous and spreading. A very old variety brought to North America around 1770. Still widely recommended but susceptible to brown rot. Better European and Japanese varieties are available for organic growers. Requires cross-pollination.
Imperial Epineuse Zones 4–9	The medium-sized to large, reddish-purple, prune-type fruits have yellow, freestone flesh and excellent flavor. Ripens early to mid-season. The trees are upright and require cross-pollination.

Italian Zones 5–9	The large, purplish-blue, prune-type fruits have greenish-yellow, freestone flesh with excellent flavor. Suitable for fresh use, processing, or drying. Ripens in late mid-season to late season. The trees are moderately vigorous, upright to somewhat spreading, and moderately productive. Widely recommended.
Kuban Comet Zones 4–9	The medium-sized, orange to red fruits have tart flesh. Resists cracking and ripens early. The trees are small and, reportedly, productive and easy to grow. A Russian introduction that is adapted to colder climates. Self-fruitful.
Mount Royal Zones 4–8	The small to medium-sized, bluish-black, Imperatrice-type fruits have yellow, freestone flesh of good quality. Suitable for fresh use, processing, or freezing. Ripens in mid-season to late mid-season. Trees are productive and among the most cold-hardy European plums. A Canadian introduction. Self-fruitful.
President Zones 4–8	The large, blue, Imperatrice-type fruits have yellow flesh. Suitable for fresh use, canning, or freezing. Ripens very late. The trees are vigorous and upright. Recommended in California and New Mexico. Requires cross-pollination.
Reine de Mirabelle Zones 4–8	The medium-sized to large, yellow fruits have good flavor and are suitable for fresh use and preserves. Requires cross-pollination.
Seneca Zones 4–9	The large, reddish-purple, prune-type fruits have soft, yellow, freestone flesh. Suitable for fresh use, processing, or drying. Resists cracking but is too soft for shipping. Best for home orchards. The small, spreading trees tolerate black-knot disease. A New York introduction. Requires cross-pollination.

Stanley Zones 5–7	The large, blue, prune-type fruits have yellowish-orange, freestone flesh with excellent flavor and quality. Suitable for fresh use, processing, and drying. Ripens in late mid-season to late season. The trees are moderately vigorous, upright to somewhat spreading, and productive. The most widely recommended prune-type plum but susceptible to brown rot. Will perform best for organic growers in semi-arid climates. Self-fruitful.
Valor Zones 5–8	The medium to large, blue fruits have greenish-yellow, freestone flesh with good flavor. The trees are vigorous, upright to somewhat spreading, and reliably productive. Resistant to bacterial spot but susceptible to black knot. Self-fruitful. Pollen-incompatible with 'Victory' and 'Vision' plums.

Suggested Japanese and Japanese-Hybrid Plum Varieties

Table 6.8

AU Homeside Zones 6–8	The medium-sized, red fruits have yellow flesh of good quality. Ripens in mid-season. The trees lack vigor but are resistant to plum scald. Highly recommended in the southeastern United States. Requires at least 700 chilling hours. Pollinate with 'AU-Producer', 'AU-Roadside', 'AU-Rosa', or 'AU-Rubrum'.
AU Producer Zones 6–8	The small, dark red fruits have sweet, red flesh suitable for fresh use and processing. Ripens in early mid-season. The trees are moderately vigorous to vigorous, require heavy pruning, and bear early. Resistant to black knot disease and plum leaf scald. Highly recommended in the southeastern United States. Requires at least 700 chilling hours. Pollinate with 'AU-Homeside', 'AU-Roadside', 'AU-Rosa', or 'AU-Rubrum'.

AU Roadside Zones 6–8	The small, dark red fruits have red flesh of very good quality. Suitable for fresh use and processing, but fruits tend to be too soft for shipping. Ripens in early mid-season. Among the most disease-resistant plums. Highly resistant to black knot, bacterial canker, bacterial fruit and leaf spot, and plum leaf scald. Best for home use and local markets. The trees are vigorous. Highly recommended in the southeastern United States. Requires at least 700 chilling hours. Pollinate with 'AU-Homeside', 'AU-Producer', 'AU-Rosa', or 'AU-Rubrum'.
AU Rubrum Zones 6–8	The medium-sized to large, red fruits have red flesh with good flavor. Excellent for fresh use and acceptable for canning. Ripens early and stores well. Highly resistant to bacterial canker, bacterial fruit spot, bacterial leaf spot, and black knot, and tolerant of plum leaf scald. Requires at least 700 chilling hours. Perhaps the best of the AU plum series and highly recommended in the southeastern United States. Self-fruitful and a good pollinizer for other AU plums.
Burbank Zones 5–8	The medium-sized to large, red fruits have amber flesh. Ripens in mid-season. Productive and bears early.
Burbank Elephant Heart Zones 5–8	The large to very large, purple fruits have red, freestone flesh. Excellent fresh or for canning. Ripens late. Best for arid and semiarid areas. Pollinate with another Japanese plum.
Byron Gold (Byrongold) Zones 5–8	The medium-sized, yellow, red-blushed fruits have firm, yellow flesh with good quality. Ripens early season to early mid-season. Somewhat susceptible to plum leaf scald. Requires about 450 hours of chilling. Widely recommended. Pollinate with 'Segundo'.

Early Golden Zones 5–8	The small to medium-sized, yellow, red-blushed fruits have amber, clingstone flesh and good flavor. Ripens early season to early mid-season. Somewhat resistant to bacterial spot. Productive and vigorous. Especially popular in the eastern United States. Pollinate with 'Methley' or another early-blooming Japanese plum.
Fortune Zones 5–7	The large to very large, yellow, reddish-purple-blushed fruits have firm, yellow, clingstone flesh with good flavor. Ripens late. The trees are upright to somewhat spreading, vigorous, and consistent producers. Reportedly grows better on heavier soils than some other plums. Requires cross-pollination.
Gulfbeauty Zones 8–9	The small, reddish-purple fruits have soft, yellowish-green, clingstone flesh. Good quality for an early plum and suitable for fresh use. Ripens early, before 'Gulfblaze'. The trees are vigorous and somewhat spreading. Very resistant to bacterial spot and moderately resistant to plum leaf scald. A low-chilling variety (150–250 hours) released by the University of Florida. Pollinate with another 'Gulf' series plum.
Gulfblaze Zones 8–9	The small, dark reddish-purple fruits have orange, semifreestone flesh of good quality. Ripens in mid-season, 1–2 weeks after 'Gulfbeauty'. The trees are less vigorous than 'Gulfbeauty' and may not live as long. Highly resistant to bacterial spot and plum leaf scald. A low-chilling variety (150–250 hours) released by the University of Florida. Pollinate with another 'Gulf' series plum.

Gulfrose **Zones 8–9**	The medium-sized to large, red fruits have red, semifreestone flesh of high quality with good firmness and storage characteristics. Sweet and aromatic without a bitter aftertaste. Ripens with or up to a week after 'Gulfblaze'. The trees are moderately vigorous, somewhat spreading, and bear as early as their second season. Resistant to plum leaf scald and bacterial spot. A low-chilling variety (150–250 hours) released by the University of Florida. Pollinate with another 'Gulf' series plum.
Methley **Zones 5–8**	The small, reddish-purple fruits have soft, red, clingstone flesh with fair to good flavor. Suitable for fresh use or preserves. Ripens early and does not store well. A medium-chilling variety (about 450 hours). An older Japanese × American hybrid plum that is still widely recommended across the United States. Susceptible to plum leaf scald and other bacterial diseases. Among the more cold-hardy Japanese plums, but better-quality, more disease-resistant varieties are now available for organic growers. Reportedly self-fruitful but benefits from cross-pollination.
Morris **Zones 5–8**	The large, red fruits are sweet and have high quality, particularly for baking and processing. A medium- to high-chilling variety (800 hours) often recommended in the southeastern United States from South Carolina to Texas. Ripens in early to mid-season. The trees are naturally small. Texas fruit specialists recommend this variety for the Hill Country and recommend pollinating with 'Ozark Premier'.

Ozark Premier Zones 5–8	The medium-sized to large, purplish-red fruits have yellow to red, clingstone flesh with a tart flavor. Suitable for fresh use or processing. Ripens in early to mid-season and does not store well. Medium- to high-chilling (800 hours). One of the most widely recommended Japanese plums but susceptible to plum leaf scald and moderately susceptible to bacterial spot. Best grown in arid and semiarid locations. The trees are vigorous, spreading, and moderately productive. Organic growers have better choices, especially for humid areas. Pollinate with another Japanese plum.
Redheart Zones 5–8	The medium-sized, red fruits have firm, crisp, red, semifreestone flesh. Ripens in mid-season to late mid-season. The trees are vigorous, upright to somewhat spreading, and moderately productive. Among the most widely recommended Japanese plums. Requires cross-pollination from another Japanese plum and is a good pollinizing variety.
Ruby Queen Zones 5–8	The large, reddish-purple fruits have firm, red, clingstone flesh with good flavor. Ripens in mid-season to late season. A Georgia introduction that is moderately susceptible to bacterial spot but often recommended for the southeastern United States because it tolerates humid conditions well. Requires cross-pollination with another Japanese variety.
Santa Rosa Zones 5–8	The medium-sized, reddish-purple fruits have firm, rather tart, yellow flesh. Ripens in mid-season. The trees are vigorous, rather spreading, and not consistently productive. Susceptible to plum leaf scald and moderately susceptible to bacterial spot. Among the most often recommended Japanese plum varieties but not the best choice for organic growers, especially in humid areas. Best suited to arid and semiarid locations. Pollinate with another Japanese plum.

Satsuma Zones 5–8	The small, red fruits have red flesh best suited for home canning and desserts. Among the more cold-hardy Japanese plums. Recommended in arid and semiarid regions of western North America. Requires cross-pollination with another Japanese plum.
Segundo Zones 6–8	The yellowish-orange fruits have soft, red, clingstone flesh with good flavor. Ripens in early mid-season. A Japanese × American hybrid, medium-chilling variety (400–500 hours) that is well suited to the southeastern United States. Pollinate with 'Byron Gold'.
Shiro Zones 5–8	The medium-sized to large, yellow fruits have soft, juicy, yellow, clingstone flesh with good quality. Ripens early. Among the most widely recommended Japanese plums in Canada and the United States. Has proven reliable in such diverse areas as British Columbia, California, western Washington, Kansas, upstate New York, New Hampshire, and Virginia. Moderately vigorous and moderately resistant to bacterial spot. One of the better Japanese plum varieties for organic growers in colder areas. Well suited to U-pick and roadside stands. Requires cross-pollination with another Japanese variety.
Vanier Zones 5–8	The small to medium-sized, purplish-red fruits have yellow, clingstone flesh with good flavor. Ripens in mid-season. An Ontario introduction and among the more cold-hardy Japanese plums. Quite resistant to bacterial spot. The trees are vigorous, somewhat spreading, and moderately productive. The small fruit size limits commercial use. Requires cross-pollination.

Suggested Hybrid Plum Varieties

Table 6.9

Alderman Zones 4–8	The large, burgundy-red fruits have yellow, clingstone flesh. Suitable for fresh use and processing. Ripens late. The trees are vigorous and spreading. Often recommended in the Central Plains and Northcentral states but not the most cold-hardy variety. Pollinate with another variety in this table.
LaCrescent Zones 4–8	The small to medium-sized fruits have freestone flesh with apricot-like flavor. Ripens in mid-season to late mid-season. Not suitable as a pollinizer due to nonviable pollen. Pollinate with another variety in this table.
Pembina Zones 3–8	The large, red fruits have yellow, clingstone flesh. Ripens in late mid-season. A South Dakota introduction and among the most cold-hardy plums. Pollinate with another variety in this table.
Pipestone Zones 3–8	The large, red fruits have amber, clingstone flesh suitable for fresh use and preserves. Among the most cold-hardy plums. Widely recommended across the northern United States. Pollinate with another variety in this table.
South Dakota Zones 4–8	The small to medium-sized, yellow, red-blushed fruits have yellow, freestone flesh with good flavor. Ripens late. The trees are very vigorous, spreading, and productive. Not as widely recommended as other hybrid plums and can be hard to find in nurseries. A good pollinizing variety. Pollinate with another variety in this table.
Superior Zones 4–8	The large to very large, red fruits have yellow, clingstone flesh with good quality for processing. Requires thinning to prevent overbearing. Widely recommended across the northern United States. An excellent pollinizing variety. Pollinate with another variety in this table.

Toka Zones 4–8	The small to medium-sized red fruits have yellow flesh with a spicy flavor. Widely recommended across the northern United States, primarily as a pollinizing variety for other hybrid plums. Pollinate with another variety in this table.
Underwood Zones 4–8	The medium-sized to large, red fruits have yellow, clingstone flesh suitable for fresh use and processing. Ripens early. Widely recommended across the northern United States. Not suitable as a pollinizer due to nonviable pollen. Pollinate with another variety in this table.
Waneta Zones 4–8	The red fruits have yellow flesh. Ripens in mid-season. A South Dakota introduction. Not as widely recommended as other hybrid plums. Pollinate with another variety in this table.
PLUOTS	
Dapple Dandy Zones 5–9	The large, greenish to yellowish, red-spotted fruits have firm, juicy, pink to red, freestone flesh. Very sweet and ranked highly for flavor. A plum-apricot hybrid that ripens in late mid-season. Requires 400–500 chilling hours. Pollinize with 'Flavor Supreme' or 'Santa Rosa' plum.
Flavor King Zones 5–9	The medium-sized, burgundy fruits have very sweet, juicy, red flesh. Requires about 400 chilling hours, perhaps less. Pollinize with 'Flavor Supreme' or 'Santa Rosa' plum.
Flavor Supreme Zones 5–9	The medium-sized to large, greenish-red fruits have mottled skins and sweet, juicy, red flesh that is noted for good flavor. Requires 500–600 chilling hours. Pollinize with 'Dapple Dandy', 'Flavor King', or 'Santa Rosa' plum.

7

CHAPTER

OBTAINING AND PLANTING YOUR TREES

- Purchasing and Handling Your Trees | 230
- Preparing for Planting | 235
- Planting and Training | 243
- First Year Care and Management | 245

We have reached the stage where it is time to order trees for planting. By now, you have carefully evaluated your site, corrected drainage and nutrient problems, designed and laid out your orchard or evaluated your existing orchard layout, and picked out crops and varieties. The next steps involve purchasing high-quality plants, caring for them before planting, and setting them into the ground.

Purchasing and Handling Your Trees

IF YOU HAVE A LARGE ORCHARD, you will likely buy your trees from specialized wholesale nurseries. Ask other fruit growers, fruit-growing organizations, fruit specialists at universities, and Cooperative Extension and Provincial offices for referrals. You want to make sure that you deal with nurseries that you can trust, not only to provide quality trees, but also expertise in fruit tree care during and after the purchase.

Homeowners have the options of buying from local retail nurseries or purchasing from mail-order companies. Here I cannot overemphasize the importance of dealing with reputable nurseries experienced in working with fruit crops. Ask other fruit growers, Master Gardeners, or local fruit specialists for referrals of nurseries you can trust. This goes for local nurseries as well as mail-order companies.

Avoid the temptation to save a few dollars by buying at big chain stores. At such places, you will rarely find salespeople skilled with tree fruits, and chances are great that you will end up with varieties and rootstocks unsuitable for your area. Instead, try to purchase only from well-known nurseries that specialize in growing the fruits you are interested in and that work for your climate. You are generally better off avoiding nurseries that offer everything from vegetables to ornamentals to fruits to herbs. Fruit growers in Florida are likely to deal with different suppliers than those in Manitoba, simply because of dramatically different climates. Avoid nurseries that sensationalize or exaggerate plant descriptions. If it sounds too good to be true, it probably is.

Before purchasing your trees, you should have a clear idea of the varieties and rootstocks that you want. Be firm and shop around, if necessary. Do not allow yourself to be pressured into buying varieties or rootstocks that you do not want. You may find it necessary to make some compromises, as varieties come and go in the nursery trade and not all the varieties and rootstocks that you want may be available. When you select your alternate trees, however, make sure you purchase something you like and that will perform well in your orchard. Don't buy plants simply because a nursery is overstocked on that item or is sold out of what you want.

Bare-Root Trees

Fruit tree planting stock for commercial orchards and from mail-order nurseries is usually sold as unbranched (whip) or branched (feathered), bare-root trees that have been dug from fields and had the soil shaken from the roots (see figure 7.1). Depending on the crop, these trees are about 3 to 6 feet tall. They are usually dug in the fall, stored in refrigeration over the winter, and shipped for spring planting. Particularly in warmer climates and for some crops, the trees or bushes may be planted in the fall or dug and shipped in the spring.

Figure 7.1

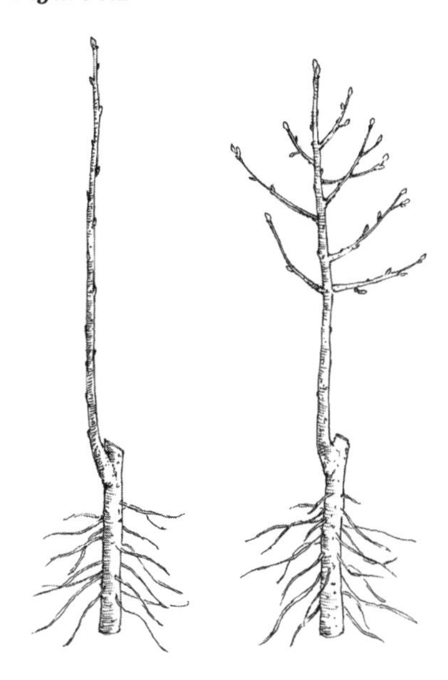

A) Whip

B) Feathered planting stock for orchards

Whips were once favored by commercial growers who wanted to choose the locations of the main scaffold branches for themselves. Particularly for apples and pears, feathered planting stock has become much more popular and works well with high-density plantings and axis training systems that avoid permanent scaffold branches in favor of small, short-lived fruiting wood that is replaced frequently. Not having to head back a whip to stimulate lateral branch growth can save at least a full year in getting to market with the new training systems.

When ordering bare-root trees, arrange to have the nursery deliver them as close to planting time as possible. Bare-root trees are highly perishable unless stored at temperatures around 32 to 38°F (0 to 3°C) in high humidity. By carefully timing the shipment and being ready to plant as soon as the stock arrives, you will save yourself the many headaches involved in storing planting stock.

If you cannot plant immediately upon delivery, arrange to have the planting stock placed into cold storage. The roots will probably be packed with damp shredded paper, sphagnum moss, or shredded wood fibers and wrapped in plastic film. Be sure to keep the materials and wrappings in place to prevent the roots from drying out.

If cold storage is not available, remove the plastic film and bury the roots and several inches of the lower trunks in damp sawdust or compost. This process is called "heeling" and prevents the roots from drying out. If the trees have broken dormancy and the buds are swelling or leaves are developing, protect the trees from freezing while they are heeled in.

Trees in Containers

Planting stock in local retail nurseries is often sold in containers. In many cases, the planting stock is purchased by the nursery as bare root and stuck into containers of potting soil for resale. You pay a premium for containerized trees, their chief advantage being convenience. You can allow the trees to remain in the pots for several months or more before planting, if necessary, without having to provide refrigerated storage. Containerized trees are also generally larger than the bare-root trees favored for commercial planting. Smaller orchard crops, such as bush cherries or saskatoons, are typically grown in pots and shipped with or without the containers enclosing the root balls. Again, the container-grown plants give you flexibility in deciding when to plant.

Containerized trees are generally easier to store than bare-root trees. If the trees have been kept outdoors in their containers at a nursery near your site, you can probably continue to leave the plants outdoors until planting. If, however, your site is still experiencing freezing temperatures and the trees are coming from a milder climate and have already broken dormancy, protect them against freezing temperatures. Even for the most cold-hardy varieties, buds that are close to breaking and emerged leaves and flowers can be killed at temperatures of about 29°F (–2°C) or less.

For example, your orchard might be located in Missoula, Montana, and the planting stock comes from a nursery in a coastal or southern region. You may still have snow on the ground with temperatures below freezing when the trees arrive, already leafed out. Exposing those trees to your subfreezing temperatures could damage or kill them. In such a case, protect the trees against freezing. Generally, they can be kept in cold storage at 34 to 40°F (1 to 4°C) for several weeks without serious problems. For orders of just a few trees, consider potting the trees up and leaving them inside an unheated garage or porch. For all kinds of planting stock, keep the roots moist but not soaking in or saturated with water.

Root-Bag Trees

Some planting stock is now grown in fabric root bags with either soil-filled root bags buried in the ground or bags filled with potting soil supported above the ground. These bags are made of the same material used for weed barrier fabrics and confine the roots within the bags. Some types of bags effectively prune at least some of the roots where they contact the fabric, thereby creating a dense, rather fibrous root system that will establish well once planted. Be aware, however, that root circling within root bags is not uncommon. Sometimes one or two roots penetrate the bags and most of the root system develops outside of the root bag, where it is lost when the plant stock is lifted from the nursery.

Be sure to examine each root ball at the time of planting. You should see an evenly distributed network of fine roots all around the outside of the root ball. Reject those with large, circling roots that will provide poor anchorage and possibly cause girdling later on. Also reject those with large roots that have penetrated the bag and been cut off.

When planting containerized trees, remove plastic containers completely before setting the trees into the ground. While this might sound absurdly simple, my colleagues and I are often called out to examine dead or dying plants only to discover that the plants were set into the ground pots and all. With no way for the root systems to grow, the trees soon die. For fiber pots, the situation is less clear. Some fiber pots quickly degrade in the soil and allow the roots to penetrate, while others do not. I recommend removing fiber pots completely, if at all possible. At the least, remove four wide, vertical strips of the fiber pot material around the sides of the pot and make an X-shaped cut across the bottom of the pot, penetrating into the root ball. When planting root-bag trees, always slit the root bags on several sides and strip away and remove all of the fabric as the trees are set into the holes. This practice is different from that used for ball-and-burlap landscape tree planting stock, where the burlap is usually left partially or completely in place. Tree roots can penetrate burlap to form healthy, well-distributed root systems. The roots cannot penetrate root-bag fabrics. If your trees come planted in plastic film bags, be sure to remove all of the film before you plant. Take care to disturb the root balls as little as possible when planting.

Ball-and-Burlap Trees

Fruit trees that are dug with the root balls intact and enclosed in burlap are sometimes used for small plantings where large trees are wanted quickly. This technique is called ball-and-burlap, and the burlap-wrapped root balls may or may not be supported with wire baskets.

Keep the root balls buried in moist compost, sawdust, or wood chips for short-term storage and plant as soon as possible. While they are in storage, keep the root balls moist and apply a liquid fertilizer weekly if the trees have broken dormancy and begun forming leaves and flowers.

Handle the trees very carefully, and do not allow the root balls to be dropped, shaken, or otherwise damaged. Lift the plants by the bottoms of the root balls and never by the trunk or stem. Pulling on or twisting a trunk will break many small roots within the ball and can stunt or kill a plant. Once the root ball is gently placed in the planting hole, remove any twine or other ties around the trunk and peel the fabric back away from the trunk for at least several inches. Slit the fabric vertically at least four times around the root ball. Do not remove the burlap fabric or wire basket; trying to do so often damages the root balls and can stunt or kill the trees. Research has shown that the roots easily penetrate the burlap and grow through the wire basket openings.

Irrigation Systems, Trellises, and Fencing

Before planting, you should have your irrigation system in place and operational, as well as trellis poles and anchors (see chapter 12). If your site is susceptible to pressure from moose, white-tailed deer, or other herbivores, strongly consider installing an herbivore fence before planting your trees or bushes (see chapter 11). This is the only effective method of keeping herbivores out of an orchard in areas where the animals are abundant.

Preparing for Planting

In chapter 4, we went into detail on how to prepare the field and lay out planting blocks. Now we need to mark the planting rows within those blocks. Two methods are commonly used to mark planting rows: you can stake out rows by hand, or use a tractor. For small orchards, choose one side of the planting block that will be parallel to the tree rows. Using this line as your reference, measure along the adjacent ends of the planting blocks and place stakes at what will be the ends of each row. For example, say you want a 12-foot row spacing. Starting at the corners of the block and following the marked end lines, place a stake every 12 feet along both end lines. Depending on your design, you might want to locate the first row 6 feet from the side of the block and the second and all other rows at 12-foot spacings. Stretch a measuring tape or marked rope between corresponding stakes and set stakes or flags to mark where you want the trees within rows (see figure 7.2). Tying knots or flags in a rope as long as the orchard block simplifies this process (figure 7.2). Ensure the knots or flags are the same distance apart as you want your trees.

For larger orchards or where the tree lines follow a curve, obtain a plastic pipe or metal electrical conduit a few feet longer than the desired distance between rows and fasten a short pipe at a right angle to one end (see figure 7.3). Fasten the other end of the pipe to the front end of a tractor. The pipe should extend the length of the desired between-row

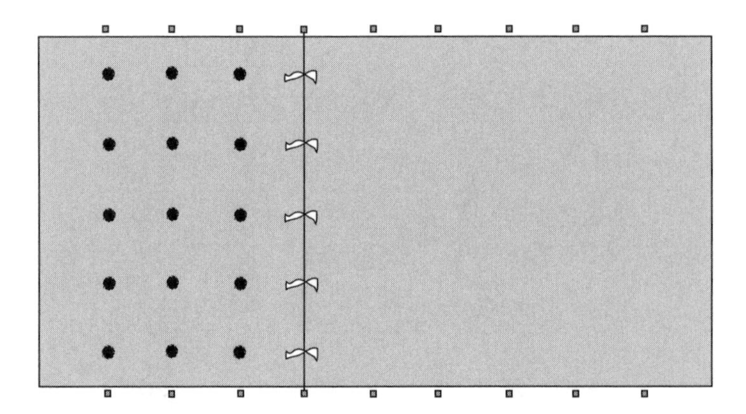

Figure 7.2 · **Laying Out Planting Rows with a Marked Rope**

Use stakes or flags to mark the ends of the tree rows along the sides of the planting block. Using a rope longer than the length of the tree rows, tie knots or flags the distance apart that you want your trees planted. Pull the rope tightly between corresponding stakes along the ends of the planting block, and use stakes to mark the location for each tree in that row. The trees can be set in a square grid (as shown here) or offset (figure 3.7).

spacing. Equip the tractor with a small plow blade, shank, or coulter wheel attached to a tool bar and centered on the back of the tractor. By centering the outer end of the pole over the reference line for the block, you create a furrow in the field marking the first tree row. Repeat the procedure using the furrows you make as reference lines.

You can use a measuring tape or flagged rope, as described above, to mark planting hole locations within the rows. A wooden pole or section of PVC pipe cut as long as you want the trees apart is also a convenient way to measure between planting holes. This technique works particularly well in high-density plantings where trees are closely spaced within rows.

When the trees are spaced more than about 12 feet apart, you can use the tractor and reference bar method to locate the planting holes. Set the bar length to mark one-half the desired in-row spacing, place the reference point of the bar over a block-end reference line, and plow at right angles to the planting rows (see figure 7.4). Mark every other furrow intersection point in each row with a flag or stake. This design creates a triangular planting that maximizes the distance between trees in adjacent rows.

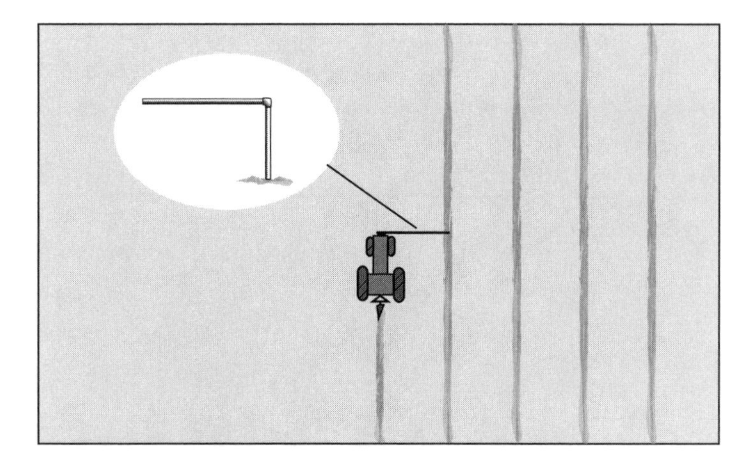

Figure 7.3 · **Laying Out Planting Rows with Tractor**
Install a chisel or narrow-furrow plow to the toolbar on the tractor. Attach a spacing rod that extends from the front center of the tractor the distance that you want the tree rows apart. Lightweight metal electrical conduit or rigid PVC pipe works well. A 90-degree extension (see inset) that reaches nearly to the ground at the outer end of the spacing bar is helpful in keeping the rod centered over reference lines. Keep the end of the spacing rod centered over the reference line you established earlier along the edge of the planting block and plow along the tree rows. Then use the tree row you have made as a reference line and repeat.

Digging the Planting Holes

When planting only a small number of trees, a hand shovel works very well. For larger numbers of trees, a tractor-mounted auger will be highly desirable for most sites, except those too rocky for augers to be used effectively. In extreme cases, a small backhoe may be used. In all cases, dig the holes on the centerline of the planting row and be sure the holes are evenly spaced within the row. As the trees mature, uneven rows and poorly spaced trees will make access difficult and increase management problems.

However you dig them, make sure the holes are large enough to place the roots inside without bending any of the roots or crowding them together against the sides of the hole. In most cases, planting holes 18 inches in diameter and 18 inches deep should suffice. Ball-and-burlap trees will require larger holes. If in doubt, measure the root balls

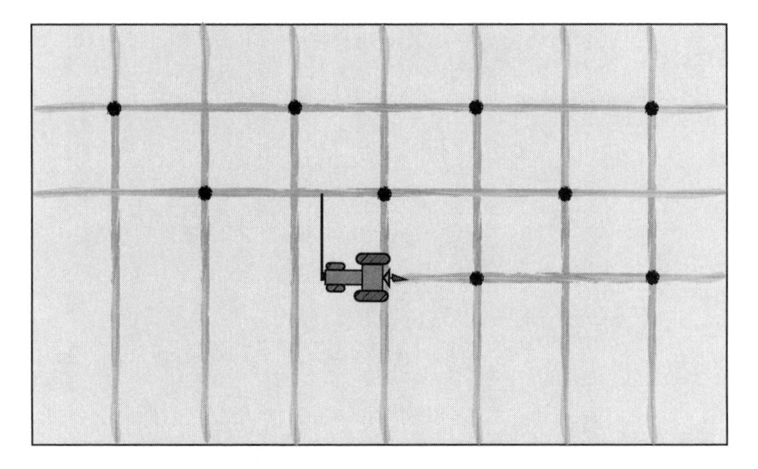

Figure 7.4 · **Locating Planting Holes Using a Tractor and Shank**

Lay out the tree crop rows as described in figure 7.3, then adjust the spacing bar for one-half the distance between trees within rows. Use one edge line of the planting block as a reference line and plow a furrow. Every other intersection of this new furrow with the tree crop rows is the location for planting a tree. Staggering the planting hole locations in adjacent rows creates a triangular tree layout that maximizes the distance between trees. Repeat using each new furrow as a reference line.

or spread of the bare roots, or contact the nursery you are buying from before digging the holes. Have a shovel on hand during planting to make any necessary adjustments to the hole dimensions. Do not prune roots or cut down root balls in order to fit them into a hole. Instead, make the hole wider. When planting large numbers of trees, you may find it helpful to have the holes dug before the trees arrive from the nursery.

Use a hand shovel to break away any smooth, compacted layers on the sides of the holes (glazing) before setting in the trees. Glazing is most likely to occur when soils are wet, contain large amounts of clay or silt, and when an auger is used to dig the holes. The dense, smooth glazing can inhibit root growth and interfere with water movement.

Make the holes deep enough to set the plants at the same depth as they grew in the nursery. Planting too deep or too shallow can stunt or kill plants. Because the soil and the trees will settle somewhat in the planting holes, you may choose to plant trees slightly shallower than they grew in the nursery. In some cases, trees with interstems are planted deeply enough to place the lower graft union at or slightly below the soil surface. When using this strategy, use only trees that have had the interstem grafted very low on the rootstock so as not to plant too deeply.

A common recommendation is to form a firm, cone-shaped mound of soil in the bottom of the hole to drape the tree roots over. This strategy works well for bare root trees and small plantings. The idea is to set the tree roots on the cone and backfill the planting hole, evenly distributing the roots around the hole and providing maximum support. When planting hundreds or thousands of trees, however, the method is usually too slow and labor intensive to be practicable. Holes for containerized or ball-and-burlap trees need to be just deep enough to receive the root ball and maintain the height at the same depth the tree grew in the nursery. You can make the holes wider, however, so that you can make adjustments to the plant's location.

Prepping the Trees

Years ago, fruit specialists sometimes recommended pruning the roots and tops of newly planted trees to "keep them in balance." We no longer make that recommendation. Prune bare roots only if they have been broken during digging or shipping. Never prune roots simply so that they will fit into a planting hole.

Dealing with girdling roots. For container-grown plants and some plants grown in root bags, circling roots can form around the outsides of the root ball on plants left too long in the containers. A few roots circling partway around the outside of a root ball do not cause any problems. In such cases, use a very sharp utility knife to make two or three slices down the sides and a cross on the bottom of the root ball. If in doubt, wash the soil gently away from the roots of a plant or two. I have found plants where the roots developed first in coils about 4 inches in diameter and later in coils about 6 inches in diameter. In these severe cases, the plants had been rootbound in 4-inch starter containers before being transplanted to 1-gallon containers and again allowed to become rootbound. If severe root circling is present, document what you have found with photos and reject the shipment. Trees with circling root systems will never anchor properly and may eventually die due to the roots girdling one another or the main trunk.

You may have heard of a practice called "butterflying," where the planter attempts to salvage rootbound container-grown plants by splitting the root ball with a shovel. At best, rootbound plants treated this way will be slow to establish and may well be stunted their entire lives; in many cases, plants will simply die. If your plants are so rootbound as to need the root balls split, discard the plants and obtain healthy planting stock.

Checking for pests and diseases. If you see evidence of diseases or pests on the roots, collar, or trunks, stop immediately. Disease and pest symptoms include tumor-like swellings, cankers, brownish or black discoloration, stubby roots, swollen root tips, borer holes on trunks and stems, and the pests themselves. *Do not plant infected or infested trees.* Contact the nursery immediately and provide them with photos, if you can, of the affected tissues. Planting infected or infested stock not only ensures that you will have at least one problem tree; it can also introduce very serious pests and diseases into your orchard. Plant only healthy, pest- and disease-free trees.

Keeping roots moist. During the planting process, keep bare roots and root balls moist at all times. For bare-root trees, bring out of storage no more trees than you can plant in an hour or so and keep the roots covered with wet burlap sacks or similar materials. People sometimes recommend placing the roots inside a bucket of water before planting. If you have kept the roots properly moist, this step should not be necessary, and extended immersion can damage or kill roots.

Adding Beneficial Microorganisms

Mycorrhizal fungi (mycorrhizae) are naturally occurring, soil-dwelling fungi that form associations with plants by colonizing the roots and spreading fungal strands (mycelia) into the soil. The fungi benefit from receiving carbohydrates and other biochemicals from the roots. In exchange, the mycorrhizae greatly increase the effective root surface area and increase uptake of water and nutrients. By colonizing the root surface, the mycorrhizae also help exclude and protect against harmful microbes, including certain root rot fungi. Trees with mycorrhizal associations will generally be more drought-tolerant and require less fertilizer than trees without mycorrhizae.

Types of microorganisms. Many different mycorrhizal types and species exist and your soils probably already contain healthy populations. If your soil has been damaged by fumigation, neglect, or poor management, however, your trees may benefit from adding mycorrhizae. Otherwise healthy sandy soils can also be deficient in mycorrhizae.

Endomycorrhizae are the most useful for fruit trees and many other crop species. These species of fungi penetrate into the root cells of the host plants. Mycorrhizal products sold for fruit growers and other organic producers typically include one or more of the following endomycorrhizae (*Glomus* species are the most common endomycorrhizae found in commercial products):

- *Gigaspora margarita*
- *Glomus aggregatum*
- *Glomus clarum*
- *Glomus deserticola*
- *Glomus intraradices*
- *Glomus monosporum*
- *Glomus mosseae*
- *Paraglomus brasilianum*

Ectomycorrhizal fungi live on the outside surfaces of host plant roots. While ectomycorrhizae have a much smaller range of plant hosts than endomycorrhizae and are not generally considered beneficial for fruit trees, they may be included in blends with the endomycorrhizae species listed above. Ectomycorrhizal species found in commercial mycorrhizal blends include:

- *Pisolithus tinctorius*
- *Rhizopogon amylopogon*
- *Rhizopogon fulvigleba*
- *Rhizopogon luteolus*
- *Rhizopogon villosullus*

Ericoid mycorrhizae are beneficial for blueberries, huckleberries, lingonberries, rhododendrons, and other acid-loving plants. They are not useful for the crops described in this book.

Choosing products. When purchasing a mycorrhizae product for your orchard, keep the following two strategies in mind. First, deal with companies that specialize only in microbial products, rather than general garden products. There are many commercial mycorrhizal products sold, not all of which are effective; deal with someone who really knows mycorrhizae. Second, don't buy any mycorrhizal product that does not list the species included and the spore count. If you do not know what is in the product, you have no way of knowing if it will benefit your orchard.

The same caution applies to spore count. The higher the spore count, the more concentrated and effective the product will be. If the species included in the product and spore count are not specified, be very skeptical of the product. Don't be lulled by advertising "testimonials." There are very good products on the market; it's worth taking the time to find them. Before using any microbial product, ensure that it can be used by certified organic growers in your area.

A product that blends at least several species of endomycorrhizae is usually preferable to a product with a single fungal species. More species means that it will be adaptable to a wider range of soils, host plants, and growing conditions. Again, endomycorrhizae are much more likely to benefit fruit trees than ectomycorrhizae. For root dips and root ball drenches, purchase only mycorrhizal inoculants. Products that include both mycorrhizae and fertilizers can damage the roots and are not recommended at the time of planting. These combination products are best applied after the trees are planted.

Treating the plants. The best way to treat bare-root plants is to dip the roots into a slurry containing desirable mycorrhizae just before planting. For container-grown or ball-and-burlap trees, drench the root balls with water containing mycorrhizae. Follow the product label directions.

Certain bacteria can also be valuable as a root treatment at the time of planting. Crown gall is a bacterial disease that causes tumor-like galls on the roots, collars, trunks, and stems of many species of woody plants. The disease is caused by *Rhizobium radiobacter* (formerly known as *Agrobacterium tumefaciens*), which dwells in the soil and can exist without a plant host for many years. The disease is usually spread to uninfected fields on infected planting stock or on soil particles carried from an infected field on equipment, runoff water, or feet. Certain closely related but nonpathogenic bacteria can protect against crown gall by colonizing the roots and excluding the disease organism. Dipping the bare roots into a slurry containing *Rhizobium (Agrobacterium) radiobacter* strains K84 or K1026 immediately before planting can provide good protection for at least the first year and help the plants become well established. Be sure that the product you use is registered for certified organic growers and for your state or province.

Adding Amendments to Soil

As we discussed in chapter 4, if you are going to add an amendment such as compost, peat moss, or other organic materials, with the exception of phosphorus, it is best to add it to the planting row or the entire orchard rather than to the planting holes.

Adding amendments to planting holes can create very serious problems and should never be used for fruit trees and bushes. The rule of thumb is that the only thing besides the roots that goes into the hole is what came out of it in the first place.

Add amendments throughout the planting row or entire orchard and till them in 6 to 12 inches deep. For bush fruits, you may obtain satisfactory results by amending only the planting row. Fruit trees have much larger root systems that quickly grow far beyond the planting row. Once the amendments have been applied and tilled in, mark the planting rows, dig the holes, and set the trees. You gain all of the benefits of the amendments and avoid the problems. The amendments will be available to the growing root systems as the orchard crops mature and to the roots of alley crops and companion plants.

In soils where available phosphorus is deficient, you can add phosphorus directly to the planting hole before setting the trees into place. This is also a good strategy on soils with pH below 6.0. For organic growers, add one or two handfuls of steamed bonemeal to the bottom of the planting hole, cover the bonemeal with 2 inches of soil, and set the trees in place. This can help maintain the plants until the root systems become well established. Bonemeal will not damage the plants. Never add materials that contain nitrogen, potassium, or boron to the planting holes.

Planting and Training

Once the holes are dug and you have pruned off damaged roots and treated the roots with beneficial mycorrhizae and radiobacter organisms, set the roots or root balls gently into the planting holes at the same depth as or slightly shallower than the plants grew in the nursery. Start backfilling the planting hole in shallow layers, firming each layer with your hands, until the hole is filled. Remember to backfill using only the material that came out of the hole and avoid the temptation to tramp the soil into place with your feet. The unfortunately popular "death stomp" can severely damage roots, especially bare roots. Once the hole is filled, you can gently walk around the plants to firm the soil in place.

A common practice is to create shallow bowls of soil around newly planted trees to catch rain and irrigation water. On soils that are well drained to excessively well drained, the practice can be helpful. On heavier soils or where drainage is otherwise poor, the tree basins can create saturated soils around the roots and increase the chance of root disorders and diseases. Tree basins are most likely to be used for very small plantings where the trees are irrigated individually by hand. Where overhead or drip irrigation is used, tree basins are likely to cause more trouble than good.

Once the trees have been set and the soil firmed around the roots, an excellent strategy is to irrigate heavily to settle the soil around the roots. This practice helps eliminate air pockets around the roots that can cause the roots to dry out. After irrigating, check to see that the plant is at the same level at which it grew in the nursery and adjust the depth, if necessary.

Depending on crop, rootstock, and training system, you may now need to stake or trellis the trees, as we will discuss in chapter 12. Not all trees need staking or trellising, however. Unfortunately, the current trend, particularly in the landscape industry, is to stake every newly planted tree and then tie it down with enough rope for the last roundup. There seems to be a fear that the trees will somehow escape.

Larger ball-and-burlap and containerized trees can be top-heavy and may benefit from temporary staking to prevent them from falling or blowing over, even if they will eventually be freestanding. In these cases of freestanding trees, the usual procedure is to use three evenly spaced wires staked to the ground, each separated by 120 degrees. The wires are looped around the tree trunk just above a branch to keep the loop from slipping down the trunk. Rubber hose is used to line the loops and protect the trunk from damage due to rubbing. When staking trees in this fashion, keep the loops loose enough to allow the tree to move in the wind several inches in each direction. That motion forms the reaction wood that will give the trunk strength and allow it to eventually stand on its own.

Even small trees destined to be freestanding are sometimes staked in order to form straight trunks. Likewise, trees that will be trained to tall spindles and other axis systems are often trained to poles, even if only temporarily. Metal tree stakes, bamboo poles, metal electrical conduit, and PVC pipe all serve as excellent temporary training stakes, depending on the height that you need. Fasten the trees to the stakes using wide tape designed for tying trees and grapes. Avoid zip ties, wire, fishing lines, and thin cords, as these cut into the bark. Check the ties frequently and loosen them, as necessary, to prevent girdling the trunk.

For trees that will be trained to trellis wires, some growers prefer to avoid the use of stakes or poles, largely due to cost, and tie the trees directly to the trellis wires. In such cases, the lowest trellis wire is usually no more than 2 to 3 feet above the ground. Special wire clips that surround the trunk and clip to the trellis wire are available from nursery supply companies. These clips allow you to quickly and securely support the trees on the trellis wires and are a marked improvement over tying.

For trees that are to be grown on poles or trellises, whether to install the support posts and wires now or later depends on the training system and size of the trees. In general, it is best to train the trees early. Straightening a bent trunk or developing a new scaffold limb on an older trunk can be challenging. Within reason, the sooner you can get the trees trained, the quicker the tree will produce and the fewer problems you will have correcting badly trained trees.

If you are training the trees to posts, such as with the slender spindle system described in chapters 3 and 12, you can plant the trees and set the posts at the same time. When the tree support posts are metal pipes or T posts that will be driven into the ground, rather than buried, the posts can be driven in after the trees are planted, taking care not to drive the posts through the root balls.

First Year Care and Management

FOR MOST ORCHARD CROPS, the critical steps following planting are to keep the soil moist, control vegetation around the trees and bushes, and control pests and diseases.

Depending on the crop and training system, you may also head back newly planted trees to encourage scaffold limbs at a particular height and may begin summer pinching on whips and feathered trees as you select and develop scaffold limbs or fruiting wood. For slender spindle and axis systems, you will need to begin tying trunks and lateral branches to trellis wires and poles. For all tree crops, use clothespins, weights, and/or spreader bars to bend down branches and develop wide branch angles. The latter step, alone, greatly adds to a tree's strength and health. These steps are discussed in detail in chapter 12.

Irrigation

Regardless of the irrigation system used (see chapter 2), the goal is to keep the soil moist, but not saturated, throughout the root zone. A common tendency is to overirrigate newly planted trees and bushes. As long as the soil is moist, no irrigation is needed. Use this strategy throughout the planting year and the year following planting, while the root systems are establishing. By the third year, the root systems should be extensive enough that irrigation can be reduced somewhat.

Weed Management

Plan on keeping the planting rows free of vegetation for at least the planting year and following year (we'll go into orchard floor management in detail in chapter 9). Eliminating competition during the critical establishment stage greatly increases tree survival and growth.

Mulches and weed fabrics. Now that your trees are in the ground, you are faced with complicated decisions on whether and how to apply organic mulches and weed barrier fabrics, or to keep the soil surface bare. Each strategy has its advantages and challenges.

The first factor to consider is how well the soil has been prepared before planting. If you are confident that the pH is both correct and stable, all nutrient deficiencies have been corrected, the soil organic matter concentration is in the range you want, and annual and perennial weeds are under control, then you have great flexibility in your choices. On the other hand, if the pH is still too high or low, weed barrier fabrics and deep organic mulches are not good choices. This caution is especially true when pH is too low and you need to incorporate a liming material into the planting zone. Weed barrier fabrics and organic mulches, such as bark or sawdust, make incorporating lime and immobile fertilizers and soil amendments very difficult or impossible. Organic mulches are also largely ineffective against established perennial weeds.

Leaving the soil bare, however, greatly increases weed problems within the planting rows. With few organic herbicides available and none particularly effective against perennial weeds, organic fruit growers must rely on mechanical cultivation, hand weeding, thermal weeders, and spot spraying organic contact herbicides.

A compromise that provides good weed control while allowing you to continue applying soil amendments is to place a piece of weed barrier fabric around individual trees or bushes (see figure 7.5). Depending on your in-row spacing, cut out weed fabric squares that are 24 to 36 inches per side. Make a slit from one side to the center of each square and make a hole in the center just large enough for the trunk or collar to fit through. The fabric squares can be staked into place with wooden pegs or aluminum or steel wire staples. Alternatively, you can bury the fabric edges with soil to hold the squares in place. These fabric squares keep a bare zone of soil around the orchard crop plants while allowing cultivation and other cultural activities. Using small squares also greatly reduces the cost of weed fabric and reduces disposal problems later.

Figure 7.5 · Use small squares (2×2 feet to 3×3 feet) of weed barrier fabric to maintain a vegetation-free zone around individual trees.

Fertilization

Fruit trees and bushes seldom need fertilization at the time of planting, particularly if you have done a good job of preparing the site. Depending on soil tests and plant size, you may want to fertilize later in the year following spring planting. We will go into detail on fertilization in chapter 8. For now, do not fertilize newly planted trees and bushes.

Perennial alley crops. Bare alleyways create problems in the form of soil erosion, soil compaction, dust, loss of organic matter, and poor access during wet weather. For mature orchards where irrigation water and/or rain are adequate for both the orchard crops and alley crop, permanent alley crops can do a good job of preventing erosion, soil compaction, and loss of soil organic matter. The negative side is that alley crops compete with your orchard crops for soil moisture and nutrients and can reduce fruit yields. There are several approaches to permanent alley crops, two approaches being especially popular at the moment: 1) The alley crops can be made up of a sod-forming grass or blend of grasses, often supplemented with clovers; or 2) the alley crop can be made up of legumes, usually alfalfa or alfalfa blended with clovers, such as white Dutch clover. In areas where soil frost heaving is not severe, bunch-forming grasses that go dormant during the hot, dry summer months can reduce competition with the orchard crops while still reducing soil compaction and erosion and allowing access during wet conditions. Typical permanent cover crops include various fescues, wheatgrasses, alfalfa, and clovers.

Annual alley crops. For a young fruit planting where soil amendments may be required, consider using annual cover crops, such as barley, oats, ryegrass, or wheat, supplemented with peas. Annual alley crops provide many of the benefits of permanent covers, but they can be quickly and easily removed, if necessary.

Whatever strategies you choose, keep in mind that young fruit trees and bushes compete poorly with other vegetation. At a minimum, keep a 3- to 4-foot diameter, vegetation-free circle around each newly planted tree or bush.

NUTRIENT MANAGEMENT

- Essential Elements | 252
- Determining the Nutrient Status of Your Crop | 254
- Approved Materials | 265
- Applying Fertilizers | 280

Nutrient management is critically important for all orchards and is an area on which many organic fruit growers concentrate. In the early days of conventional agriculture, when an abundance of cheap nitrogen fertilizers became available, this chapter would have been titled "Fertilization." Fields long managed by organic methods or virgin lands were often already rich in macro- and micronutrients, and little was required to produce a fruit crop except to dump commercially available, nitrogen-rich fertilizers into the field. These concentrated forms of plant nutrients eliminated the need to transport and apply large quantities of fertilizers or to grow green manure crops, and they fit perfectly into the concept of modern horticulture at that time.

As our knowledge of soil environments and plant-soil interactions has grown, the concept of fertilization has also grown. We now recognize the soil as far more than an inert sponge that holds up the trees and provides a place for the roots. Soil ecosystems are incredibly complicated and diverse, and the health of those ecosystems greatly influences the health of the orchard crops. Soil ecosystems are also dynamic, changing with the seasons and from one year to the next. Every change you make in the orchard and in your management practices affects the soil and the macro- and microorganisms that dwell there, in turn affecting your fruit trees and bushes.

Unlike fertilization, which often has a quick or short-term effect, nutrient management requires a long-term approach, particularly in an organic system. Preplant applications of colloidal phosphorus, for example, will provide phosphorus to the fruit crops for decades. The green manure crop that you till in this season will supply nitrogen and micronutrients to your crops for years. Our goal is to guide the nutrient status of the soil so that the crop plants always have adequate, but not excessive, readily available supplies of all required nutrients. In some cases, quick fixes will still be required, even in the best organic orchard. That should be the exception, however, not the rule.

At the very beginning, a caution is needed: Having been a professional fruit specialist for 30 years, I have found nutrient deficiencies in market and home orchards to be quite rare and problems associated with excessive fertilization quite common. The old adage that "if a little is good, more is better" does not apply to fruit growing. Quite the contrary is true! The abundant, lush foliage created by excessive applications of nitrogen is generally accompanied by many problems that are described below. Adequate but not excessive is the rule.

A second caution: As the size of an organic orchard grows, so does the difficulty and cost of supplying adequate nitrogen in the form of fertilizers. Organic fertilizers typically have low concentrations of nitrogen, compared with industrial fertilizers, and are required in much greater quantities. Composts, manures, alfalfa pellets, alfalfa meal, soybean meal, feather meal, fish emulsions and meal, and cottonseed meal are all fine organic sources of nitrogen. They tend, however, to be expensive to purchase and ship in the large amounts needed for a big commercial orchard. Along with pest and disease management, supplying adequate nitrogen is one of the greatest challenges in operating a large organic orchard. As in the days before industrial fertilizers, fruit growers near livestock operations or other sources of inexpensive, nitrogen-rich materials have a distinct advantage over their more distant counterparts.

Newly established organic orchards and orchards transitioning from conventional to organic systems are most likely to suffer nutrient deficiencies. The problem is compounded by weed competition, which can greatly impair the survival and growth of young trees. These facts emphasize the need for careful and thorough preplant soil testing, applications of soil amendments, corrections of soil pH, and weed management.

As we will discuss in chapter 9, new variations on old cultural practices allow us to supply much of the needed nitrogen on-site using nitrogen-fixing cover crops. While we still have much to learn in selecting and managing cover crops, orchard floor management has great potential to reduce the need for off-site nitrogen, while enhancing and maintaining soil structure and biological activity.

Essential Elements

PLANTS TAKE CHEMICALS CALLED "ESSENTIAL ELEMENTS" from their environments to create living tissues. We refer to them as essential because plants must have them in order to complete their life cycles. Sixteen elements are generally considered necessary for normal plant growth and development. Among them are carbon, hydrogen, and oxygen, which come from water and air and combine to form cellulose, starches, sugars, and other carbohydrates that make up the bulk of plant tissues. The remaining 13 essential elements (mineral nutrients) come from the soil and are divided into "macronutrients" and "micronutrients." Although both macro- and micronutrients are equally important, plants need larger amounts of macronutrients, which are normally measured as percentages, by weight, of plant tissues. Micronutrients are needed in tiny amounts, ranging from about 5 to 400 parts of micronutrients per million parts of plant tissues (ppm). Tables 8.2 and 8.3 list macro- and micronutrients, along with their recommended ranges for leaf tissues in orchard crops.

Nitrogen. The nutrient most often deficient in orchards is nitrogen, for three reasons: First, nitrogen is needed in greater amounts than any other mineral nutrient. Second, nitrogen is not derived from the mineral portion of the soil, coming instead from the atmosphere by way of nitrogen-fixing plants and microorganisms. Third, nitrogen is negatively charged, as are soil and organic matter particles. Rather than binding to these particles like Ca^{++} or K^{+}, nitrogen is repelled and washes from the root zone or volatizes into the atmosphere.

Although nitrogen is essential and needed in large amounts, too much nitrogen forces excessive, lush growth that is highly susceptible to damage from pests and diseases and creates dense canopies that reduce air movement, further increasing disease problems. Excess nitrogen also delays fruit crop bearing; reduces flower bud and fruit formation; results in large, soft, poorly colored fruits; and creates excessive, erect shoots, thereby increasing pruning requirements and complicating training.

In orchards that are transitioning to organic and in newly planted organic orchards, nitrogen deficiency can be severe and often limits tree survival and health. Proper site preparation, combined with careful orchard floor management and supplemental nitrogen fertilizers, should prevent most nitrogen deficiencies from developing.

Research has demonstrated that organic apple orchards with low, although acceptable, amounts of nitrogen have better crops and healthier trees than those orchards with high, although still acceptable, amounts of nitrogen. The amount of nitrogen needed depends on the crop and variety. In apples, for example, 'Stayman', 'Turley', 'McIntosh', 'Jerseyred', 'Gravenstein', 'Starr', 'Summer Pippin', and 'Britemac' require less nitrogen than 'Red Delicious', 'Rome Beauty', 'Golden Delicious', and 'Jonathan'.

Potassium. Potassium is abundant in many soils across North America. Deficiencies can arise in orchard production because some fruits are relatively rich in potassium and large amounts of the nutrient are transported off-site in the harvested fruits. Potassium, although often considered rather immobile in the soil, can leach out of the root zone in poorly managed orchards. Remember: When applying potassium fertilizers, base your rates on the K_2O analysis for your fertilizer (see page 28).

Other nutrients. While other macronutrients can be deficient in orchard soils, careful preplant soil testing and site preparation should prevent significant problems from developing. Boron, iron, and zinc are micronutrients that are occasionally deficient in orchards. In many cases, the nutrients are actually available in adequate amounts in the soil, but their availability to plants is limited by pH values that are too high or low, excessive amounts of other nutrients, and soils that are wet and cold. Depending on soil conditions, boron can leach from the root zone. These elements are easily added to the orchard by applying compost, boric acid, chelated foliar sprays, and other materials. Tissue testing, as discussed below, is the most effective method for tracking macro- and micronutrient status in orchard crops.

It bears repeating: In planning and carrying out your site preparation and soil nutrition program, avoid the temptation to add excessive amounts of nutrients to the soil. Adding too much calcium or phosphorus, for example, can trigger the onset of iron chlorosis and apparent deficiencies of copper and zinc, even though these nutrients are plentiful in the soil. Excessive potassium can trigger or exacerbate apparent calcium and magnesium deficiencies by interfering with plant uptake of these nutrients. Strive to develop and maintain balanced soil and plant nutrition profiles.

Determining the Nutrient Status of Your Crop

YOU NEED TO KNOW THE NUTRIENT STATUS of your crop to make effective decisions about fertilization and orchard floor management. You can find recommendations to apply 60 to 90 pounds of nitrogen per acre per year, but such recommendations do not take into account your orchard's soil types, precipitation and irrigation, climate, training systems, fruit varieties, and management practices. In short, they are not particularly useful, especially for a commercial orchard. Several methods can help you judge the nutrient status of your crops.

Observing the Foliage and Fruit

Regular scouting is a critical part of maintaining an orchard. While scouting for pests and diseases, take the time to observe the growth and appearance of trunks, foliage, and fruit. Our goals are to maintain moderate vegetative growth and sustainable yields of high-quality fruit. Excessive nutrients can cause excessive vegetative growth or toxicity to your crop. Inadequate nutrition results in spindly trees and poor fruit crops.

To judge nutritional status, you will need to know what a healthy tree or bush should look like, which comes through experience. Appearance varies from one crop to another and even from one variety to another, and plant responses to poor pruning and other cultural practices can mimic nutritional disorders. To assist your learning, keep a journal of your observations. Over a period of a few years, you will discover patterns of annual growth and development and correlate these with fruit yields and quality. You will learn how your orchard system responds to fertilization, pruning and training, and orchard floor management. Keep your observations separate for each crop and each variety. To help get you started, here are some general guidelines:

Leader growth. A tree or bush with adequate nutrition will produce average leader growth. In a young apple tree, for example, you should see the leader grow 18 to 30 inches annually. As trees mature, the leader's elongation slows. In all crops, there should be enough lateral branches to fill the canopy. Short leaders and few or short laterals indicate problems.

Most newly developed lateral branches should form at wide-enough angles to the trunks to avoid bark inclusions.

Shoot growth. Excessively vigorous trees develop too many shoots and shoots that are too long and upright. The result is a dense, columnar tree that requires much effort to prune out undesirable branches and spread branches that you want to keep. In a vicious cycle, heavy pruning stimulates more vigorous, vertical growth.

Fruit color and size. Overly vigorous, upright fruit trees are slow to come into bearing and typically bear light crops of large, soft, poorly colored fruit. There are far greater pest and disease pressures on an excessively vigorous tree than on a healthy tree exhibiting moderate growth. In such a case, your approach might be to stop applying nitrogen, bend branches down to form 45- to 90-degree angles with the trunks, and prune as little as possible, but enough to keep the canopy open and allow light exposure to the trunk and interior branches.

Leaf color. Normally, leaves on most fruit trees should be green, exceptions being ornamental varieties selected for purple or other exotic leaf colors. Leaves on nitrogen-deficient trees are yellowish or pale green and may be smaller than normal. Nitrogen is mobile in plants and deficiencies often show up first in older leaves. Very dark green leaves usually indicate too much nitrogen. Sulfur deficiency also causes leaf yellowing, but it is much less common than nitrogen deficiency.

Potassium-deficient leaves show browning (scorching) around the leaf margins, with the leaf margins on stone fruits often curling upward. Magnesium-deficient leaves show yellowing between the veins of older leaves, forming a herringbone pattern on the leaves. Iron chlorosis shows up as yellow to white areas between dark green veins and is often associated with stunted shoot growth. Iron chlorosis is more often due to high soil pH and/or excessive soil moisture.

Leaf size. Leaves should be of normal size for that particular variety. Very large leaves might suggest too much nitrogen and/or shading caused by a canopy that is too dense. Very small, elongated leaves, especially when they are clustered at the tips of branches (rosetting) often indicate zinc or copper deficiencies, although glyphosate herbicide drift can cause the same symptoms.

Typical symptoms associated with nutrient deficiencies and excesses are described in table 8.1.

Typical Symptoms and Treatments of Nutrient Deficiencies and Excesses

Table 8.1

NUTRIENT DEFICIENCY	NUTRIENT EXCESS
Nitrogen (N)	
Symptoms. Appear on older leaves first. Symptoms include small, light green to yellowish leaves, possibly with dead tissues at the leaf tips. New shoots are stunted and contain little vigor. Low fruit set is common. **Treatment.** Apply blood meal or other rapidly available nitrogen in two or three split applications during May through July.	**Symptoms.** The numbers and growth of new shoots are excessive, and growth tends to be unusually upright. Leaves are dark green, and the foliage is lush and succulent. Fruit set is poor, and the fruits are often large, soft, and poorly colored. Susceptibility to pests and diseases (particularly fire blight) increases greatly. Growth fails to harden off in the fall and winter injury to the branches and trunk increases. **Treatment.** Reduce N fertilization and the use of N-fixing alley and in-row cover crops.
Phosphorus (P)	
Symptoms. Visual symptoms of P deficiency are uncommon in tree fruits. Leaves may be reduced in size and bluish-green in color, possibly with purplish veins and undersides of the leaves. Symptoms appear first on older leaves. Flower bud formation can be reduced, and older leaves may drop early. **Treatment.** Apply 250 lb/A (9 oz/100 square feet) of bonemeal.	**Symptoms.** Excessive amounts of P can trigger apparent Cu, Fe, and Zn deficiency symptoms. **Treatment.** Keep soil pH adjusted correctly. Avoid excessive P applications.
Potassium (K)	
Symptoms. Potassium deficiency in tree fruits typically appears as necrotic (dead) leaf margins and brownish-yellow spots on the leaf blades, starting with older leaves. Leaves may take on a bronzed appearance. Spurs and new shoots may exhibit weak growth and early defoliation. Fruits may be small, poorly colored, and lacking in acidity. Potassium-deficient trees fail to properly develop cold hardiness. **Treatment.** Apply 400 lb/A (15 oz/100 square feet) of sul-po-mag or 160 lb/A (6 oz/100 square feet) of potassium sulfate. Apply when symptoms develop. Potassium should normally be applied during early spring or late fall.	**Symptoms.** Excess K can trigger or exacerbate apparent Ca and Mg deficiency symptoms. **Treatment.** Avoid excessive K applications.

NUTRIENT DEFICIENCY	NUTRIENT EXCESS
Calcium (Ca)	
Symptoms. Calcium deficiencies seldom appear in leaves, although root and shoot growth can be reduced and leaf tips scorched. Symptoms are more often seen as fruit disorders, such as bitter pit, cork spot, and internal breakdown of apples. **Treatment.** Keep soil pH correctly adjusted. Add 1,000 lb/A (2.3 pounds per 100 square feet) of gypsum in spring or whenever symptoms develop. Foliar calcium sprays during fruit development may help reduce bitter pit and cork spot in apples.	**Symptoms.** Excess C does not produce visual symptoms but can trigger phosphorus deficiency. **Treatment.** Avoid excessive Ca applications.
Magnesium (Mg)	
Symptoms. Yellowish-brown necrotic areas appear, generally first between the veins on older leaves. Spurs and new shoots may be thin, weak, and brittle. Fruiting spurs may fail to develop (blind wood). Leaves and fruits may drop prematurely. **Treatment.** Apply 500 lb/A (18 oz/100 square feet) of magnesium sulfate (Epsom salts) or potassium magnesium sulfate (sul-po-mag) when symptoms develop or in early spring. Apply to the soil, not to the foliage.	**Symptoms.** Excess Mg can hinder Ca uptake and make Ca deficiency symptoms more prevalent when soil Ca supplies are low. **Treatment.** Avoid use of dolomitic limestone or other Mg-containing materials when soil Mg concentrations are high.
Boron (B)	
Symptoms. The most characteristic of B deficiency symptoms include delayed bud break, poor flower development, and poor fruit set. Shoot terminals and spurs may abort, and leaves are often small, narrow, and elongated. Apple fruits may develop corky spots within the flesh and wrinkled skins. **Treatment.** Apply a soluble boron-containing product as a foliar spray during the growing season or to the soil in early spring. CAUTION — boron becomes toxic to plants at very low concentrations. Never band boron fertilizers within crop rows. Use soil and foliar tests to determine the need for boron. Do not apply more than 1–2 pounds of actual boron per acre. Follow label directions when applying foliar spray materials.	**Symptoms.** Symptoms of excess B are similar to deficiency symptoms. Leaves may show yellowing (chlorosis) along the midrib, and shoots may defoliate early, beginning at the tips. Fruits may crack and drop early. **Treatment.** Avoid excessive applications of boron-containing materials.
Copper (Cu)	
Symptoms. Copper is not mobile in plants, and deficiency symptoms usually appear on younger leaves first. The leaves are stunted or misshapen, and the margins may be irregular. Yellowish or white mottling may appear between the leaf veins. Fruits may be small, poorly colored, and of poor quality. **Treatment.** Seldom a problem in organic orchards where copper fungicides are used to control diseases.	**Symptoms.** Excess copper may damage or kill roots and can trigger symptoms of other micronutrient imbalances. **Treatment.** Use copper-containing fungicides and bactericides carefully and according to label directions to avoid excessive accumulation of copper in the soil under the trees.

Typical Symptoms and Treatments of Nutrient Deficiencies and Excesses *continued*

NUTRIENT DEFICIENCY	NUTRIENT EXCESS
Iron (Fe)	
Symptoms. Iron deficiency is among the easiest nutritional disorders to spot. Leaf blades turn yellow to white but the veins remain dark green. Symptoms appear first on younger leaves and are generally worse in early spring when soils are cold and wet. Necrotic areas can develop along the leaf margins and between the veins. New shoots are stunted and may die back. Flowering and fruiting are decreased. **Treatment.** Foliar sprays of approved iron-containing products provide quick but temporary relief. Follow label directions. Iron chlorosis is seldom due to Fe being deficient in the soil, and is more often due to high soil pH, wet soils, cold soils, and/or excess Ca and P. It is often caused by overirrigation or poor drainage, especially on alkaline soils. Ensure that irrigation water is evenly distributed throughout the field. Drain or do not plant on wet sites. Use sulfur to lower soil pH to 6.5 on alkaline sites.	**Symptoms.** None
Manganese (Mn)	
Symptoms. Manganese deficiency resembles iron chlorosis somewhat but tends to appear as dull, chlorotic herringbone patterns on the older leaves. Young terminal leaves remain green. Shoot tips may die back, and flowering and fruit set are reduced. **Treatment.** Apply a foliar spray of an approved Mn-containing product as soon as leaves are well developed. Follow label directions. High soil pH can create manganese deficiency. Use sulfur to lower soil pH to 6.5 on alkaline sites.	**Symptoms.** Symptoms of excess manganese are only seen on 'Red Delicious' apple in the form of measles-like pimpling of the bark. **Treatment.** Avoid excessive applications of manganese-containing materials.
Zinc (Zn)	
Symptoms. Zinc deficiency produces stunted shoots. Leaves are small and narrow with a striped, irregular chlorotic pattern between green veins. Rosettes of leaves form at the tips of shoots, with bare wood below the rosettes in a condition called "little leaf disease." Fruit set, size, and quality are reduced. **Treatment.** Apply an approved Zn-containing product as a foliar spray beginning in early spring or as soon as symptoms develop. Follow label directions.	**Symptoms.** Excess Zn can trigger symptoms of apparent Cu deficiency. **Treatment.** Avoid excessive applications of zinc-containing materials.

Adapted, in part, from Cornell University tree fruit guidelines and Garcia, E. *Orchard Nutrition*. University of Vermont Extension.

Soil Testing

As we discussed in earlier chapters, testing the soil is a very important part of preparing an orchard for planting. Once the trees or bushes are established, soil tests for nutrient concentrations become much less useful, for several reasons: Trees and fruit bushes have spreading root systems that gather nutrients from large volumes of soil. These plants are also very efficient at concentrating mineral nutrients in their tissues at much higher concentrations than are found in the soil. And unlike annual crops that grow and die in a single season and release the nutrients back into the soil, woody plants recycle and store nutrients from year to year. For these reasons, soil nutrient tests on established orchards often correlate poorly with fruit crop performance.

It is important to test your soil's pH levels throughout the life of the orchard. Remember that soil pH has a strong impact on nutrient availability, as shown in figure 4.6. An excellent goal is to maintain the pH between 6.0 and 7.0 and as near 6.5 as possible, recognizing that soil pH varies throughout the year. Soil pH also varies across the orchard. Research has shown that soil pH in the planting row often differs significantly from the pH within the alley, just a few feet away.

Plan on testing the soil within crop rows every 2 to 3 years. While the time of sampling is not critical, you should sample at the same time each year because soil pH varies with the seasons. I normally sample the soil in mid-spring, when plant growth is most rapid and the need for plant nutrients is the greatest. See box on page 260 for how to collect a sample.

When gathering samples, an easy approach is to walk diagonally from one corner of the orchard or planting block to the opposite corner, sampling as you cross tree rows. Repeat the process on the other diagonal to obtain a representative sample for that block. Mix the samples inside the plastic pail and take out enough to fill the sample bag provided by your testing laboratory. If you are concerned about the pH effects on alley crops, you may want to repeat the procedure along the centers of the alleys.

Many universities with agriculture programs offer soil testing, and there are many good private laboratories in the United States and Canada. Choose a laboratory as close as possible to your site. Soil testing procedures vary greatly from one region to another, based on different soil types. Always use a laboratory that is familiar with your soils and stick with that lab. A common complaint from fruit growers is that testing

How to Collect a Soil Sample

Collect samples within tree rows throughout the orchard. In large orchards, sample each planting block separately. For bush fruits and high-density tree plantings, collect samples midway between plants.

1. Scrape away any vegetation, mulch, or organic litter on the soil surface.
2. Using a soil sampling tube or a trowel, gather samples to a depth of 8 inches midway between the trunk and drip line of the tree.
3. Dump each sample into a clean, plastic pail. Collect about 20 to 30 samples per acre for larger orchards and at least 1 quart of soil in small orchards. Thoroughly mix the soil in the bucket to create a composite sample for the laboratory.

results from different laboratories are not the same, even for the same sample. Use consistent sampling procedures and the same testing facility. Cooperative Extension offices and Ministry of Agriculture offices can usually provide recommendations for a reliable testing laboratory in your area.

Tissue Testing

By far the most reliable method of determining the nutrient status of perennial fruit crops is to analyze the leaves and petioles for mineral nutrient concentrations, a process called tissue testing or foliar analysis. As mentioned above, the results of soil analyses of established orchards are not good indicators of tree health and productivity. Visual observations of growth, leaf color, and fruit characteristics are valuable and necessary, but they do not reveal problems until damage has already occurred and fruit quality and yields have been reduced. Tissue testing, on the other hand, tells you precisely how much of each essential mineral nutrient is present in the plant.

While foliar analysis is seldom needed for home orchards, it is standard for commercial plantings. Foliar analyses help guide your nutrition and orchard floor management programs and provide early warnings of nutrient deficiencies and toxicities. Knowing, with accuracy, your crop's nutrient status takes the guesswork out of fertilization programs, saving time and money on unneeded fertilizer applications and helping ensure healthy plants and high-quality fruit.

In commercial orchards, foliar analyses are usually conducted yearly. Nutrient concentrations fluctuate greatly during a growing season, and it is best to sample after shoot and leaf growth have ceased but before the trees begin senescing (leaves changing color or dropping). At that time, the nutrient concentrations in the leaves are fairly stable and allow comparison from one year to another. In USDA Plant Hardiness Zones 4 to 6, the last week of July through the first week of August is the most popular time for sampling. In warmer or cooler climates, you may need to adjust the timing for leaf collection. Other fruit growers, state and provincial fruit specialists, and your analytical laboratory can give you guidance for your particular region.

Sample collecting procedures. Collecting leaves for testing is easy and straightforward. As with soil tests, you are looking for average values across many trees, if possible. Sample and test each variety separately because different fruit varieties of a single crop often show differences

Tissue Testing Tips

Here are some general guidelines for collecting leaves to be sent away for tissue testing.

- Collect both the leaf blades and petioles (leaf stems).
- Collect leaves from typical trees of the same age.
- Collect clean leaves that are typical of the tree.
- Collect leaves from mid-height of trees or bushes.
- Collect no more than two leaves per shoot and from several shoots per tree.
- Shoots should be of average vigor.
- Shoots should receive abundant sunlight.
- Collect roughly 60 to 100 total leaves from at least 5 to 10 trees, if possible, spread throughout the orchard.
- Place leaves inside a paper bag and allow to air dry until brittle.

in foliar nutrient concentrations. Do not sample soon after a cover spray of macro- or micronutrients or materials such as kelp or fish emulsion.

Collect only from typical trees. If most trees in a planting appear healthy and one or a few isolated trees do not, it is unlikely that the problem is being caused by nutrient deficiencies or excesses and is more likely to be pests, diseases, or localized irrigation or soil problems. All of the trees sampled should be the same age.

Sample from the mid-height of the trees or bushes, typically about 5 to 7 feet above the ground for mature fruit trees and 3 to 6 feet for young trees. Collect samples from the middle portion of shoots, taking no more than about two leaves per shoot, and from several shoots per tree. The shoots should be of average vigor for the trees. Do not collect spur leaves or leaves from excessively vigorous or weak shoots, such as those near pruning cuts.

Shoots should be well exposed to sunlight. If collecting more than 60 to 70 days after petal fall, collect the first and second fully mature leaves below the shoot tip. Collect only clean leaves that are typical of those on the tree. Avoid those damaged by pests, diseases, weather, or mechanical injury, and do not collect leaves showing unusual symptoms, unless they are typical of the trees being sampled. Collect both the leaf blades and petioles by grasping the petioles (leaf stems) firmly and pulling them downward and away from the shoot.

Repeat the leaf collecting process across the orchard or planting block, much as you did for soil samples, collecting roughly 60 to 100 leaves in total for each variety. If possible, collect from at least 5 to 10 trees and do not collect all of your samples from a few trees planted closely together. Repeat for different varieties and crops, if necessary.

Place the leaves inside a paper lunch sack and allow them to air dry at room temperature until they are brittle. Transfer the leaves to the sample bag provided by the analytical laboratory and send the samples to the laboratory.

Analyzing test results. Interpreting foliar analyses is generally straightforward. Much research has been conducted on commercial orchard crops, and tables showing recommended foliar nutrient concentrations are available. Most tables list ranges of concentrations, from deficient to adequate to excessive. In general, you want to be in the adequate range. Because organic orchard systems perform best with less nitrogen than in conventional orchards, you will want the nitrogen values to fall toward the lower end of the adequate range. Tables 8.2 and 8.3 provide suggested nutrient values. Because published standards are not available for mayhaw, saskatoon, quince, medlar, and bush cherries, the suggested values are extrapolated from similar crops. Consider these values as starting points, and adjust, as necessary, for your crops, varieties, and cultural practices.

To Wash, or Not

Growers often ask whether they should wash the leaves before drying them to remove dust or other contaminants. In general, most analytical laboratories prefer that you do not wash the leaves. If the leaves are dusty, covered with kaolin clay (Surround), or have recently received a cover spray, you may want to wash them. The following guidelines are courtesy of Cornell University's Agro-One Soils Laboratory.

Wash the leaf samples while they are still fresh and before they wilt. If a large number of samples are involved, they may be stored overnight in a refrigerator or ice chest to keep them from drying out.

Clean one sample at a time, taking care not to mix different samples together. Use distilled water, available at most drug stores, for washing and rinsing the samples. Change the water if it becomes dirty or after 8 to 10 samples (whichever occurs first).

1. Gently and lightly scrub the leaves together in a mild detergent solution (most dishwashing detergents are satisfactory).
2. Shake the leaves to remove excess water, and immediately rinse them in clean distilled water. Shake the leaves again to remove excess water.
3. Immediately rinse the sample again in clean distilled water. Shake the leaves to remove excess water.
4. Spread out the leaves on clean paper towels until the leaf surfaces are dry.
5. Transfer each sample to its own, individual paper bag. Leave the bag tops open, and dry the samples at room temperature until the leaves are brittle.

Suggested Nutrients in Pome Fruits

Table 8.2

Nutrient	Apple	Mayhaw	Medlar	Pear	Quince	Saskatoon
MACRONUTRIENTS (%)						
Nitrogen (N)	1.75–2.25	1.75–2.25	2.2–2.8	2.2–2.8	2.2–2.8	1.75–2.25
Phosphorus (P)	0.2–0.3	0.2–0.3	0.11–0.25	0.11–0.25	0.11–0.25	0.2–0.3
Potassium (K)	1.25–1.75	1.25–1.75	1.0–2.0	1.0–2.0	1.0–2.0	1.25–1.75
Calcium (Ca)	1.2–1.6	1.2–1.6	1.0–1.5	1.0–1.5	1.0–1.5	1.2–1.6
Magnesium (Mg)	0.25–0.4	0.25–0.4	0.25–0.5	0.25–0.5	0.25–0.5	0.25–0.4
MICRONUTRIENTS PPM						
Boron (B)	25–50	25–50	20–70	20–70	20–70	25–50
Copper (Cu)	5–20	5–20	5–20	5–20	5–20	5–20
Iron (Fe)	100–300	100–300	60–250	60–250	60–250	100–300
Manganese (Mn)	40–100	40–100	30–100	30–100	30–100	40–100
Molybdenum (Mo)	20–100	20–100	20–100	20–100	20–100	20–100
Zinc (Zn)	20–50	20–50	25–200	25–200	25–200	20–50

Adapted from *Leaf Analysis for Fruit Trees.* Rutgers NJAES Cooperative Extension.

Suggested Nutrients in Stone Fruits

Table 8.3

Nutrient	Apricot	Cherry (sweet)	Cherry (tart)	Cherry (bush and hybrid)	Peach	Plum
MACRONUTRIENTS (%)						
Nitrogen (N)	2.0–2.5	2.1–3.0	2.6–3.0	2.6–3.0	3.25–4.0	2.4–3.0
Phosphorus (P)	0.13–0.35	0.16–0.5	0.16–0.22	0.16–0.22	0.2–0.4	0.14–0.25
Potassium (K)	2.5–3.0	2.5–3.0	1.6–2.1	1.6–2.1	1.5–2.0	1.6–3.0
Calcium (Ca)	1.6–2.5	2.0–3.0	1.5–2.6	1.5–2.6	1.5–2.25	1.5–3.0
Magnesium (Mg)	0.3–1.2	0.3–0.8	0.3–0.75	0.3–0.75	0.3–0.6	0.3–0.8
MICRONUTRIENTS PPM						
Boron (B)	25–70	20–100	20–55	20–55	25–75	25–60
Copper (Cu)	5–25	5–50	8–28	8–28	5–20	6–16
Iron (Fe)	70–150	100–200	100–200	100–200	100–200	100–250
Manganese (Mn)	25–100	40–200	40–60	40–60	50–150	40–160
Molybdenum (Mo)	20–100	20–100	20–100	20–100	20–100	20–200
Zinc (Zn)	20–100	20–50	20–50	20–50	20–60	20–50

Adapted from *Leaf Analysis for Fruit Trees.* Rutgers Rutgers NJAES Cooperative Extension.

Approved Materials

IN DEVELOPING ORGANIC CERTIFICATION PROGRAMS for the United States, Canada, and other countries, organic growers, fruit specialists, legislators, and others worked together to identify materials suitable for sustainable organic crop production. The following list of fertilizers, growth promoters, and soil amendments is approved for use in organic crop production by the U.S. National Organic Program. Similar lists have been developed for Canadian certified organic growers. Even if you are not interested in organic certification, the approved lists provide excellent guidance for developing effective and environmentally friendly orchard management practices.

Some approved materials have certain restrictions regarding their use in organic crop production, and some organic certifying organizations follow standards more restrictive than those detailed in the following list. If certification is important to you, become thoroughly familiar with your certifying organization's rules. Always carefully read labels and any other documentation. Be sure that the materials you purchase meet organic certification standards.

NOP-Approved Nonsynthetic Materials

- Aquatic plant extracts (not hydrolyzed). The extraction process is limited to the use of potassium hydroxide or sodium hydroxide; the solvent amount used is limited to that amount necessary for extraction.
- Elemental sulfur
- Humic acids. Includes only naturally occurring deposits, water, and alkali extracts.
- Lignin sulfonate. Used as a chelating agent, dust suppressant, flotation agent.
- Magnesium sulfate. This is allowed with a documented soil deficiency.

- Micronutrients. These cannot be used as a defoliant, herbicide, or desiccant. Those made from nitrates or chlorides are not allowed. Soil deficiency must be documented by testing.
- Soluble boron products
- Sulfates, carbonates, oxides, or silicates of zinc, copper, iron, manganese, molybdenum, selenium, and cobalt
- Liquid fish products. The pH can be adjusted with sulfuric, citric, or phosphoric acid. The amount of acid used shall not exceed the minimum needed to lower the pH to 3.5.
- Vitamins B1, C, and E.
- Sulfurous acid (CAS # 7782-99-2). For on-farm generation of substance utilizing 99 percent purity elemental sulfur.

NOP-Prohibited Nonsynthetic Materials

- Ash from manure burning
- Arsenic
- Calcium chloride, natural brine process. Prohibited for use except as a foliar spray to treat a physiological disorder associated with calcium uptake.
- Lead salts
- Potassium chloride, unless derived from a mined source and applied in a manner that minimizes chloride accumulation in the soil
- Sodium fluoaluminate (mined)
- Sodium nitrate, unless use is restricted to no more than 20 percent of the crop's total nitrogen requirement
- Strychnine
- Tobacco dust (nicotine sulfate)

Certification standards allow for the use of many naturally occurring and a few synthetic materials for crop production. In the following section, we will cover some common soil amendments and fertilizers used in organic fruit production. Because organic certification standards vary between Canada and the United States and some U.S. certifying organizations enforce stricter rules than the NOP, be sure you know your certifying organization's rules if you are to be certified. Table 8.4 provides information on the nutritional values of selected organic fertilizers.

Algae. Algae are incredibly diverse, from giant kelps along ocean coastlines to one-celled organisms floating in a pond. Perhaps the most diverse are the blue-green algae (also called cyanobacteria) that are found almost everywhere life exists, as free-floating blooms, strands, and sheets in lakes and rivers to the symbiotes living inside lichens on desert rocks. For growers, the primary benefits of algae are as sources of plant nutrients.

One problem with using algae is that most algae has little dry matter and is primarily water, making harvesting and processing very labor intensive. You can harvest sheets, strands, and balls of algae from a farm pond and dump the materials around your fruit trees. You may also choose to wait until the pond has dried and scrape off the surface crust and algal remains to apply as a soil amendment. For a very small orchard, this approach might yield some benefits, depending on the soil conditions. You will add very little organic matter, and the primary benefit will be the addition of micronutrients. Any microorganisms you add will be adapted to watery environments — certainly not what you want in an orchard soil.

Probably most useful to organic orchardists are kelps (brown algae), when used in small quantities and as commercially available products. If you live close to kelp forests along a coastline, you might economically obtain large enough quantities to use as a source of nitrogen and other macronutrients. However, the energy and labor required to harvest the material and process it into a form suitable for orchard applications would make kelp a very expensive option for most growers. When using large quantities of kelp, be aware that it contains large quantities of sodium that can damage orchard soils by displacing calcium, phosphorus, and other plant nutrients from the soil and organic matter particles. Commercially available kelp meals and extracts are available and can serve as good sources of micronutrients, depending on how the kelp was processed. Some blue-green algae products may also be available. The greatest value of these kelp and blue-green algae products will probably be as foliar sprays to provide micronutrients.

Approximate Percentages of Nutrients in Fertilizer Materials

Table 8.4

Material	N	P_2O_5	K_2O	Ca	Mg	S	Mn	Fe	B	Rate of Availability
Alfalfa meal	3	1	2	—	—	—	—	—	—	ms
Blood meal	12	1–2	0.6	—	—	—	—	—	—	mr
Bonemeal (steamed)	0.7–4	11–34	—	23	tr	—	tr	tr	tr	s–m
Borax	—	—	—	—	—	—	—	—	11	r
Cocoa shell meal	2–3	1	2–3	—	—	—	—	—	—	s
Coffee grounds (dry)	2	0.4	0.7	—	—	—	—	—	—	s
Colloidal phosphate	—	18–24	—	—	—	—	—	—	—	s
Compost (unfortified)	1–2	1	1–2	—	—	—	—	—	—	s
Corn gluten meal	9	—	—	—	—	—	—	—	—	m–r
Cottonseed meal (dry)	6–9	2–3	1–2	—	—	—	—	—	—	s–m
Dolomite	—	—	—	22	13	—	—	—	—	s
Feather meal	11–15	—	—	—	—	—	—	—	—	s
Fish emulsion	5	2	2	—	—	—	—	—	—	m–r
Fish meal	8–10	4–9	2–4	—	—	—	—	—	—	s
Granite dust	—	—	6	—	—	—	—	—	—	vs
Greensand	—	1–2	6	—	—	—	—	—	—	s
Guano (bat)	5–6	8–9	2	—	—	—	—	—	—	m
Guano (Peruvian)	12–3	11–12	2–3	—	—	—	—	—	—	m
Gypsum	—	—	—	22	—	19	—	—	—	m
Hoof and horn meal	12–14	2	—	—	—	—	—	—	—	s
Iron sulfate	—	—	—	—	—	9–12	—	16–20	—	r
Kelp	1	0.5	1–4	—	—	—	—	—	—	m
Limestone	—	—	—	40	—	—	—	—	—	s
MANURE (FRESH)										
Cattle	0.3	0.2	0.3	—	—	—	—	—	—	m
Horse	0.3	0.2	0.5	—	—	—	—	—	—	m
Sheep and goat	0.6	0.3	0.8	—	—	—	—	—	—	m
Swine	0.3	0.3	0.3	—	—	—	—	—	—	m
MANURE (DRY)										
Cattle	0.7–2	0.4	0.6–2	—	—	—	—	—	—	m
Horse	0.7	0.3	0.5	—	—	—	—	—	—	m
Rabbit	2	1.3	1.2	—	—	—	—	—	—	m
Sheep and goat	2–3	1–2	2–3	—	—	—	—	—	—	m
Swine	1	0.7	0.8	—	—	—	—	—	—	m
Marl	—	2	4–5	—	—	—	—	—	—	vs
Mushroom compost	0.7	0.9	0.6	—	—	—	—	—	—	s

Material	N	P_2O_5	K_2O	Ca	Mg	S	Mn	Fe	B	Rate of Availability
Potassium chloride	—	—	60	—	—	—	—	—	—	r
Potassium sulfate	—	—	54	—	—	18	—	—	—	r
Soybean meal	6–7	1–2	2–3	—	—	—	—	—	—	s
Sulfate of potash magnesium	—	—	22	—	11	23	—	—	—	m–r
Rock phosphate	—	5	—	—	—	—	—	—	—	vs
Wood ashes	—	1–2	3–7	—	—	—	—	—	—	r

vs = very slow; ms = moderately slow; s = slow; m = moderate; r = rapid; mr = moderately rapid; vr = very rapid
To convert between % P and % P_2O_5: % P × 2.29 = % P_2O_5, % P_2O_5 × 0.44 = % P
To convert between % K and % K_2O: % K × 1.2 = % K_2O, % K_2O × 0.83 = % K

Animal manures. Historically, animal manures have been a valuable source of organic matter and macro- and micronutrients, and they remain so today. Economics and governmental rules, however, have changed how we obtain and use manures. Large, integrated orchard and livestock operations are quite rare today. Even 60 years ago, operators of such enterprises often found they were making money on the fruit but just breaking even on the livestock and were essentially working very hard for manure. Larger growers usually find it most effective to specialize on their orchards and let someone else produce the livestock and manure. Separating crop and livestock operations creates transportation problems, however, and raises the cost of manure products unless the livestock operations are very close. Integrated market and home orchards and livestock production are still common.

The nutrient content of fresh manures varies according to the kind of animal and what the animal has been eating. As manures dry and age, leach, or are composted, nutrient concentrations are reduced, but so are the salt concentrations and the chances of burning plants. Fresh manures also contain large amounts of water, making them difficult to handle and expensive to transport. They can damage plants by providing too much nitrogen too quickly, and the high concentrations of salts in fresh fertilizers can build up in soils.

Raw (fresh or dried) manures can contain human pathogens, and NOP organic certification regulations do not allow raw animal manures to be applied to tree orchards within 90 days of harvest, or small fruit orchards within 120 days of harvest. The NOP states that, for crops intended for human consumption, raw manures must be *incorporated into the soils* at least 90 or 120 days, respectively, before harvest.

Both bat and bird forms of guano are allowed under the U.S. National Organic Program. Use these materials as you would other manures. Nitrogen contents range from about 5 percent for bat guano to 13 percent for bird products. Guanos also contain 8 to 12 percent P_2O_5 and about 2 percent K_2O. These nutrients are available to plants at moderate rates and can be applied as late as the spring for the coming growing season.

Composting solves many of the problems with manures, providing a stable, relatively light (if bulky) material from which excessive salts have been leached out. Composting various mineral products, such as colloidal phosphate, with animal manures can make the mineral nutrients more available to plants. Proper composting also helps destroy potential human pathogens, such as *E. coli* bacteria. Manures can be contaminated with pesticides used to control pests in and around barns, corrals, and feedlots. When obtaining composts or materials for making compost, ensure that they are pesticide-free. Note that the U.S. National Organic Program prohibits the use of ash from burned manures.

If you are meeting certification standards, be aware that the NOP has specific requirements for composting manure so that it can be applied to an orchard within the 90 or 120 days-to-harvest window mentioned above. Composted plant and animal materials must be produced though a process that establishes an initial C:N ratio of between 25:1 and 40:1, respectively. The temperature must be maintained between 131 and 170°F (55–77°C) for 3 days using an in-vessel or static aerated pile system, or you must maintain the temperature between 131 and 170°F for 15 days if you are using a windrow composting system. During the 15-day period, you must turn the materials a minimum of five times.

Biodynamic preparations. Biodynamic agriculture was one of the cornerstones of today's organic agriculture, beginning in 1924 with Rudolf Steiner in Germany and quickly spreading to the UK, Australia, and North America. Biodynamic farming emphasizes an integrated, holistic approach, where the soil and all organisms on a farm are managed as a single, integrated organism. Many of the practices we discuss in this book fit well into a biodynamic approach.

The term "biodynamic" is trademarked by Demeter-International, which represents an association of biodynamic farmers. Some biodynamic farmers bring religious, occult, and astrological considerations into farming, which is beyond the scope and intent of this book. Scientific studies have shown that biodynamic farming operations have better soil health and structure than similar conventional farms and are more energy efficient. Biodynamic farming has not produced results that are significantly different from similar organic farms. Whether to follow strict biodynamic practices or the organic practices we cover here is a personal choice. Both approaches are very similar in most respects and produce similar results.

Blood meal. Blood meal is a by-product of slaughterhouse operations and is one of the few relatively high-concentration, rapidly available sources of nitrogen approved for organic farmers. Blood meal also supplies iron and some other micronutrients. It is most useful when establishing an orchard or transitioning from conventional to organic, when soil organic matter is low and soil nitrogen is not yet available in the concentrations needed by the orchard crops. As the orchard matures and soil organic matter and slow-release sources of soil nitrogen build, blood meal becomes less necessary.

Due to its relatively low cost, rapid availability, and ease of application, blood meal is also useful in established orchards where a quick fix of nitrogen is needed. Follow recommended rates carefully, as you can damage plants with excessive rates of blood meal.

Bonemeal. Like blood meal, bonemeal is a by-product of slaughterhouse operations. It is a rich source of calcium (23 percent) and P_2O_5 (12 to 14 percent) and contains trace amounts of micronutrients. Depending on its source and preparation, it may contain 1 to 4 percent nitrogen.

Alfalfa Meal, Corn Gluten Meal, and Soybean Meal

While they provide nitrogen and organic matter, these products are usually expensive sources of both and are best used in very small orchards. While advertised as a natural preemergence herbicide, research trials on corn gluten generally show it to be ineffective as an herbicide for fruit crops.

Bonemeal is particularly valuable when establishing an orchard. Use steamed bonemeal preparations to reduce the likelihood of human pathogens and increase the availability of phosphorus. Because it contains very little nitrogen or potassium, excess amounts of bonemeal will not "burn" plants, although excess amounts can create nutrient imbalances in young orchard crops, as we discussed earlier.

Boron products. Boron deficiency can cause poor flowering, fruit set, fruit development, and shoot development. Boron, however, becomes toxic to plants at very low concentrations. Boron products can be applied to the soil or as foliar sprays when tissue analyses show boron deficiencies. Use only organic-approved boron products and do not apply more than 1 to 2 pounds of actual boron per acre. Follow label directions very carefully. See page 107 for allowable sources of boron.

Chelates. Chelated micronutrient sprays may be used when soil and/or plant tissue analyses show nutrient deficiencies. Chelates are used to protect nutrient molecules from becoming chemically bound up in the soil and unavailable to plants. Some chelates may also facilitate the uptake of nutrients by plants, which normally release chelates from their roots to make soil nutrients more available. Amino acids, lignosulphate, citric acid, malic acid, tartaric acid, and other diacid and triacid chelates are acceptable chelates for organic orchards.

Chelated products may be applied to the soil, but they are most useful in orchards as foliar sprays to correct micronutrient deficiencies. Use only products approved for organic use.

Cocoa bean hulls. Cocoa bean hulls are the shells of cocoa beans and are removed from the beans during roasting. The hulls are generally applied as mulches and slowly decompose. They contain about 2 to 3 percent nitrogen and small amounts of phosphorus and potassium. While the mulch is weed-free, organic growers need to ensure and document that the products they use have been tested for pesticide residues and found to be clean. Cocoa bean hull mulch is best used for home orchards, as its cost is prohibitive for larger, commercial operations.

Compost. Compost is an outstanding amendment for orchards, providing both organic matter and nutrients. Because it is derived, in large part, from plant materials, compost normally has most or all of the nutrients plants need and in the correct proportions.

The quality of compost and its nutrient content depend on what goes into it and how it is prepared. It can be made strictly from plant materials or include animal manures and/or amendments such as colloidal phosphate. Properly managed compost has temperatures high enough to kill most weed seeds and pathogenic microorganisms. Many beneficial microorganisms tolerate the composting. Storey Publishing carries several guides for composting. The *On-Farm Composting Handbook*, available online from the Natural Resource, Agriculture, and Engineering Service (see Resources), is an excellent guide for large farm-scale composting.

When purchasing or making compost, take great care to ensure that none of the components have pesticide residues that can damage your crops or cost you certification. Several years ago, farmers and nursery growers in Washington State obtained compost from a municipal source. This compost contained high concentrations of the herbicide clopyralid, once used to control dandelions and other broadleaf weeds in lawns. Clopyralid is not destroyed during the composting process, and many crops were damaged or destroyed by the residues remaining in the compost. Clopyralid is, reportedly, still registered for use in some forage and grain crops.

When making your own compost (highly recommended), do not include diseased or pest-infested plant materials or weed seeds. Burn these materials and add the ashes to your compost, if you wish. Be very careful not to include materials that may contain pesticide residues. Grass clippings, leaves, wood chips, and other plant debris from municipal sources should generally not be used to produce compost for organic orchards unless you are absolutely sure they are free of pesticides.

Cottonseed meal. Cottonseed meal is the dry matter left after cotton has been ginned and the oil extracted from the seeds. This high-protein product is often used for animal feed in the southern and western United States and is used as an organic fertilizer. It provides 6 to 9 percent of slow- to moderately rapid-release nitrogen and lesser amounts of phosphorus and potassium. If you have access to a cost-effective source of this product, it can be a good orchard amendment. Ensure that any product you use has been tested for pesticide residues and you have documentation that it is free from such contamination.

Dolomite. Dolomite (also known as dolomitic limestone) is primarily mined from sedimentary deposits and consists largely of calcium magnesium carbonate ($CaMg(CO_3)_2$). Its primary use in horticulture is to raise soil pH levels while supplying magnesium. Because this is a natural product, the percentages and chemical compositions of calcium and magnesium vary, as does its calcium carbonate equivalent value (liming ability). As we discussed in chapter 4, do not apply dolomite unless soil magnesium concentrations and pH are both low. Excessive amounts of magnesium can interfere with calcium uptake and increase bitter pit problems in apples.

Enzymes. Various organic products containing enzymes, often combined with supposedly beneficial microbes and plant nutrients, are marketed as fertilizers and soil amendments. According to the U.S. National Organic Program, enzymes are acceptable if derived microbiologically from natural materials and not fortified with synthetic plant nutrients. In actual practice, it is extremely difficult to understand how applying enzymes to the soil would benefit orchard crops.

Enzymes are chemically complex biochemicals that carry out or regulate very specific chemical reactions. Microbes, soil macroorganisms, plants, and all other living things produce the enzymes they need for their life processes. While adding enzymes to the soil or spraying them on your fruit crops is not likely to cause any damage, you will earn much greater returns on your investments of time and money by focusing on creating the conditions needed for a biologically rich and active soil containing adequate amounts of plant nutrients. We have covered many such practices, such as correcting soil drainage problems, adjusting soil pH, incorporating green manure crops, applying compost, and adding optimal amounts of plant nutrients.

Epsom salts or magnesium sulfate. Epsom salts are a rich source of magnesium and are best applied before planting orchard crops. They are much more rapidly available than dolomite, however. According to the U.S. National Organic Program, they may be used when you can document a soil magnesium deficiency. Base your application rates on soil or tissue analyses. Do not apply to apples when soil calcium concentrations are also low. Kieserite is a hydrated form of magnesium sulfate and used in the production of Epsom salts.

Feather meal. This is a by-product of poultry slaughterhouses and contains 7 to 12 percent nitrogen, depending on how it is processed. The nitrogen is released slowly over 4 months or more. Nitrogen from feather meal generally costs about 10 to 15 percent more than that from the more rapidly available nitrogen in blood meal. If you can obtain feather meal inexpensively, it can be an effective slow-release fertilizer. Be sure to obtain a product that is free from pesticides.

Fish emulsions. Emulsions contain only about half of the nitrogen, phosphorus, and potassium found in fish meal, but the nutrients are available much more rapidly. According to the U.S. National Organic Program, forms that are fortified with urea or other synthetic plant nutrients are prohibited. Phosphoric acid can be used as a stabilizer, but the resulting product may not exceed 1 percent, by weight, of P_2O_3.

While some people recommend applying fish emulsion to the foliage in order to correct nitrogen deficiencies, the practice is not effective. Nitrogen and the other macronutrients generally cannot be taken up through the foliage in large enough quantities to significantly improve a plant's nutritional status. Studies have shown that, in cases where foliar feeding increased the macronutrient concentrations in orchard trees, the fertilizer solution dripping from the leaves onto the soil provided the benefits. For nitrogen, phosphorus, and potassium deficiencies, apply fish emulsions to the soil. Fish emulsion can be applied as a foliar spray, however, to supply micronutrients; these can be taken up in significant quantities through the leaves.

Fish meal. Fish meal is a rich source of nitrogen (8 to 9 percent) and also provides phosphorus and potassium. Depending on the preparation, fish meal can also provide micronutrients. Nitrogen from fish meal is available rather slowly and is best used as part of a soil-building program, not as a quick fix for a nitrogen deficiency.

Fruit pomaces. Grape, apple, and other fruit pomaces (spelled incorrectly as "pomades" in the U.S. National Organic Program) are the pulpy materials left after the juice has been extracted from the fruits. Pomaces may be nearly dry or wet, depending on the press. They can be applied directly to the soil, although the relatively large amounts of water make handling and applying the materials somewhat difficult and transportation over great distances expensive. The fruit odor and sugar will also attract flies and other insects, possibly including fruit pests. Where facilities and equipment are available, composting pomaces along with other organic materials can be an effective practice to recycle orchard wastes and create nutrient-rich compost. In an orchard operation where cider or juice is extracted, an excellent practice is to apply the pomace waste to the orchard floor, preferably after composting.

Granite dust. This is sometimes used as a source of potassium (see page 103).

Green manure crops. These crops are not grown to be harvested for human consumption but are incorporated into the soil as part of a soil-building program. As we discussed in earlier chapters, growing and incorporating green manure crops before planting an orchard is an excellent strategy for adding organic matter and nitrogen to the soil. Used properly, green manure crops are also important in reducing weed problems. Green manure crops are at the heart of organic annual vegetable and grain production. In vegetable or grain systems, once a cash crop is harvested, you can plant and turn in a green manure crop before replanting the cash crop. The use of green manure crops in established, long-lived perennial crops is more challenging because you are not removing and replanting your cash crop frequently. In some orchard systems, green manure crops can be used as annual alley crops or as in-row companion plantings. We will discuss these strategies in chapter 9.

Greensand. This is a source of potassium (see page 104).

Gypsum. Also known as calcium sulfate, gypsum adds both calcium and sulfur to the soil without changing the soil pH (see page 105 for more information). When calcium is needed and the soil is already at pH 6.5 or greater, gypsum is an excellent source of calcium. Likewise, when sulfur is needed and the soil is pH 6.5 or lower, gypsum is the material of choice.

Gypsum is often touted as a "soil conditioner." That claim is true, but only in very specific cases. Gypsum is used to reclaim saline-sodic soils that are found most often in the arid regions of western North America. These soils have both high pH values and high sodium contents. They are very compacted and drain poorly, often forming what appear to be oily patches, earning them the common name of "slick soils." High concentrations of sodium in such soils displace calcium and other nutrients, destroying soil aggregates and creating dense, virtually lifeless soils. When gypsum is combined with tillage and water to flush the sodium soils out of the root zone, the calcium in gypsum occupies the negatively charged sites on soil particles, flocculating (fluffing up) the soil and forcing the sodium out. Aside from its effectiveness on saline-sodic soils, however, gypsum's "soil conditioner" properties are highly questionable. Although it is a valuable tool for orchardists, be careful not to give gypsum more credit than it deserves.

In conventional cropping systems, borated gypsum is often used to supply boron. This material typically contains about 1 percent boron and is applied at a rate of about 175 pounds per acre, broadcast across the entire field. There are naturally occurring sources of boron-containing gypsum, but not many. If you choose to apply borated gypsum, ensure that it is approved for organic use and ensure that the boron concentration is not high enough to damage your crops. Limit boron applications to no more than 1 to 2 pounds of actual boron per acre and only when soil or tissue analyses document a boron deficiency. Never band boron-containing fertilizers into crop rows.

Hoof and horn meal. Containing 12 to 14 percent nitrogen and 2 percent K_2O, hoof and horn meal are suitable sources of slow-release nitrogen for use as soil builders. Like blood meal, they are by-products of the slaughterhouse industry.

Humates and humic acids. Humic materials are derived from plant materials, such as lignin, and have weathered to the point that they are very resistant to further degradation. They are important components of the soil because they help bind the soil particles together, creating aggregations and structure.

The term "humic acid" refers to a large group of related compounds. Although they can be created synthetically, humic materials are generally mined from lignite coal or an oxidized clay-like form of lignite called leonardite. The latter is particularly rich in humic acids. While humic acids are certainly critical components in a biologically active soil, the benefits of adding them to the soil are less clear. You would need to add vast amounts to increase soil organic matter even slightly. Mined humates are chemically different from those found in the soil. From a fruit grower's perspective, commercial humate products appear to have little to offer.

Kelp meal and extracts. See algae entry, page 267.

K-mag, sul-po-mag, or sulfate of potassium magnesia. This naturally occurring material is mined for use as a fertilizer. It is moderately to rapidly available to plants and contains 23 percent sulfur, 22 percent K_2O, and 11 percent magnesium. This is a valuable product when you are faced with potassium, magnesium, and sulfur deficiencies. It has no effect on soil pH. Make your applications based on recommendations from a soil analysis.

Limestone. Limestone is a naturally occurring material comprising mostly various concentrations of calcium carbonate. A typical limestone

Leather Meal

Leather meal is a by-product of the tanning industry. Although a relatively rich source of nitrogen (10 percent), it is often contaminated with chromium and cannot be used in organic orchards.

contains about 40 percent calcium, which is slowly available to plants. Its greatest value in orchards is its ability to raise the soil pH. Dolomitic limestone, or dolomite, is a magnesium-containing form of limestone.

Micronutrients. Boron, copper, iron, manganese, molybdenum, and zinc are occasionally deficient in orchards. While sulfates of these nutrients are available, your best results will probably come from applying foliar sprays of chelated products (see page 272), described above. Apply them only when there is a documented deficiency of the nutrient in the soil or tissues.

Mushroom compost. Mushroom compost is usually comprised mostly of sawdust, rice hulls, straw, or similar woody organic materials that have been used to grow mushrooms for human consumption or medicinal purposes. It typically contains less than 1 percent each of nitrogen, phosphorus, and potassium and is most valuable as a source of organic material for mulching. Test the material for pesticides and document that it is pesticide-free before applying it to your soil.

Peat moss. Peat moss is extremely valuable for containerized plant production, but it is normally prohibitively expensive for all but a small home orchard. Its value as an amendment is to increase the soil's water-holding capacity and is most useful on sandy soils. It is also acidic and somewhat useful in temporarily lowering soil pH. Ensure that any peat moss you use has not been amended with synthetic fertilizers or wetting agents (surfactants).

Potassium sulfate. Also called potash of sulfate, potassium sulfate is a valuable fertilizer in conventional farming systems. Unfortunately, it is usually produced in factories and naturally occurring forms that can be mined are rare. As described in chapter 4, potassium sulfate can also be derived from evaporating saline lake water. Not all sources of potassium sulfate are suitable for organic production and not all certifying organizations allow its use. Sul-po-mag is a good alternative product. If you use potassium sulfate, ensure that it is allowed under your certification guidelines.

Rock phosphates and colloidal phosphates. See page 102.

Shells. Ground oyster, clam, lobster, and crab shells are similar to bonemeal in some ways. They are rich in calcium but rather low in phosphorus. A typical crab shell–based fertilizer might contain 2 to 3 percent nitrogen, 3 percent P_2O_5, 0.5 percent K_2O, and 23 percent calcium. Nutrients from shell products become available slowly. Shell products are probably best used as slow-release forms of calcium.

Sodium nitrate. Also known as Chilean nitrate, this is a nitrogen-rich, rapidly available fertilizer. Because of its high sodium content, most organic certification programs discourage or prohibit its use. Some certification programs allow Chilean nitrate to be used, but only enough to provide 20 percent of the total amount of nitrogen applied. This material is best used sparingly during the early stages of establishing an orchard or as an occasional supplement when a quick fix of nitrogen is needed. Ensure your organic certification organization allows its use before applying it to your crops.

Sugar beet lime. In processing sugar beets, limestone is heated to produce calcium oxide and carbon dioxide, which are then mixed with the beet juice to absorb impurities. During the process, the calcium oxide and carbon dioxide recombine to form calcium carbonate, which must be disposed of. Sugar beet lime can be useful as a soil amendment to raise pH, but the material often contains large amounts of water, making it difficult and expensive to transport and apply in an orchard. Obtaining uniform applications of the wet product is also difficult, and uneven applications can create pH and nutritional problems for your orchard crops. Be sure you have an accurate measurement of the calcium equivalent of sugar beet lime before applying it to your soil.

If you are in an area where sugar beet lime is readily available, it can be a good choice as a soil amendment. Ensure that any product you use has been tested to ensure it is pesticide-free, and check with your certification organization to ensure it is allowed for your operation.

Sulfur. We discussed the use of elemental sulfur in chapter 4 as a means of lowering the pH in alkaline soils. Elemental sulfur can also be used to add sulfur to the soil when pH is high and soil sulfur concentrations are low. Elemental sulfur is available as pellets, prills, and powders. The smaller the product particles are (the finer it is ground), the more rapidly the sulfur reacts in the soil to lower pH and the more rapidly available it is to plants. If the soil pH is already in the desired range but sulfur concentrations are low, add sulfur in the form of gypsum or sul-po-mag.

Wood ashes. Wood ashes are allowed in organic production but can be problematic. They contain 1 to 2 percent P_2O_5 and 3 to 7 percent K_2O, both of which are rapidly available to plants. Wood ash is highly alkaline and reacts very quickly in the soil to raise the soil pH. Do not apply ashes to soils that are already at or above the desired pH range and do not add more than 20 pounds of wood ash per 1,000 square feet per year. Use only ashes from untreated and unpainted wood. They must not be created using colored paper, plastic, or other prohibited substances. Excessive applications of ash can cause pH and nutrient imbalances. Never use coal ash.

Worm castings. Worm castings are literally the manure formed by worms. This organically rich material contains variable amounts of nutrients, based on the worms' diets. Few commercially available worm casting fertilizers provide a fertilizer analysis, making it impossible to determine application rates. At best, worm castings would be a very expensive fertilizer and source of organic matter for any but the smallest orchards. A better strategy is to create a biologically active soil where your own worms produce the castings in the soil where they are needed by your orchard crops.

Applying Fertilizers

Soil tests, tissue analyses, and rule-of-thumb guides generally include estimates of how much nitrogen, phosphorus, and other nutrients are needed. Recommendations are usually given in pounds per acre or pounds per 1,000 square feet in the United States or kilograms per hectare or kilograms per 100 square meters in Canada. The question then becomes how much fertilizer is needed to provide those amounts of nutrients. Commercial fertilizers, organic and nonorganic, are analyzed for the concentrations of nutrients in them, and the fertilizer analyses are printed on the containers. At least three numbers are shown, representing nitrogen, phosphorus, and potassium, in that order. Other macro- and micronutrients, if present, may be listed in a table.

Ammonium sulfate, a popular conventional fertilizer, has an analysis of 21-0-0, indicating 21 percent N and no P_2O_5 or K_2O. A typical conventional garden fertilizer might be labeled as 16-16-16, indicating that it has at least 16 percent each of N, P_2O_5, and K_2O. Most fertilizers contain at least trace amounts of micronutrients. Manufacturers are not required to list these nutrients unless they advertise that fertilizer as a source of those particular nutrients. If they do list micronutrients, the claim is usually that the fertilizer contains "at least" the amounts listed on the label.

Unfortunately, some manufacturers of commercial organic fertilizers choose not to include the analyses and sell the products as "soil amendments" or "soil conditioners" without specifying exactly what is in them. When buying such products, you have no clear idea of the benefits, if any, that you are receiving. You are far better off using materials that are clearly and accurately labeled. Treat testimonials in advertisements with a huge grain of salt. If a material is intended to be used as a source of plant nutrients and no fertilizer analysis is printed on the label, there is usually a very good reason why it is missing.

There are, however, some good commercial organic fertilizers, including some naturally occurring materials, that do not come with fertilizer analyses. These products include such things as feather meal, greensand, and soybean meal. Table 8.4 provides approximate concentrations of plant nutrients for many common organic fertilizers. Like conventional fertilizer labels, phosphorus values are given as P_2O_5 and potassium as K_2O equivalents.

Calculating Application Rates

For an example, say that a foliar analysis of your young apple trees showed low levels of nitrogen and the laboratory recommended applying 60 pounds of nitrogen per acre (fruit crops often need about 60 to 90 pounds of nitrogen per acre [68 to 103 kg/ha] per year). We will further assume that this is a newly established orchard and that we do not yet have a nitrogen-fixing alley crop in place, nor much decomposing organic matter in the soil. All nitrogen must come from the applied fertilizer.

Because the need for nitrogen is acute and our soil's organic matter is not yet releasing much nitrogen, we will use a fairly rapid-release fertilizer in the form of blood meal. Blood meal contains about 12 percent nitrogen. If you were to buy a commercial blood meal fertilizer, it might have a label something like 12-2-0.6, indicating 12 percent N, 2 percent P_2O_5, and 0.6 percent K_2O.

Amount of fertilizer per acre. To calculate how much fertilizer to apply, divide the amount of nitrogen needed by the percentage of nitrogen in the fertilizer. The same procedure is used to calculate the amount of phosphorus, potassium, or other nutrient to add. In this case:

60 pounds of N per acre ÷ 0.12 = 500 pounds
of blood meal fertilizer per acre

67 kilograms per hectare ÷ 0.12 = 558 kilograms
of blood meal fertilizer per hectare

In this example, we decide to split the fertilization into two applications in order to reduce the amount of nutrients that are lost from the soil by leaching and volatilization. We will make the first application as the buds begin swelling and the second 1 month later. It is common practice to split the application of rapidly available fertilizer materials, such as blood meal. For materials that are available to plants at moderate or slow rates, the entire amount of fertilizer is usually applied at once, allowing plenty of time for it to become available to the crop.

Amount of fertilizer for less than one acre. To calculate the amount of fertilizer needed for orchards smaller than one acre, you can easily convert the rate per acre to the rate per 1,000 square feet or the rate per tree.

One acre contains 43,560 square feet. **To calculate the amount of nitrogen per 1,000 square feet** (an area 10 feet wide by 100 feet long), divide the recommended rate per acre by 43.56.

500 ÷ 43.56 = 11.5 pounds of blood meal per 1,000 square feet
(11.5 lbs per 1,000 ft^2)

To calculate a per-tree rate, divide the number of trees per acre, based on your planting density. Say, for example, that your apple trees are 6 feet apart in rows 12 feet apart.

6 feet in rows × 12 feet between rows = 72 square feet per tree

43,560 square feet per acre ÷ 72 square feet per tree =
605 trees per acre

500 pounds of blood meal per acre ÷ 605 trees per acre =
0.82 pounds of blood meal per tree

For small orchards, you may need small amounts of fertilizer and it may be more convenient to convert pounds to ounces (dry) of fertilizer.

0.82 pounds of blood meal × 16 ounces per pound =
13.2 ounces of blood meal per tree

Amount of fertilizer for metric areas. Exactly the same procedures apply when calculating kilograms of fertilizer per hectare or 100 square meters.

Soil and tissue test recommendations in Canada are likely to be given in kilograms of nutrient per hectare. If you are converting between U.S. and metric units, remember that kg/ha = 1.12 × lbs/acre. Say that your analyses show potassium and magnesium to be deficient and you plan to correct the problem using sul-po-mag. From table 8.4, you see that sul-po-mag contains about 22 percent K_2O and 11 percent magnesium. The laboratory's recommendation is to apply 60 kg of K_2O and 30 kg of Mg per hectare.

60 kg of K_2O ÷ 0.22 = 273 kg of sul-po-mag per hectare

30 kg Mg ÷ 0.11 = 273 kg of sul-po-mag per hectare

One hectare = 10,000 square meters. To calculate the amount of sul-po-mag needed per 100 square meters:

273 kg per 10,000 square meters ÷ 100 = 2.73 kg of sul-po-mag
per 100 square meters

If your tree density is 1,400 trees per hectare:

273 kg of sul-po-mag per hectare ÷ 1,400 trees = 0.195 kg =
195 g of sul-po-mag per tree

Amount of fertilizer in mature orchards. As our organic orchard matures and we develop organically rich soils and nitrogen-fixing alley and/or companion crops, determining how much nitrogen fertilizer to apply becomes much more complicated. We need to account for the nitrogen that is fixed in the alley and companion crops and available to our fruit crops. We also need to account for the estimated amount of nitrogen that becomes available to plants each year as the soil organic matter decomposes.

The most effective strategies are to 1) carefully observe and document tree growth and productivity each year, and 2) use annual foliar tissue analyses and compare them with your observations. Be consistent each year in where you sample and from which varieties. Keep the samples for different crops and varieties separate. Keep accurate records of alley crops, companion crops, and types and amounts of compost, fertilizers, mulches, and other materials added to the orchard. At the end of each growing season, try to determine how these practices affected your crops.

In the long term, we want to greatly reduce or eliminate the need for off-farm nitrogen fertilizers. Our focus will be on building abundant supplies of soil organic matter and developing an orchard floor management system that helps provide or conserve nitrogen and make it available to our fruit trees.

How and Where to Apply

Because we are covering many different crops, ranging from a few trees to large commercial plantings, it is hard to make specific recommendations on how to apply fertilizers. It is most important to make accurate applications. For large orchards, dry, granular fertilizers and soil amendments are often broadcast using cone spreaders mounted on the backs of tractors. Some fertilizer spreader designs allow the product to be applied in narrow bands along the tree rows. Smaller cone spreaders are available that work with garden tractors and ATVs (all-terrain vehicles). For smaller orchards, handheld spreaders (belly grinders) can be used to broadcast fertilizers across the orchard floor. For a few trees, hand spreading fertilizer from a small pail works well.

Whether you are using a large commercial fertilizer spreader or are sprinkling fertilizer out of a pail, be sure you know how much material you are applying in a given area. Mark off an area on bare ground or on a parking lot, measuring it so that you know exactly how much area it covers. Carefully weigh the fertilizer and apply it across the area, then measure the amount of fertilizer you have left and calculate how much you applied to that known area. Keep practicing until you know very closely how much fertilizer you are applying to a given area in your orchard.

Be Cautious with Boron

Because boron becomes toxic to plants at very low concentrations, boron-containing fertilizers should never be applied to the soil in bands within crop rows. Broadcast these materials evenly across the orchard floor. For foliar boron sprays, carefully follow the label directions to avoid damaging the trees.

In an orchard, fertilizers can be broadcast across the entire orchard floor or banded within the tree rows. While both methods have advantages and disadvantages, the fact is that both work. I prefer to broadcast fertilizers across the orchard, knowing that the tree roots spread far beyond the rows and into and beyond the alleys. This strategy is particularly effective in high-density plantings with narrow alleys. Other fruit growers enjoy good success concentrating the fertilizers within tree rows. This strategy is particularly effective when the trees are planted farther apart within and between rows.

To a large degree, the orchard floor management system that you choose will determine your fertilization practices. In areas with abundant precipitation or ample overhead irrigation, you may choose to plant permanent alley crops. In drier areas, these permanent alley crops might consist of grasses that go dormant during the summer to reduce competition with the trees for moisture. Annual crops, such as cereal grains, can be planted as temporary alley crops. In arid regions where water for alley crops is unavailable or prohibitively expensive, you might maintain bare ground between the trees. With annual or perennial alley crops, you need to address their nutritional needs, as well as those of the fruit crops. In such cases, broadcasting fertilizers can be a good strategy. In a bare ground situation, fertilizers might best be kept within the drip lines or planting rows of the trees.

In summary, ensure that your trees are well fed—but not too well fed. Our goals are to establish and maintain healthy trees that produce moderate growth and sustainable harvests of high-quality fruit.

ORCHARD FLOOR MANAGEMENT

- History of Orchard Management | 287
- Soil Organic Matter | 289
- Controlling Weeds | 291
- Cover Crops | 300
- Managing Fruit Crop Rows | 311

Orchard floor management lies at the heart of organic fruit production and is the key difference between conventional and organic growers. It includes all aspects of weed control, fertilization, pH management, management of ground- and soil-dwelling beneficial organisms, and companion plantings. Simply exchanging conventional pesticides for those approved for certified organic growers does not make an orchard organic and seldom gives great results.

History of Orchard Management

IN THE NOT-DISTANT PAST, this chapter would have been titled "Weed Management" and may have contained a few short sections on soil structure and erosion. Today the situation is far more complex and integrated.

Prior to the development of tractors and herbicides, orchards were complex ecosystems that included fruit trees, often a wide assortment of understory plants, and an even greater assortment of micro- and macro-organisms. Cultivation in the form of tillage was limited to horse-drawn implements and hand weeding. Trees were often planted in square grids with a tree at each corner of the square, allowing cultivation in all directions. Given reasonable care, soil compaction was minimal and biodiversity within the orchard high.

Advent of the Tractor and Herbicides

With the introduction of these two factors, the situation changed. As recently as the early 1980s, many commercial orchards were kept bare of all vegetation except trees by using a combination of mechanical cultivation and herbicides. While the scorched earth approach to orchard floors eliminated weed competition, serious problems began developing. In many orchards, soil compaction and erosion became severe. With the compaction and loss of soil structure, water infiltration decreased and root damage from diseases and physiological disorders became common. Root damage from improper mechanical cultivation was sometimes serious. Erosion became even more severe, and concentrations of soil organic matter decreased, as did populations of micro- and macroorganisms.

By the mid-1980s, much research on the use of alley crops was under way. The focus was initially on reducing soil compaction and erosion. With the introduction of size-controlling rootstocks for apples and pears, tree density began increasing — we went from about 100 trees per acre or fewer to around 200 or more trees per acre planted relatively closely together in rows separated by distinct alleys. The use of permanent alley crops, narrower alleys, and more closely spaced trees altered mechanical soil cultivation practices.

Most of the alley crops originally tested were monocultures or blends of sod-forming grasses. Tree rows were kept weed-free with preemergence and contact herbicides. Erosion and soil compaction were reduced, but some of the early results showed disappointing fruit yields and rates of tree growth. Some turf crops appeared to reduce tree growth rates, not only due to competition for water and nutrients, but also due to alleopathic effects of the grasses on the trees. Rodent problems also increased as alley crops were reintroduced to orchards. Providing for the nutritional needs of the tree fruits and alley crops became more complicated than it was for a scorched earth monoculture.

As tree density continued to increase with improved rootstocks and training methods, closely spaced fruit trees in well-defined rows became standard practices. In most cases, the alleys were covered with grasses, sometimes with clovers mixed in. Herbicides were used to create 2- to 4-foot-wide weed-free strips along the tree rows. These practices remain very common for nonorganic fruit growers. In arid regions with limited irrigation water, both alleys and tree rows (again, except for the fruit trees) may be left bare for at least part of the year.

Organic Production Gains Traction

In the late 1900s and early 2000s, the viability and importance of organic fruit production gained notice, and research on developing environmentally sound and economically productive orchard floor practices evolved.

Much progress was made from about 2002 to 2010, but there is much we still do not understand about orchard floor management.

What we know is that everything we do in an orchard affects everything else. Imagine your orchard as a net or web. Pull on one small string and it pulls on others, which pull on others, which pull on yet others until the entire net moves. Planting an alley crop increases soil organic matter and helps create soil structure, which influences water infiltration and soil aeration, which affects nutrient uptake and micro- and macro-organism populations in the soil, which affects root health and habitat for beneficial organisms, which affects tree health and vigor, which affects tree root growth, which affects the soil structure, and so on.

Soil Organic Matter

EARLY BIODYNAMIC AND ORGANIC FARMING supporters recognized the importance of soil organic matter (SOM), despite lacking the technology and ecological science backgrounds to fully understand why it is important. Greatly improved research equipment and practices are showing that soil organic matter is even more vital than its early advocates recognized.

Soil organic matter is made up of plant and animal remains that have been partially disintegrated and decomposed, as well as organic compounds that have been synthesized by soil microorganisms. About 60 to 80 percent of soil organic matter consists of humus (humic material), a complex mixture of organic compounds formed by microbial activities. Humus, which is quite resistant to further decomposition, is made up of fulvic acid, humic acid, and humin. Fulvic acid is the most susceptible to decay by microbes and breaks down over 15 to 50 years. Humic acid is more stable, lasting at least 100 years. Humin is the most stable and highly resistant to degradation.

Soil organic matter is important to organic fruit growers for a variety of reasons:

Improves CEC and nutrient uptake. Humus, one of the components of soil organic matter, increases the soil's cation exchange capacity (CEC), which refers to the soil's ability to bind and hold positively charged plant nutrients. Humus has a much greater cation exchange capacity than the mineral portions of the soil. As humus concentrations increase, so does the soil's ability to store positively charged plant nutrients. Further, humus enhances mineral breakdown, making nutrients more available to plants as organometal complexes (specialized organic compounds that include metallic elements) that form with ions, such as Fe^{3+}, Cu^{2+}, Zn^{2+}, and Mn^{2+}. These organometal complexes are more available for plant uptake and use than the mineral form of the elements.

Increases soil aggregates. Soil organic matter is also critical in binding mineral soil particles into stable aggregates. These aggregates combine to form the macro- and micropores in the soil that are critical for water infiltration and drainage. The presence of aggregate-building organic matter improves the drainage of fine-textured soils, yet also improves the water-holding capacity of sandy soils. Aggregated soils are less susceptible to erosion and compaction than those low in organic matter. Aggregate formation is also important for gas exchange, root penetration, and providing suitable habitats for micro- and macroorganisms. The degree to which a soil is composed of aggregates is a measure of the soil's tilth or friability and is a vital key to the soil's productivity.

Adds nitrogen and sulfur. Soil organic matter is a source of nitrogen and sulfur. As organic matter is broken down by micro- and macroorganisms, nitrogen and sulfur are released into the soil and made available for plant uptake. By some estimates, decomposing soil organic matter releases about 10 to 20 pounds of nitrogen per acre each year for each 1 percent of soil organic matter present. The actual amount of nitrogen released, however, is quite variable and heavily influenced by many factors, including temperatures.

Increases biological activity. While helping to support plant growth, soil organic matter is also critically important for biological activity in the soil. It is, literally, the base of the food chain for soil-dwelling micro- and macroorganisms. Soil biological activity, in turn, is a key component of healthy and productive soils that are needed for successful organic fruit production.

Soil organic matter concentrations vary greatly across different soil types, from less than 1 percent in sandy soils to 6 percent or more under grasslands and up to 80 percent in some wetland soils. How much organic matter is needed to produce a high-quality, sustainable crop, however, is very hard to say. Commercial organic orchard crops are produced successfully in Washington State on sandy soils with as little as 1.5 percent organic matter. Unlike pH and nutrient concentrations, we have no recommendations for soil organic matter concentrations. We also lack a standard method of measuring soil organic matter, and testing methods vary across regions and between laboratories.

Rather than worrying about the soil organic matter concentration in your orchard, use tests every 3 to 5 years to monitor the effects of your soil-building cultural practices. Such practices might include the use of short-term green manure or permanent alley crops, in-row cover crops, and applications of composts or mulches. Because laboratory tests measure the decomposed portion of organic matter, and not pieces of roots or other visible plant materials, organic matter concentrations in the soil change slowly. Given effective orchard floor management practices, losses of organic matter should stop within a year or so of transitioning to organic or starting a new orchard, and levels should gradually increase over 10 to 15 years. On sandy soils in arid climates, 3 percent SOM is a good target, while 5 percent is reasonable on loamy soils. The goal is not to increase soil organic matter concentrations for the sake of increasing them but rather to develop and maintain a biologically diverse and active soil with good structure and adequate, but not excessive, plant nutrient concentrations. Monitoring soil organic matter, as well as tree health and productivity, will help you track your progress. Because testing procedures (and therefore the results) for soil organic matter differ between laboratories, find a good lab and stick with it. If you switch from one laboratory to another between years, you will not be able to accurately compare results for different years or track your progress.

How we manage an orchard floor has great impact on organic matter. Cultivation and bare-ground treatments cause soil organic matter concentrations to decrease. Green manure crops, alley and in-row companion crops, and mulching increase soil organic matter.

Controlling Weeds

SOME ORGANIC PRACTITIONERS OBJECT TO THE USE OF the word "weed," believing that all vegetation in an orchard is desirable. For our purposes, we will consider any plant growing where we do not want it to be a weed. Regardless of semantics, vegetation management on the orchard floor is critically important while we are establishing our trees and remains important throughout the life of many orchards.

It is important to understand that any time we create a vegetation-free area, we have created a vacuum that nature will strive to fill. As with any ecosystem, highly aggressive, invasive, and competitive plant species have the advantage over our nonaggressive and relatively slow-growing fruit crops. On the positive side, the required amount of vegetation-free space around the fruit trees is smaller than most conventional herbicide strips. The need for aggressive weed control also diminishes during the course of the growing season and as the trees mature.

In orchards consisting of large, established trees that essentially form an unbroken canopy across the entire orchard floor, weed control is often a rather minor concern. The trees have extensive root systems capable of obtaining water and nutrients far beyond the reach of shallow-rooted understory vegetation. The dense shade produced by the trees also greatly reduces the establishment and growth of other vegetation in the orchard.

While weed problems are often minor in a mature orchard with large trees, it is important to manage vegetation throughout the life of high-density orchards if you want to produce high-quality fruit and high, sustainable yields. Orchard systems that confine the trees to narrow, trellised rows have a particular need for careful vegetation management for two reasons: First, the trees are kept small, often with weak trunks and relatively little foliage. These trees are designed and maintained to produce high yields of fruit, not structural wood and extensive root systems that provide abundant nutrient and carbohydrate reserves. Large trees are better able to compete with orchard floor vegetation for nutrients and water. Second, an abundant amount of sunlight penetrates the orchard floor, greatly increasing the amount, diversity, and growth of competing vegetation.

For young trees just getting started, in any type of orchard, weed control is a major concern. One of the greatest challenges in establishing an organic orchard is weed pressure. Perhaps 90 percent of the young trees' root systems are in the top foot of soil, placing them in direct competition with "weeds." Most young fruit trees and bush fruit plants are rather poor competitors.

In Washington State trials, trees in organic plots where weed pressure was high had far lower survival and growth rates than those in conventional plots with few weeds. The situation is worst in areas with ample summer rains, where vegetation is naturally abundant, varied, and vigorous. An old axiom in horticulture is that you can double the size of a tree during its first 3 years by keeping a vegetation-free zone of several feet around the tree.

The challenge is to minimize competition with the trees while maintaining biologically diverse and active orchard floors and soil environments, as well as minimal soil compaction and erosion. Unfortunately, there is no perfect strategy and no single strategy fits all situations. The vegetation management practices of a West Texas apple grower, a Louisiana mayhaw grower, a British Columbia cherry grower, and a Quebec plum grower are likely to be quite different. The strategies below will guide you in developing a system that works for your crops and growing region.

Methods of Control

A key strategy in weed management is preventing the weeds from forming viable seed. Whatever weed suppression practices you employ, strive to prevent the weeds from setting seed. After a time, the weed pressure will diminish. Weed seeds can be very persistent, however. The seeds of common purslane can survive in the soil for 20 years, while black mustard seeds can remain viable in the soil seed bank for 40 years or more.

Thermal Weeders

Thermal weeders use heat produced by propane-fired flames, infrared radiation from a red-hot metal or ceramic plate, or steam to kill or stunt unwanted plants. Thermal weeders work by disrupting the cells in the plants, causing the affected plant parts to wilt and die. Although not effective against established perennial grasses or many perennial broadleaf weeds, these weeders are especially effective against annual weeds and young perennial weed seedlings and offer a chemical-free way to manage orchard floor vegetation without damaging the soil. Thermal weeders can be adapted for maintaining the centers of tree rows in some orchard designs and in maintaining the narrow bare ground strip in the "sandwich" system (see page 317).

Flame Weeders

Using fire as a weed control strategy in crop production has been practiced for centuries, if not millennia. Although the thought of fire raging through your orchard can be unsettling, there are applications for flame weeders. Flame weeders are the simplest type of thermal weeder, usually consisting of a handheld, pushcart-mounted, or tractor-mounted nozzle attached by a flexible hose to a propane tank. Flame weeders were first used in North America for cotton in the 1930s. In orchard situations, they are most useful for managing vegetation along fencerows and for spot-spraying within tree rows on bare ground. They should not be used on or near combustible mulches such as bark, straw, or alleyway clippings.

Large, tractor-mounted flamers consist of metal beds up to about 8 feet wide with flame nozzles underneath the beds that create sheets of flames. Bed flamers are designed primarily for preparing planting beds for field crops or for burning off plant residues after harvest. In an orchard situation, bed flamers might be useful for preplant site preparation. Of the three types of thermal weeders discussed here, flame weeders create the greatest risk of damaging trees or of starting unintended fires. Use these with extreme care.

Infrared Weeders

Infrared weeders prevent direct flames from contacting the plants and soil. Instead, the flame is directed onto a ceramic or metal plate that radiates heat at temperatures around 1,800 to 2,000°F (982–1,093°C). As with flame weeders, units range from small handheld to large tractor-mounted devices. Infrared weeders are safer to use than open-flame weeders and can be effective in some orchard situations when used carefully.

Infrared weeders are presently manufactured in Sweden and available in North America. They generally cost about three times more than comparable flame weeders. A commercially available unit presently being marketed in North America uses a handheld wand that resembles an open flamer and is primarily designed for spot treatments. The unit costs about $300 USD, not including the cart and propane tank.

Steam Weeders

Steam weeders can be useful in an orchard floor management program. Early steamer models were extremely expensive tractor-mounted units. Pushcart models with handheld wands are now available in North America for around $4,600 USD, compared with a comparable flame weeder for around $300 or less. Steam weeders have promise for fruit growers, but their high initial cost, slow speed of travel, and large propane requirements limit their usefulness for many growers at this time. Expect improvements in design that will lower equipment costs and make the devices more attractive to fruit growers.

Infrared and steam weeders are less likely to cause unintended fires than flame weeders, but they can still damage fruit trees if used incorrectly. Be sure that the equipment's design fits your application.

Organic Herbicides

While conventional growers enjoy a wide range of herbicides registered for fruit crops, organic fruit growers have very little to choose from. Worse, herbicides that are accepted by organic programs do not work very well for anything but young seedlings. Herbicides can be effective for spot treatments in an organic orchard. Five basic types of organic herbicides are available to organic fruit growers: acetic acid (vinegar), citrus products, clove oil, corn gluten, and herbicidal soaps. Except for corn gluten, all appear to be effective against young seedlings, although none control established perennial weeds well. Combining soap and citrus oil products may prove more effective than either product alone.

Acetic Acid

Acetic acid can be used as an organic herbicide. Acetic acid is the active component of vinegar, and it chemically burns delicate plant tissues. The vinegar sold at grocery stores contains about 5 percent acetic acid; several commercially available herbicides contain higher concentrations. In tests conducted by Cornell University, household vinegar had some short-term effectiveness in suppressing weeds. The commercial products marketed as organic herbicides performed better, and a solution containing 20 percent acetic acid provided the greatest and longest-lasting control. Young annual weeds were killed readily. Quack grass was suppressed, but it continued to regrow, and the 20 percent acetic acid had to be applied three times to effectively suppress the quack grass. In contrast, a single application of glyphosate (not allowed in organic production) provided effective control of even quack grass throughout the season.

When comparing products, the authors of the study concluded that organically acceptable acetic acid products cost around $40 per 1,000 square feet of area treated. If three applications are needed, the cost rises to around $120 per year. Conventional growers can achieve equivalent or better control by using a single application of glyphosate herbicide for about $13.

Even household vinegar (5 percent acetic acid) can irritate eyes and respiratory systems, so products containing 20 to 25 percent acetic acid must be used with great care. Like soap and citrus oil products, acetic acid herbicides kill succulent green tissues and are probably best used for spot treatments within an orchard.

Citrus Products

Several citrus products are approved for use as herbicides in organic crops. These products, some of which are on OMRI's approval list, are based on various forms of limonene (citrus oil). These materials are effective degreasers. They act by dissolving oils and waxes in the epidermal layers of plants, causing those plants to desiccate (dry out) and die. To be effective, the materials must be sprayed on succulent green tissues. They can quickly kill green tissues (within about two hours of application) and are most effective against young annual and perennial plant seedlings.

Citrus products are not effective against established perennial weeds, such as Canada thistle or quack grass, because the oils do not kill the roots, rhizomes, or other underground organs that produce new shoots when the tops are killed back. These products can damage young trunks and stems, foliage, flowers, and developing fruit on fruit trees and bushes, but they should be safe to use in an orchard, given careful application.

Their greatest value would appear to be as spot treatments within tree and "sandwich" rows. At least one citrus oil product now approved for organic use also contains a soap or surfactant component similar to the herbicidal soaps described in this section. Surfactants help herbicides and other pesticides cover and stick to plant surfaces and can make some contact herbicides more effective.

Clove Oil

Clove oil (eugenol) is found in several commercially available products, at least one of which is registered with OMRI. In some cases, the clove oil is blended with vinegar and other ingredients to increase the herbicidal activity. While these products show promise for ornamental landscape and garden use, more testing is needed to determine how well they work in commercial orchards.

Corn Gluten Meal

Corn gluten meal is a by-product of the wet milling process for corn. It was originally patented and introduced by Iowa State University as a natural herbicide in 1991. According to the developers, corn gluten works when applied before weed seeds germinate, when it can inhibit the root formation of germinating plants. Unfortunately, after 2 years of trials in Oregon, researchers concluded that corn gluten meal did not control any of the weeds under any circumstances. It would appear that, for organic fruit growers, corn gluten might best be used as a fertilizer rather than an herbicide. It has a fertilizer analysis of 9-0-0. OMRI presently lists two corn gluten products, one of which is advertised as a fertilizer and soil amendment.

Herbicidal Soaps

Several herbicidal soap products are sold for use in North America. The materials are quite similar to insecticidal soaps that have been used for many years, and they are marketed primarily for use in vegetable and flower gardens, landscapes, lawns, and areas such as driveways. Chemically, herbicidal soap products are salts of long-chain fatty acids. They act by damaging the outer layers of green, succulent plant tissues, causing the affected plant parts to wilt and die. They are not effective against woody stems or trunks and do not damage roots or other underground organs. The following discussion of one particular herbicidal soap, ammonium nonanoate, provides a good description of this type of product.

The following excerpts on the use of ammonium nonanoate are taken from the *Ammonium Nonanoate (031802) Fact Sheet Related Information*, issued by the U.S. Environmental Protection Agency (EPA) in November of 2006. This fact sheet applies to a particular herbicidal soap product presently approved by the Organic Materials Review Institute (OMRI).

> Ammonium nonanoate is closely related to other salts of fatty acids known as soap salts. The active ingredient is a C9 saturated-chain fatty acid soap salt. It . . . is a clear, colorless to pale yellow liquid with a slight fatty acid odor. Ammonium nonanoate is a non-systemic, broad-spectrum contact herbicide that has no soil activity. . . . Ammonium nonanoate is to be used for the suppression and control of weeds including: grasses, vines, underbrush, annual/perennial plants, including moss, saplings, and tree suckers.
>
> Ammonium nonanoate can be applied using all standard methods of liquid herbicide application, including hand-held, boom, pressure, and hose-end sprayers. For use, the concentrate is diluted with water to a specified concentration.[. . .] For the active ingredient to be effective, the leaves of undesirable vegetation must be uniformly sprayed and thoroughly wetted. Application can be repeated as often as necessary to obtain the desired control.
>
> The Agency concludes that no risks to human health will be expected from the use of ammonium salts of higher fatty acids (C8–C18 saturated and C18 unsaturated), based on their low toxicity and the fact that residues from pesticide use are not likely to exceed the levels of naturally occurring or intentionally added fatty acids in commonly eaten foods. Since ammonium nonanoate has the potential for eye, skin, and mucosal irritation [mucous membranes], the Agency is requiring stringent precautionary labeling. Exposure and attendant risks are expected to be negligible for applicators when they follow the directions for use by wearing the appropriate personal protective equipment.
>
> Ammonium nonanoate is expected to degrade rapidly, primarily via microbial action, with a half-life of perhaps less than one day. Although ammonium nonanoate is slightly toxic to both warm water and cold water fish species and highly toxic to aquatic invertebrates, the agency believes that ammonium nonanoate, when used as directed, will not persist in the environment. If ammonium nonanoate is used according to the directions on the label, it should not seriously impact aquatic invertebrates because it is not applied directly to water and it undergoes very rapid microbial degradation in the soil.

Application rates of herbicidal soaps used for organic production are high. One product recommends using 26 fluid ounces of their product per 1,000 square feet of area covered. The same amount of glyphosate herbicide, which is very popular in conventional horticulture, would cover an area nearly 24 times as large and kill both annual and many perennial weeds.

While herbicidal soaps may have a place in organic orchards, they appear to be best suited for spot treatments of young perennial and annual seedlings. They are not effective in controlling established perennial weeds.

Biological Management of Vegetation

Chickens and weeder geese are used to control weeds in gardens and orchards. You will need to use pens to confine the birds to the areas that you want weeded. Young geese are more effective at weeding than older geese, and some people prefer Chinese strains over other breeds.

Unfortunately, organic certification standards greatly limit the use of livestock in orchards. The U.S. National Organic Program requires that livestock, including birds, be removed from a tree fruit orchard 90 days before harvest. For bush fruits, the livestock must be removed at least 120 days before harvest. For most organic orchards in North America, these rules effectively prevent the use of livestock for preharvest vegetation management. The animals can be returned to the orchard after harvest. (Chickens can also serve a valuable pest control role in orchards by devouring apple maggot pupae, as discussed in chapter 11.)

Insects are also used in biological weed control, although their usefulness in orchards is still very limited. Beetles, for example, are used to manage yellow starthistle and knapweeds in western North America, but neither starthistle nor knapweeds are typically problems in orchards.

Non-fruit Crops

Non-fruit plants in and around orchards provide many benefits. Clover and other legumes provide habitat for beneficial organisms and add nitrogen to the soil. Perennial grasses and alfalfa in the alleys reduce soil compaction and erosion while providing easy access in wet weather and reducing dust in dry weather. Alley and in-row crops also help prevent nutrients from leaching out of the root zone and release them slowly to the fruit crops. Cover crops add organic matter, play key roles in the formation of soil aggregates, and help create biologically diverse and active soil. Insectary crops provide habitat for beneficial insects and mites.

Cover Crops

Several types of non-fruit crops can be used in orchards, including alley crops, in-row cover crops, and insectary crops. Depending on where and how the non-fruit plants are grown in the orchard, benefits include reduced soil compaction and erosion, improved soil structure, increased biological diversity and activity, increased nutrient cycling, and habitat for beneficial organisms.

Including an annual or perennial alley crop in your orchard is highly recommended when climate and soil moisture allow. Suggested annual and perennial cover crops are listed in table 9.1. These crops work well in most fruit-growing areas of the United States and Canada. Check with Cooperative Extension and provincial fruit specialists in your area for additional recommendations. In-row cover crops, also called living mulches, have great potential for organic orchards as well.

Alley Crops

In general, alley crops are highly recommended for many orchard situations. Alley crops are critically important in reducing soil compaction caused by tractor and vehicle wheels, reducing wind and water erosion, maintaining or increasing soil organic matter, providing improved habitat for soil micro- and macroorganisms, reducing dust contamination on fruit, and improving access during wet weather. Depending on the alley crop, they can provide an attractive appearance, and grass cover crops

Criteria for Understory Crops

In choosing understory crops, try to select those that:

- Do not compete heavily with fruit trees for moisture and nutrients
- Provide poor habitat for rodents and rabbits
- Provide good habitat for beneficial organisms
- Improve soil quality

provide a pleasant surface to walk on. These latter two benefits are especially important in home and U-pick orchards.

There are also disadvantages to alley crops, however, and you need to take precautions when using them. The two greatest concerns with alley crops are 1) competition with the trees for nutrients and moisture and 2) the crops can provide habitat for rodents that may damage fruit crops by girdling trunks and stems.

There are other potential drawbacks associated with alley crops. Besides rodents, they can harbor other pests and diseases that attack fruit crops. They increase the humidity low in the tree canopy, which can increase disease problems, particularly in low-growing crops. In arid climates, lack of precipitation and limited or prohibitively expensive irrigation water may make alley crops less attractive. Invasive weeds can also be troublesome in cover crops: In the arid Southwest, sandbur, johnsongrass, and Bermuda grass can all be invasive problems; Canada thistles, quack grass, and hawkweeds can make life difficult for fruit growers farther north. For home and U-pick orchards, cereal crop and alfalfa stubble are less attractive and less pleasant to walk on than grass. These cover crops may be better suited to commercial grower-pick orchards.

Each growing region faces its own particular challenges in orchard floor cover crop management. We will discuss strategies to overcome some of these potential problems.

Annual Alley Crops

For a newly planted orchard where the tree root systems are small, tilling under the alley crop to make it a green manure crop can provide some soil-building benefits. This practice, however, makes access to the orchard difficult during the growing season when the trees need care. A better strategy is to grow your green manure crops during the site preparation phase, before planting your trees.

Annual crops are often cereal grains, such as wheat, barley, oats, or annual rye. Austrian winter peas and other legumes are sometimes used as annual cover crops, and grains and legumes can be blended together in a single crop. Some annual cover mixes might include 80 pounds of wheat, barley, or oats per acre; 30 pounds of common buckwheat mixed with 4 pounds of clover; or 100 pounds of winter or spring peas per acre.

Buckwheat is well adapted to cool, moist climates, short growing seasons, and acid soils, but it is not winter-hardy and is killed by mild freezing temperatures. Buckwheat is particularly effective at taking up otherwise insoluble phosphorus and making it available to the fruit crop. Buckwheat can become a troublesome weed, so be sure to till it very shallowly into the soil before the seeds ripen. Mowing is not an effective way to prevent reseeding because some buckwheat plants are short enough to escape the mower blades, leaving some seed production.

When to plant. The best time to plant an annual cover crop depends on your climate and the availability of water. Where spring rainfall is abundant or water is readily and economically available for sprinkler irrigation, you can plant early in the growing season. This strategy is good for spring wheat, spring barley, oats, peas, and buckwheat.

In areas where precipitation during the growing season and irrigation water are limited, another strategy is to plant late in the season, after harvest in bearing orchards. This practice works well when using winter wheat, winter rye, or winter barley as an alley crop. No further care of the cover crop should be needed until the following summer. As with spring-planted alley crops, you may need to mow from one to several times. Leave the stubble in place until it is time to plant the annual cover crop again. In Washington State University trials, winter rye proved particularly effective in managing grassy weeds in orchard alleyways.

How to plant. In a new orchard, when tree roots are still confined to the tree rows, there are few problems with plowing in preparation for planting a cover crop. As the tree root systems expand, however, cultivating the alleys can damage the tree root systems. Remember that many of the tree feeder roots are located within the top six inches of soil.

Many years ago, some authorities recommended cultivating deeply around fruit crops to drive the roots deeper into the soil. Apparently they thought the fruit crops would be more drought-tolerant and have better access to soil nutrients if their roots were deeper. Deep cultivation, however, does not drive the root systems deeper into the soil. It simply cuts off the vital shallow roots that lie in the most nutrient-rich and biologically active portion of the soil. If you must cultivate in your orchard, keep the cultivation as shallow as possible.

When planting an annual cover crop through the stubble of the preceding cover crop, use a no-till planter, if possible. Set it at the shallowest planting depth that will serve for your cover crops. Regardless of the planter used, keep tillage as shallow as possible to reduce damage to tree roots. Suggested annual cover crops are listed in table 9.1.

How to manage. Mowing annual alley crops once to several times during the growing season prevents them from becoming overgrown and producing seed. Cereal grain crops provide excellent rodent habitat. Reduce this habitat by mowing the crops before they set seed. Mow by early- to mid-fall to leave a short stubble going into winter, when most tree girdling by rodents occurs.

Blowing the clippings into the tree rows during mowing helps reduce weed problems and adds organic matter to the tree rows. In areas with heavy rainfall or where the soil moisture levels are already naturally high, keeping the soil in the crop rows covered with mulch can increase root rot problems in some crops. Mulching has aggravated root rot problems in some New York apple orchards, for example.

Despite their advantages, especially for young orchards, annual alley crops have some drawbacks. Most importantly, the stubble provides little weed control and leaves the orchard floor open for invasion during much of the growing season. Wheat and barley stubble provide reasonable access for equipment, but they are not overly pleasant to walk on and are less desirable for U-pick and home orchards than smooth, turf-covered alleys. Before using annual alley crops, be sure that you have serious weeds under control, particularly established perennial weeds.

Perennial Alley Crops

Perennial alley crops have many of the advantages of annual cover crops, as well as a few more. The greatest advantage is not having to replant each year. Permanent (or at least long-term) alley crops eliminate or reduce the need to cultivate the alleys, protecting the shallow tree roots.

Permanent cover crops should be very durable and resistant to wear from machinery and foot traffic. Particularly in organic orchards, where there are few approved herbicides, the cover crops must be noninvasive or they can move into crop rows and become weed problems themselves.

Permanent alley crops can be sod-forming or bunch-forming grasses, alfalfa, clovers, or mixes of these and other crops. You need to match the alley crops to your climate and irrigation practices.

Low-Growing, Bunch-Forming Grasses

In northern Idaho trials, I had success with a variety of sod-forming and bunch-forming grasses as orchard alley crops. A combination of sheep fescue, hard fescue, and white clover worked well as a low-maintenance alley crop, although this combination was slow to establish and benefited from sprinkler irrigation during the planting season. The advantage of this type of blend is that the fescues are naturally low-growing and become dormant during the dry summers, minimizing competition with the trees for moisture and reducing the need to mow the alleys. Various other bunch-forming grasses are available that are adapted to different climates.

A problem with low-growing, bunch-forming grasses is that they are not particularly good at competing with weeds. Especially troublesome are deep-rooted perennial weeds, such as Canada thistle, and spreading rhizomatous weeds such as quack grass. Make sure you have perennial weeds well under control before planting slow-to-establish or otherwise poorly competitive cover crops. Another drawback to the low-growing covers is that they produce very small amounts of clippings that can be blown into the tree rows to serve as mulch.

In areas with cold winters and particular types of soil, bunch-forming alley crops can suffer severe damage due to frost heaving. This damage makes weed invasion likely and occurred during my Idaho trials. Despite that, the fescues persisted and continued to provide a reasonably good working surface for 20 years. On soils that experience moderate to severe frost heaving, sod-forming grasses are probably better suited as perennial alley crops if sufficient soil moisture or irrigation is available for their establishment and maintenance.

White Clover

White clovers can make good orchard cover crops. They establish quickly, provide good weed suppression, and add nitrogen to the soil. White clover stands are often fairly short-lived, however, and typically need to be reseeded every few years. In my Idaho trials, white clover thinned out to a few scattered clumps within 3 years, leaving mostly the sheep and hard fescue grasses. Depending on your location, certain other clovers can also be used as orchard cover crops and are longer-lived. In practice, blends of several different clovers are more effective than monocultures of a single species. Table 9.1 suggests clovers that can be used as orchard cover crops. Avoid using white or yellow sweet clover (*Melilotus* sp.), which are highly aggressive, invasive, and hard to control in orchards. These crops are best used for forage production and are classified as invasive species in some areas of North America.

One drawback with all clovers is that their roots, rhizomes, and seeds are attractive to rodents. Should you choose to include clovers in your orchard understory, also include an aggressive rodent control program (see page 372).

Orchard Grass and Tall Fescues

These are used successfully as perennial alley crops in home and commercial orchards. They require more frequent mowing than low-growing fescues, but they are more aggressive at competing with weeds and tolerate frost heaving in the soil much better. Orchard grass and tall fescues produce large amounts of clippings that can be blown into the fruit crop rows as mulch. A commercially available blend of perennial ryegrass and creeping red fescue, known as Companion grass, has proven popular in many different kinds of fruit plantings. This crop is relatively low-growing. While it does not provide much in the way of clippings to mulch the rows, it forms a dense sod and requires less mowing than taller-growing crops.

Drought-Tolerant Grasses

In arid climates, drought-tolerant perennial grasses can provide advantages as alley crops, although some are problematic. Bermuda grass is often planted as lawn turf in warm, arid regions. Native to Africa and Asia, Bermuda grass gained its name because it became a serious invasive weed there, and it can also become an invasive weed in orchards. Texas fruit specialists suggest managing Bermuda grass on the orchard floor if the grass is already present, but not to introduce it into an orchard.

Buffalo grass is less competitive than Bermuda grass and tends to be expensive to establish in orchards in arid regions. Buffalo grass has not proven hardy in colder areas. King Ranch (K.R.) Bluestem is a clump-forming grass that shows good promise as an orchard crop for warm, arid regions. Kleingrass, another clump-forming grass, makes a good alley crop for the Southwest, but it is more difficult to establish than K.R. Bluestem. Fescues work well in northern and southeastern U.S. plantings, but they tend to die out in Texas orchards.

Alfalfa

In Washington State apple orchard trials, alfalfa proved to be an effective perennial alley crop. In that climate, alfalfa crops normally persist for about 5 years before they need to be replanted. In these particular trials, the alfalfa created an effective alley crop and the large amount of top growth that was produced provided good weed control when blown into the tree rows as mulch. In related trials, however, alleyways maintained in alfalfa became heavily weed-infested.

Managing Alley Crops

For both annual and perennial alley crops, a good practice is to mow adjacent rows alternately, several weeks apart, if possible. Alternate mowing leaves the predators and other beneficial organisms in every other row undisturbed for a time, allowing them to repopulate the mowed rows.

Alfalfa, clover, peas, and other legumes on the orchard floor will fix nitrogen, provided they are inoculated with the proper *Rhizobium* before planting. During establishment of the legumes, they compete with the fruit crop and other cover crops for nitrogen. When they mature or die, the legumes begin adding nitrogen to the soil. When alfalfa and clover are cut and the clippings blown into the fruit crop rows, the clippings add nitrogen to the soil there, even though the legumes in the alley remain alive.

Be cautious when using legumes in your cover crops. While nitrogen is necessary for plant growth, too much creates excessively lush, disease-susceptible foliage, reduced flowering, and poor fruit quality. This problem was observed in Washington State University orchard trials with alfalfa alley crops. Carefully monitor foliage nutrient concentrations and fruit quality. Be prepared to change the types of plants in your alley crop to adjust the nitrogen availability, if necessary.

Insectary Crops

Sometimes called nectary crops, the primary purpose of these plantings is to provide habitat for beneficial organisms, such as green lacewings, that are predators of orchard pests. Many of these beneficial insects depend on nectar, pollen, and other plant materials during certain stages of their life cycles. Insectary crops provide the habitat needed to complete their life cycles, and they also provide habitat for sufficient prey to support the predator populations. The goal in organic pest management is seldom to eradicate the pests. Doing so eliminates the food for beneficial predators that then leave or die, leaving our orchards open to rapid reinvasion by new pests. Our goal is to maintain enough pests to support the predators but not enough to do serious damage to our fruit crops.

Insectary crops can be annual, perennial, or a combination of both. They include such plants as dill, chamomile, hairy vetch, spearmint, Queen Anne's lace, buckwheat, yarrow, white clover, cowpea, and cosmos. Other plants also serve as effective nectary crops, and a blend of different plants is better than a monoculture.

If you adopt some variation of in-row living mulches (see page 316), include insectary crops within the planting. Another option is to include insectary plants in the alleyway cover crop, although it can be difficult to manage an insectary crop in alleys due to such activities as mowing, and they may not be as effective as separate plantings. For example, if you maintain a bare strip within the tree fruit rows, blowing the seed heads of insectary plants into the rows as a mulch could create weed problems. Alley crops of all kinds provide some habitat for beneficial organisms. If you choose to include insectary crops in your alleys, you will have to actively manage them.

A third option is to plant rows of insectary crops within the orchard, spaced about 50 feet apart. From a commercial standpoint, this reduces

Orchard Access

In a desert orchard, roadways and alleys are often dry and firm most of the year. In areas that receive abundant rain and/or snow, access during wet soil conditions can be challenging. Pruning, mulching, fertilizing, mowing, spraying, and harvesting can become difficult or impossible. The worst situation involves bare alleyways.

In general, sod-forming grasses for roads and alleyways provide the best year-round access. They may be supplemented with clovers or other nitrogen-fixing plants. For alleys, well-established alfalfa also provides reasonably good access. In arid locations where perennial sod-forming grasses are not well adapted, bunch-forming fescues, buffalo grass, and other arid-adapted crops can be used. Even stubble from an annual alley crop of wheat, rye, or barley can reduce soil problems while improving access.

yields per acre and can reduce profits. It might be possible for commercial fruit growers to recoup a return on the investment by planting herbal or ornamental cash crops within the insectary crop rows. Regardless of where you plant the insectary crops, try to keep them within 50 feet of your fruit crops and preferably closer.

Suggested Cover Crops

Table 9.1

Common Name	Seeding Rate lb/A (kg/ha)	Establishment Rate	Nitrogen Req'd lb/A (kg/ha)
ANNUAL COVER CROPS			
Barley (spring)	120 (135)	FAST	30–50 (34–56)
Barley (winter)	120 (135)	MEDIUM-FAST	30–50 (34–56)
Buckwheat (common)	35–60 (39–68)	MEDIUM-FAST	10–20 (22–23)
Buckwheat (tartary)	25 (28)	MEDIUM-FAST	10–20 (11–23)
Oats (spring or winter)	120 (135)	MEDIUM-FAST	30–50 (34–56)
Peas (spring or winter)	120 (135)	MEDIUM-FAST	*by soil test*
Ryegrass (annual)	30 (34)	FAST	30–50 (34–56)
Wheat (spring)	120 (135)	MEDIUM-FAST	20–50 (23–56)
Wheat (winter)	120 (135)	MEDIUM	30–50 (34–56)
Grain/Pea	80/100 (90/112)	FAST	20–30 (23–34)
PERMANENT ALLEY CROPS			
Alfalfa	13 (14.6)	MEDIUM	0
Fescue (hard)	20 (23)	SLOW	20 (23)
Fescue (sheep)	20 (23)	VERY SLOW	20 (23)
Fescue (tall)	25 (28)	MEDIUM	20 (23)
Perennial rye	25 (28)	FAST	30 (34)
Russian wild rye	30 (34)	SLOW	20 (23)
Siberian crested wheatgrass	35 (39)	MEDIUM	20 (23)
Standard crested wheatgrass	25 (28)	MEDIUM	20 (23)
Strawberry clover	4 (4.5)	MEDIUM	0
White clover	4 (4.5)	MEDIUM	0

While intuitively it seems beneficial to increase the diversity and abundance of flowering plants in an orchard, either alone or as parts of alley and in-row cover crops, research has seldom shown measurable improvements in pest management with such an approach. Washington State University researchers report they are having better success targeting specific pests and their predators or parasites. Alfalfa, for example, has proven to be a promising insectary crop in controlling leaf rollers. Unfortunately, we are just at the beginning of this type of research. For now, strive for a biologically diverse understory and carefully monitor the understory and fruit canopy for pests and beneficials. Insectary crops provide excellent habitat for rodents, so employ aggressive rodent management practices (see page 372). Keep careful records of the understory crops that are present, your management practices (such as mowing), the types and abundance of pests and beneficials, and the fruit yields and quality.

Nutrient Management

In chapter 8, we discussed crop nutrition and fertilization in detail. Here we will integrate nutritional goals with orchard floor management practices.

Integrating tree fruit nutrition practices with vegetation management in the alleys and under the trees is far more complicated than simply adding fertilizer to the soil or foliage. Some alley crops add nitrogen to the system during one phase of their growth and take it away during another phase. Other alley crops provide no nitrogen to the soil and are always competing with the trees for nutrients, but even then they are of benefit to the fruit crops by preventing the nutrients from leaching out of the root zone. Alley and companion crops also add organic matter to the soil, increasing nutrient reserves for the fruit crops. In-row companion crops may compete for nutrients or add nitrogen, but they also alter when the nitrogen is available for the trees. Adding compost to the orchard floor provides nutrients and organic matter but can interfere with mowing, tillage, and other practices.

Although the situation is complicated, the approach is quite simple:

- Correct soil nutrient deficiencies and other soil problems before planting your trees.
- Focus on your soil-building program.
- Monitor soil pH (especially in the crop rows), and keep it within a desired range.
- Monitor tree nutrition with visual observations and/or tissue testing, as we discussed in chapter 8.
- Add fertilizers and increase the use of nitrogen-fixing cover crops when tree or fruit observations or foliar tests indicate deficiencies.
- Reduce the abundance of legumes and other nitrogen-fixing plants in the understory if the trees are taking up too much nitrogen.

Erosion Control and Soil Structure

The way the orchard floor is managed has a great impact on soil structure and erosion. Part of the soil building program required by organic certifying organizations involves protecting and enhancing soil structure and reducing erosion. Alley crops are very helpful in preventing erosion, as discussed above. Even in the most arid climates, try to maintain at least an annual cover crop in the alleys, or a perennial crop of locally adapted plants.

If possible, maintain roads in permanent sod to reduce soil compaction and erosion, provide access during wet weather, and reduce dust contamination on the fruit. These areas are somewhat removed from the fruit crops, and competition for water and nutrients is less. You have flexibility in cover crop selection and management.

Mulches, whether inorganic, organic, or living, help prevent erosion within the crop rows and some add organic matter to the soil. Bare soil is highly susceptible to compaction and erosion. Even raindrops or droplets of irrigation water striking bare soil can damage structure near the surface and lead to sheet or rill erosion. To the greatest extent possible, reduce the amount of bare soil in your orchard.

Managing Fruit Crop Rows

FOR MANY YEARS, TREE ROWS have primarily been maintained free of other vegetation — a strategy that eliminates much of the competition with the fruit crops and reduces rodent problems. For newly planted orchards, vegetation-free strips in the tree rows remain the method of choice. During their first few years in the orchard, fruit trees lack the size and vigor to compete with other vegetation. Even if they survive the competition, they will likely remain stunted and unproductive.

Many strategies are available to prevent competing vegetation from establishing within crop rows. For very small plantings, hand weeding can produce acceptable results. One of the best implements for this task is a "speed hoe" equipped with a looped U-shaped blade. This implement cuts weeds off at or just below the soil surface and does not damage the crop's roots. As the orchard grows beyond a small home planting, however, more efficient methods are needed.

Mulches

Mulches have long been used to manage weeds in fruit plantings. These materials come in many different forms, but most act to prevent light from reaching the soil surface or reduce the amount of light. Some mulches also create physical barriers that seedlings and shoots cannot penetrate.

Inorganic Mulches

Inorganic mulches include plastic films and porous weed barrier fabrics, as well as gravel and crushed stone. The latter are really only useful for small landscape plantings.

Plastic Films

Plastic films act primarily by forming a physical barrier that prevents weeds from growing. Dark-colored plastics also reduce the amount of light that reaches the weed seeds; germinating seedlings; and shoots from rhizomes, bulbs, and other underground organs. Clear plastic films are not good choices for weed barriers. In mild and cool climates, they serve as greenhouses and actually increase weed growth under the plastic. Even in mild climates, clear plastic heats up the soil, sometimes enough to damage or kill fruit tree roots. In short, avoid clear plastic film.

Black and other dark-colored plastic films can effectively eliminate most weed problems in crop rows. They do have many drawbacks, however. First, for all but the smallest orchard, they require large amounts of plastic made from a nonrenewable resource (petroleum). (So far, biodegradable, plant-based plastic films have not worked well as weed barriers.) As the plastic deteriorates, it creates significant amounts of waste that usually cannot be recycled. From an organic perspective, these characteristics make the benefits of plastic film questionable.

Impermeable plastic films also create challenges for irrigation and fertilization: You can run drip irrigation lines under the plastic, but spotting plugged emitters becomes impossible, and the films limit you to soluble fertilizers applied through the drip lines under the plastic. Plastic films also interfere with the movement of oxygen into the soil and the movement of carbon dioxide out of the soil. While impermeable plastic films have been valuable in some annual and short-lived perennial crops, their value in orchards is debatable. In warmer climates, the use of black plastic film has been problematic because of increased soil temperatures under the plastic. Reportedly, the problem has been particularly serious with highly dwarfing rootstocks. More vigorous rootstocks and larger trees have proven more tolerant of the increased soil temperatures.

Landscape Fabrics

Woven and spunbonded porous landscape fabrics (also called geotextile mulches) have proven more useful than impermeable films for orchard crops. High-quality fabrics effectively control weeds but allow water and oxygen to penetrate into the soil. Landscape fabrics allow less water to infiltrate than do organic mulches, and they can interfere with irrigation in some situations. The fabrics do prevent some pests from penetrating the soil, but they also prevent beneficial organisms from doing so. Some fertilizers can be applied over the top of weed barrier fabrics, but compost and alley row clippings will have little effect as fertilizers.

High-quality fabrics should last 5 to 10 years. By covering them with bark or other organic mulch, you can reduce deterioration due to sunlight and extend the life of the fabric. The downside here is that rodents that are repelled by the mulch tunnel under the fabric and attack the roots and collars of the trees. This has, reportedly, been a problem when wood chips were applied over fabric mulches. Although not all organic fruit growers find these weed barrier fabrics acceptable, they do have a place in organic production. Be sure your certification organization allows their use.

Rather than covering an entire row with weed barrier fabric, consider making squares, perhaps 2 to 3 feet per side. Center a square around each tree or bush at the time of planting, as we discussed in chapter 7. Leave a hole in the center of the square just large enough to allow the trunk or stems to pass through (figure 7.5). The fabric squares will greatly reduce weed growth near the young crop plants for at least 5 years if properly installed. They will not, however, interfere with irrigation, fertilization, or gas exchange. They also reduce the likelihood of root rot due to excessively moist soil under a large section of fabric.

For larger commercial orchards, plastic films and landscape fabrics are most useful in young, nonbearing fields. Harvesting operations and equipment can damage the films and fabrics in mature orchards.

Organic Mulches

Mulching crop rows with organic materials after planting can help manage weeds and is very popular for organic orchards. As with dark plastic films and fabrics, organic mulches reduce the amount of light that reaches the soil surface, thereby reducing weed populations. Seeds of many weed species must be exposed to light in order to germinate.

Types. The types of mulches vary by region, historically representing waste products from agriculture or forestry. In the northwestern United States, southeastern United States, and western Canada, bark and sawdust from lumber yards are often available locally. In orchard trials in British Columbia, researchers found that locally available wood wastes provided season-long weed control when used as mulches. In other trials across North America, wood chip and bark mulches usually provided good weed suppression for about 3 years before they needed replacing. Once free for the taking, however, these materials now have value as fuel for industry and municipalities and can be expensive to purchase, transport, and apply on a large scale.

In grain-growing regions, straw is often readily available and can be used for mulching. Straw mulches, however, provide excellent habitat for rodents and are generally not good choices for organic orchards. Many communities across North America now recycle yard wastes, either chipping them for use as mulches or composting the wastes. Be sure the straw or recycled yard wastes are free of pesticides before applying them to your orchard.

Coarse-textured products, such as bark and wood chips, generally help control weeds with fewer problems than fine-textured materials such as sawdust. In windy areas, it can be difficult to keep sawdust and fine bark from blowing away. Coarse bark proved very effective in orchard establishment trials and with young orchards in Washington State.

Shredded paper also provides good weed suppression but can be difficult to apply and must be replaced annually. Spray-on mulches made from slurries of shredded paper pulp suspended in water have provided very good weed control in orchards, but they require expensive mixing and spraying equipment and tend to be challenging to apply. When using paper mulches, be sure that the paper does not have glossy or colored inks. Spot treatments using hand cultivation, steam weeders (not flame!), and organic herbicides are typically needed to control weeds, particularly at the edges of the mulches.

Advantages. In some situations, organic mulches can be useful to orchardists by lowering soil temperatures, conserving soil moisture, and reducing the need for irrigation. Organic mulches can also add nutrients and organic matter to the soil, and some provide favorable environments for beneficial insects and microorganisms. In some cases, organic mulches have been found to increase populations of beneficial nematodes while reducing populations of plant parasitic nematodes.

Disadvantages. Organic mulches do not provide complete weed control and, when contaminated with weed seeds, insects, or diseases, can actually increase problems. Grass hays and alfalfa hay harvested after flowering, for instance, often contain large amounts of seeds and are not generally recommended for mulches. Perennial weeds, such as quack grass and Canada thistle, are especially troublesome in mulched fields because of their rhizomes. Rhizomes can spread quickly in some organic mulches, particularly sawdust or compost. Organic mulches, especially straw, can also harbor mice and voles, increasing the risk of girdling damage to orchard trees. When using agricultural and municipal wastes as mulches, contamination with pesticides can be a problem. Before using any materials, make sure they are pesticide-free.

In some regions, it's a good idea to keep orchard soils cool and moist. If your soil is already naturally cool and moist, however, mulching can slow tree growth in the spring, increase problems with nutrient uptake, and greatly increase root diseases. In New York, researchers found that mulching greatly increased *Phytophthora* root rot and the death rate of young apple trees. This does not mean mulches cannot be used in your orchard; simply use them wisely. On soils that are slow to warm in the spring and that may already have ample moisture, delay applying mulches until early summer and rake them out of the tree rows in early spring.

Excessively deep mulches encourage collar rot and inhibit the movement of oxygen into the soil. Keep your mulches no more than about 4 inches deep, and keep them several inches away from trunks and canes to reduce disease problems.

Although often touted as adding nutrients and organic matter to the soil, mulches do not always achieve these goals. In Swedish trials conducted over a 4-year period, bark chip mulch did not add organic matter or biomass to the soil and caused nitrogen deficiency in black currants, despite manure applications that provided 178 pounds of nitrogen per acre.

Although compost is a good source of plant nutrients, it also provides excellent conditions for weed seed germination and growth. Compost also does not provide control against emerging perennial weeds. Consider compost to be a soil amendment, rather than a weed-controlling mulch.

Organic mulches for large plantings can be very expensive to purchase, transport, and apply. For example, applying 4 inches of bark to 3-feet-wide strips with the tree rows on 10-foot centers requires 177 cubic yards of bark weighing 35.9 tons per acre. Be creative and look for inexpensive, locally available mulch materials that will be effective and meet organic standards.

Regardless of the mulch that you use, take great care to eliminate perennial weeds (particularly rhizomatous weeds) before applying the mulch. Once the mulches are in place, your weed control options become more limited as mechanical cultivation may no longer be practical. Mulches can also make hand cultivation more difficult. Consider mulches to be a part of your orchard floor management program, and integrate them with your other practices.

Living Mulches

Living mulches provide all of the advantages of the organic mulches described above, and they eliminate some of the disadvantages. Perhaps the greatest advantage of living mulches is that you produce them on-site in the alleyways between your trees or grow them as in-row cover crops. As part of routine alley mowing operations, the clippings can be blown into the fruit crop rows, where they serve as mulches. You avoid the expenses of purchasing and transporting the materials, and you greatly reduce labor costs associated with applying the mulches. You also eliminate concerns about pesticide residues and avoid problems with your organic certification organization.

By tailoring the alley crop, you can fine-tune how much nitrogen is available to your orchard crops. If nitrogen is too abundant, shift to grasses and other non-nitrogen-fixing alley crops. If more nitrogen is needed, increase the amounts of alfalfa, clovers, and other legumes in the alley crop mix. Both annual and perennial alley crops can be used effectively as living mulches.

Much work has been conducted on living mulches recently, and they show great promise for organic fruit growers. In Washington State trials, alfalfa proved especially valuable as an alley cover crop and source of mulch, providing good weed control in the fruit crop rows and adding nitrogen to both the alleyways and crop rows. In one orchard, an innovative fruit grower installed a front-mounted mower on his tractor and had a mechanical cultivator and brush rake behind. This arrangement allowed the grower to mow the alfalfa crop and move the clippings into and out of the tree rows, as well as to cultivate in the tree rows.

Living mulches are not without their disadvantages. You must still grow and maintain the cover crop, which must grow tall and dense enough to provide adequate amounts of mulch. You will need a mower capable of cutting the tall alley crops and blowing the clippings into the crop rows. The clippings in the tree rows provide habitat for mice and voles, and you will need an aggressive rodent control program (see page 372). Depending on the crop and time of year, the increased reflectance of light from alfalfa clipping–mulched rows can increase fruit sunburn in some climates or help color fruits where light intensity is lower.

In an innovative trial, British Columbia fruit researchers found that *Equisetum arvense* (horsetail or scouring rush) "greatly reduced the annual weed populations in the tree row without significant competition to established apples in a high-density planting." Horsetail is a common and aggressive "weed" on moist soils and is very difficult to kill. In this case, the weed itself became an effective weed management practice. Foliar testing of the apple trees showed that the *Equisetum* caused little, if any, competition with the trees for nitrogen.

In the same trials, four perennial herbs were tested as possible living mulches: *Arabis alpina*, *Cerastium tomentosum*, *Saponaria ocymoides*, and *Thymus serphyllum*. The *Saponaria ocymoides* filled in the fastest and provided good early weed suppression. The *Arabis* and *Thymus* also produced good coverage and showed promise as living mulches. The *Cerastium* gave the poorest results.

Swiss sandwich design. In this orchard floor management system, the trees are planted in rows, as usual, and either an annual or perennial alley crop is planted. Within the tree row, in what would be the weed-free herbicide strip in a conventional orchard, either legume or non-legume

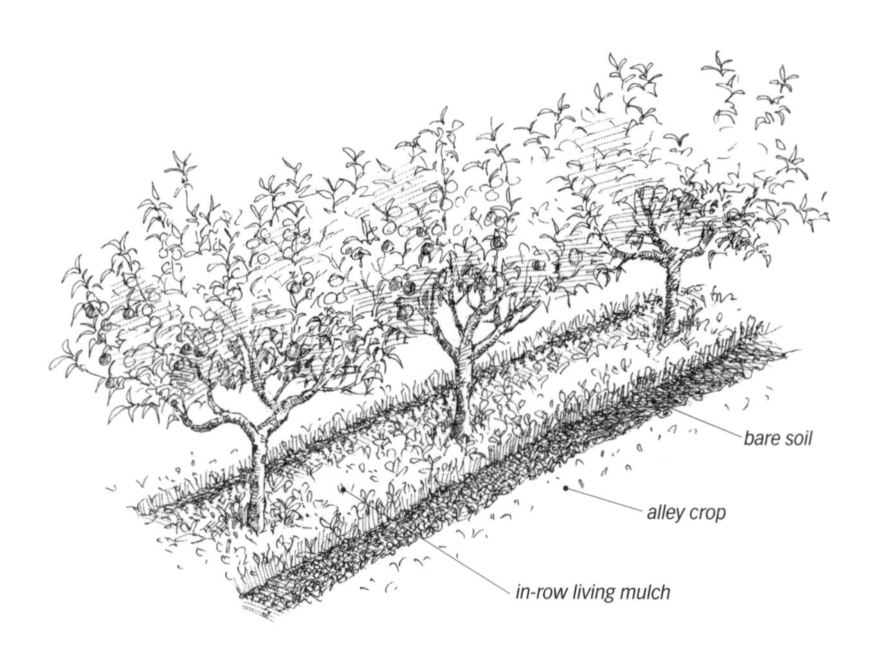

Figure 9.1 · **The Sandwich System for Orchard Floor Management**
A narrow strip of bare soil is maintained between the alley crop and in-row living mulch using a combination of thermal weeding, cultivation, and organic herbicides.

perennial crops are planted in a strip 3 to 4 feet wide. At the edges of the rows where the alley and in-row cover crops meet, a narrow strip of bare soil is maintained (see figure 9.1). If you use the sandwich system, wait until at least the third spring after planting your trees before establishing the in-row living mulches.

In extensive orchard floor management trials, Washington State University scientists examined the sandwich system in a newly planted organic apple orchard. The sandwich system was compared with clean-cultivated, wood-chip-mulched, and other living mulched tree rows. Results showed that there was less competition between weeds and young trees in the sandwich system than in a living mulch system that completely covered the orchard floor.

Although weeds were reported as being abundant in the rows mulched with wood chips, those trees produced good growth during their first 2 years in the field. The in-row clean cultivation treatment produced the lowest weed competition and also produced good tree growth, but it disrupted the tree root systems and caused the trees to lean. In the trials, both the wood chip and the clean cultivation treatments produced greater tree growth than the sandwich system.

Although the in-row living mulches (a.k.a. in-row cover crops) showed promise for controlling weeds and enhancing soil fertility and overall quality, they competed with the fruit trees, reducing tree growth. Rodents were a problem in the living mulches and represent the greatest challenge to the use of in-row living mulches. In the Washington trials, sweet woodruff (*Galium odoratum*) had significantly lower meadow vole infestation than the other crops tested, possibly due to the coumarin found in sweet woodruff. Japanese pachysandra (*Pachysandra terminalis*) has also proven somewhat repellent to rodents due to the presence of steroidal alkaloids.

How to use. We are clearly at the beginning of developing and understanding in-row living mulch cover crops. While they show great promise, as does the sandwich system, use them with caution. A leading tree fruit researcher highly experienced with living mulches advised me that rodent problems and the stunting of young trees during establishment were serious obstacles for which we do not yet have all the answers.

During the first 3 years after planting, maintain a vegetation-free area at least 4 to 9 square feet around each tree using weed barrier fabric squares, mulching, spot treatment of weeds with thermal weeders or herbicides, hand weeding, or a combination of these practices. Using blow-in mulch from a vigorously growing alley crop such as alfalfa can be beneficial. If possible, avoid in-row cultivation using tractor-mounted or other powered equipment during the establishment years.

Once the trees are well established, consider testing in-row cover crops to manage weeds and provide habitat for beneficial organisms. Protect the trunks against rodent damage using hardware cloth screen cylinders (figure 11.2), and maintain an aggressive rodent control program. If you choose to use living mulches, you might consider using slightly less dwarfing rootstocks than you would employ if there were no vegetation under the trees.

What to use. Your choice of living mulch depends on your climate and available soil moisture. Alfalfa, grasses, or grass-clover mixes can make good alley crops. In New Zealand trials, researchers tested red clover, rye grass, and an herbal crop. Which crop was used made little difference, but its management did. Blowing the mulch into the tree rows instead of leaving it in the alleys increased the nitrogen, calcium, phosphorus, and potassium in the tree row soil while substantially increasing soil organic matter.

For in-row living mulch cover crops, look for shallow-rooted herbaceous perennials that are adapted to your area, establish quickly, and form dense stands. Both legume and non-legume species can be used effectively. Legumes can fix nitrogen within the tree rows, but they are especially attractive to rodents. Annual clovers can be included with non-legume species to provide early weed suppression while the non-legume covers establish. Remember to inoculate legume seeds with *Rhizobium* to promote nitrogen-fixing nodes on the roots. Some clovers require special inoculants. Some organic growers allow grass sod to grow in the alleyways and up to the tree trunks, and manage the in-row grass by mowing before fruit thinning and harvest. Suggested in-row crops are listed in table 9.2.

Suggested In-Row Living Mulch Cover Crops

Table 9.2

These species may also serve as insectary crops and can be blended with grasses for alleyway cover crops.

NON-LEGUMES	
Alpine rockcress (*Arabis alpina*)	A perennial member of the *Brassica* (mustard) family with 16-inch flower stalks and a basal rosette of leaves. Adapted to gravelly and alkaline soils and grows best in full sun. Rated to USDA Zones 4–7.
Creeping thyme (*Thymus praecox 'Minus'*)	Produces a low-growing cover with good weed suppression and little cover for rodents. Can be slow to establish and requires supplemental weed control during establishment. May benefit from a nurse crop (annual rye or oats). Drought-tolerant in cooler climates. Does not perform as well in hot, dry areas. Rated to USDA Zones 2–9.
Mother-of-thyme (*Thymus serpyllum*)	A low-growing, drought-tolerant, evergreen perennial that seldom exceeds 3 inches in height. Moderate growth rate. Tolerates sun to light shade. Rated to USDA Zones 4–9. Produces purple flowers. Grows best on well-drained soils.
Irish moss (*Sagina subulata*)	Produces a low-growing cover with good weed suppression and little cover for rodents. Can be slow to establish and requires supplemental weed control during establishment. May benefit from a nurse crop (annual rye or oats). About as traffic tolerant as grass and rated to USDA Zones 4–9. Not drought-tolerant.
Rock soapwort (*Saponaria ocymoides*)	A fast-growing, drought-tolerant perennial that grows 6 to 8 inches tall. Produces abundant pink blossoms. Prefers full sun. Rated to USDA Zones 2–9.
Snow-in-summer (*Cerastium tomentosum*)	A hardy perennial member of the carnation family. Grows 6 to 8 inches tall and is adapted to dry, sunny sites and poor soils. Can be invasive. Tolerates mowing to about 2 inches tall. Flowers heavily. Prefers full sun. Rated to USDA Zones 1–9.
Sweet alyssum (*Lobularia maritima* L.)	Establishes quickly and provides good weed control during the establishment season. May not survive cold winters but has the potential to reseed itself. A popular ornamental plant that is usually grown as an annual because it can become scraggly by its second year. Rated to USDA Zones 4–9.
Sweet woodruff (*Galium odoratum*)	This perennial grows 12 to 20 inches long, but usually lays over and reaches heights of 6 to 12 inches. Establishes quickly and provides good weed control during the establishment season. Produces coumarin, which may have some potential for repelling rodents. Tolerates shade well and is not drought-tolerant. Rated to USDA Zones 4–8.

LEGUMES	
White clover (*Trifolium repens*)	A number of white clover varieties are available. 'White Dutch' and 'New Zealand' have been used in orchard floor systems. They emerge quickly and form dense stands 4 to 10 inches tall with good weed suppression. White clovers provided the greatest percentage of ground coverage of the legumes tested in recent university trials. Not long-lived and may need to be reseeded every few years. White clovers are best adapted to cool, moist conditions. Rated to USDA Zones 4–9.
Strawberry clover (*Trifolium fragiferum*)	Resembles white clovers in growth habit but produces stands that are less dense. More heat tolerant than white clovers but less tolerant of shade. Establishes quickly. Adapted to wet saline and alkaline soils but also reported to be drought-tolerant. Not long-lived and may need to be reseeded every few years. In California orchards, often mixed with white clovers, bird's-foot trefoil, and creeping red fescue in cover crops.
Kura or honey clover (*Trifolium ambiguum*)	Slow to establish and can benefit from high seeding rates, fall planting, and a nurse crop (annual rye or oats). Very persistent in grass stands when established but does not tolerate weed competition during establishment. Longer-lived than white clovers and produce less aboveground biomass that can attract rodents. Produces an exceptionally large amount of root biomass. Requires a special *Rhizobium* inoculant. Best adapted to cool and cold climates from latitude 40–50°N.
Subterranean clovers (*Trifolium subterraneum, T. yanninicum, and T. brachycalycinum*)	Many varieties are available of these cool-season annual legumes that grow 6 to 15 inches tall. 'Koala' and 'Clare' are tall varieties that have performed well in California orchards. Can be slow to establish and may be best when fall-planted. Low-growing and produce less biomass than white clovers, possibly attracting fewer rodents. Limited cold hardiness (5°F [–15°C]) and best grown in milder climates. May best be sown in blends with medics to provide season-long weed suppression. Although annuals, subterranean clovers aggressively reseed themselves. Survive best in hardiness Zones 7–9.
Bird's-foot trefoil (*Lotus corniculatus*)	Bird's-foot trefoil is a long-lived perennial legume that produces dense stands and grows well in mixed grass and trefoil cover crops. It establishes slowly and does not tolerate competition well during establishment. Fall plantings may be best to reduce competition during establishment. When it establishes, it provides good cover and weed suppression. Adapted to USDA Zones 2–9, but can be a bit difficult to grow in parts of Florida and other parts of the extreme southern United States.

Mechanical Cultivation

Mechanically cultivating alleys is usually straightforward and easily accomplished with various handheld or tractor-mounted rototillers, as well as tractor-mounted disk implements and harrows. Mechanically cultivating within the crop rows is much more of a challenge from an engineering perspective.

There are several tractor-mounted cultivators designed to keep a vegetation-free strip within tree fruit rows, but they are generally not useful for bush fruit crops. Some of these devices use vertical blades that rotate horizontally, essentially stirring the soil with a series of metal fingers. The cultivators can be mounted on hinged bars that swing the cultivators into and out of the tree rows. A sensor bar detects the tree trunks, automatically swinging the cultivator head away from the trees in time to avoid damaging the trunks.

These devices have proven useful in larger organic orchards, although they do have several drawbacks. First, the stirring action can damage shallow tree roots. In Washington State trials, trees in test plots where the in-row cultivators were used leaned significantly more than in other plots, due to poor anchorage caused by root damage. Adjusting the cultivator for very shallow cultivation will minimize damage to the roots. Placing metal or wooden stakes next to the trees helps protect the trees from being struck by the sensor bar or cultivator. These horizontal rotovators have proven rather slow and difficult to use in orchard systems.

A different type of in-row cultivator is presently manufactured in Washington State and appears well suited for organic orchards. This device, called a Wonder Weeder (Harris Manufacturing; see Resources), uses a series of angled, rotating cultivator tines and a scraper bar to remove vegetation within the tree rows. This device mounts to the front of a tractor and can be used at the same time as a rear mower to manage both the alley crops and tree rows with a single pass.

Intercropping Food and Nonfood Crops for Harvest

In the past, when orchards were slow to come into fruit production and all operations in the orchards were performed by hand or with animals, intercropping vegetables and low-growing fruit plants between the trees was common. Even today, you can find this practice in some parts of the world.

Intercropping is almost nonexistent today in North America, certainly in commercial orchards where trees quickly come into bearing and tractors and other equipment are used in the alleys. Trying to manage vegetable or other cash crops in the alleyways will prove very difficult for most growers and will probably not be worth the effort. Orchardists are generally better served managing fruit crops on the orchard floor and planting vegetables and other cash or subsistence crops elsewhere.

There are applications for intercropping in orchards, however. Some herbal crops can be grown as part of companion plantings within the tree rows. Likewise, flowering crops suitable for honey production might be included in the companion plantings, provided they are not allowed to flower during the fruit crop bloom.

The best choices for companion plants are perennials or annuals that reseed themselves without being so aggressive that they compete excessively with the trees for nutrients and moisture. Tilling to create planting beds in or alongside tree rows could damage the tree roots.

While intercropping is certainly not for every fruit grower, there are cases where it can be used. Should you choose to intercrop, select plants that fit well with your orchard floor management program. Even better, choose plants that also support beneficial organisms.

Bare Ground

By this point, you should know that I am not in favor of maintaining much bare ground in a mature orchard. Bare ground strategies, however, do have important roles to play, especially in newly planted and some high-density orchards.

When it comes to adding soil amendments and fertilizers, keeping the planting rows and alleys bare of any vegetation besides the fruit crop provides the greatest flexibility. This strategy also eliminates competition with weeds, companion plants, and alley crops for nutrients and moisture. Orchards with bare ground also have far fewer rodent problems than those with cover crops and mulches.

For newly established orchards, I strongly recommend maintaining a vegetation-free zone for a radius of 2 to 3 feet around each fruit tree or bush. This vegetation-free zone should be maintained for at least

3 years after planting. For high-density plantings, permanently maintaining a vegetation-free crop row can be a good strategy. The vegetation-free zone does not necessarily have to be bare ground. Various organic and inorganic mulches can be used to reduce weed pressure while protecting the soil.

Maintaining bare ground requires more labor and machine hours for mechanical and hand cultivation, raises fuel consumption if tractors are needed for tillage, increases soil erosion and compaction, and reduces soil organic matter. Bare ground can also be lacking in biological diversity and activity.

Using mechanical cultivation to maintain bare alleys does not work in all orchards. On even gentle slopes, erosion can be severe. In any orchard, tilling more than 1 or 2 inches deep can damage vital feeder roots of the fruit crops. Mechanical cultivation works best for annual weeds. It is less effective and can even increase problems with perennial weeds such as nutsedge, field bindweed, Canada thistle, quack grass, johnsongrass, and Bermuda grass. When faced with aggressive perennial weeds, you will need a combination of weed suppression practices.

In unirrigated orchards in regions that have little precipitation during the growing season, some fruit growers practice what is called dust mulching. Frequent, very shallow cultivation creates a layer of dust on the soil that helps reduce moisture loss from the underlying soil and reduces weed problems in the alley. Dust-mulched alleys are highly susceptible to wind and water erosion, and depending on the method of tilling, you can also create a shallow hardpan. Use this strategy with great care.

According to some orchard experts, the far-reaching aspect we call soil quality is strongly influenced by cover and soil management practices, as shown in figure 9.2. The take-home message is to use mechanical cultivation as little as possible and optimize the use of cover crops in established orchards.

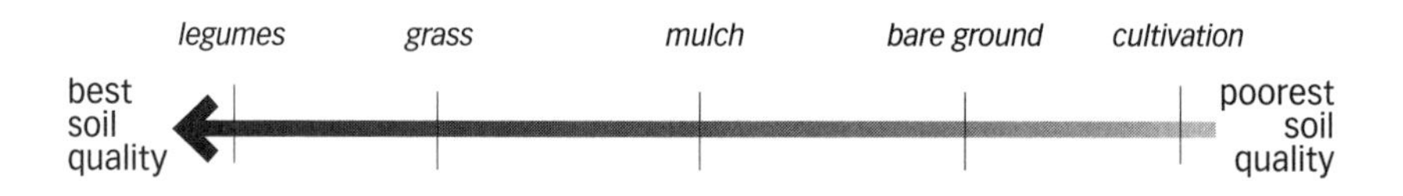

Figure 9.2 · Relative effects of orchard floor management practices on soil quality
Adapted from publications by Washington University

DISEASE MANAGEMENT

- Disease Management Strategies | 327
- Pome Fruit Diseases | 341
- Stone Fruit Diseases | 355
- Diseases of Both Pome and Stone Fruits | 363
- Root Diseases | 364

Diseases present some of the greatest challenges to organic fruit growers and are among the primary reasons much commercial fruit production is centered in arid and semiarid regions of North America, where disease pressures are relatively low. Nevertheless, improved fruit varieties, organic pesticides, and cultural practices now make organic fruit production possible throughout North America.

We have discussed many of the disease control strategies available to organic growers. It is very important to select the right site if starting a new orchard. Sites on gentle to moderate slopes above prevailing temperature inversions reduce chronic frost and freezing injuries that make fruit trees more susceptible to stem cankers and other diseases than do sites with frequent or severe frosts. The best orchard sites also allow heavy, moist air to drain away from the orchard, reducing fungal and bacterial infections. Locating your orchard on deep, well-drained soils reduces root diseases and physiological disorders.

Many readers already have located their sites and established their orchards. In these cases, adjustments to the site can help reduce disease problems. Practices such as installing drain tiles or otherwise improving water drainage in an orchard help reduce root problems. Eliminating plants that serve as alternate hosts for fruit tree diseases in and around an orchard helps reduce the amount of inoculum your trees are exposed to, thereby reducing infections. Proper orchard layout, pruning, and training create open canopies with good light exposure and air movement. Opening the canopy lowers the humidity around the fruit and facilitates rapid drying of the leaves and fruit after rain. Effective orchard floor management creates a biologically diverse and active soil that promotes healthy, vigorous root systems and tree canopies. Proper nutrition and crop load management help maintain the moderate vegetative growth and sustainable yields necessary for long-term tree health.

Disease Management Strategies

EVEN IF YOU USE EVERY CULTURAL PRACTICE and pesticide available, if you plant a highly disease-susceptible fruit variety in an area where that disease is severe you can expect serious problems. The most important disease management practice is to select crops and varieties that are resistant to the diseases in your area. In the southeastern United States, for example, plum leaf scald is a very serious disease for which European plums have no resistance. Several North American plum species and hybrid varieties, however, are resistant to leaf scald and other diseases prevalent in that warm, humid climate.

In chapters 5 and 6, we listed many disease-resistant fruit varieties and recommended growing regions. By following these guidelines, you can usually reduce disease problems to the point that they can be managed with cultural practices and organic pesticides. It is especially critical that growers in eastern North America, who typically deal with humid climates and high disease pressures, select the right crops and varieties. Organic fruit growers in more arid regions with low disease pressures have greater flexibility in what crops and varieties they grow.

If you want to grow fruit varieties not described in this book, of which there are literally thousands, do your homework! Find out if those crops and varieties are likely to do well in your location. Working hard only to watch your trees die and crops rot in the orchard is not much fun and certainly not profitable for commercial growers. Choose crops and varieties that give you every chance for success. Experiment, but do so cautiously. Try one or a few individual trees or bushes of questionable or unknown disease resistance before investing heavily in them.

While selecting the right variety for your orchard is your primary line of defense, the following strategies will also help you manage diseases.

Frequent and Regular Scouting

Even in a conventional orchard with an arsenal of highly effective fungicides and bactericides, regular and frequent scouting during the growing season is essential to produce a healthy, usable crop. Scouting is even more important for organic orchardists. If you wait until your trees or fruit are severely infected, you can do little but wait for next year's crop.

Invest in a good-quality jeweler's loupe (magnifying hand lens) that provides about 10X magnification, and obtain a journal or a three-ring binder filled with preprinted observation forms. For pest scouting, a piece of white poster board and a short stick are valuable sampling tools, as we'll discuss in the next chapter. Record weather data. You will likely find that the data you collect on-site is more accurate and valuable for your purposes than that reported regionally. Weather stations are available in many forms ranging in price from under $100 to several thousands of dollars. Even many of the inexpensive models automatically record temperature and precipitation data. If available, a computer text, spreadsheet, or database program makes your records easy to organize, search, and update.

Record dates of bud swell, green tip, bloom, and harvest, as well as information on plant vigor, flower and fruit set, fruit quality, pests and pest damage, and disease symptoms. Also include daily high and low temperatures, precipitation, and humidity. Over a few years, you will discover patterns that will allow you to predict what pest and disease problems you are likely to see and when and under what weather conditions you are likely to see them. This knowledge will allow you to take planned and timely preventative actions, rather than reacting to crises. Remember to record observations for each crop and variety separately. A sample sheet is shown in figure 10.1.

Know what to look for. Have photos and written descriptions of the common pests and diseases in your area and be able to compare them with your trees. Given the abundance of high-quality information and photos available almost instantly via the Internet, developing a good portfolio of pest and disease information for your area should be easy. Many land grant (Cooperative Extension) universities in the United States and provincial ministries of agriculture in Canada publish paper and online guides for identifying fruit pests and diseases. If you cannot find what you need in your state or province, look for the information in nearby or otherwise similar growing regions. The British Columbia Ministry of Agriculture, Oregon State University, Pennsylvania State University, the University of Kentucky, and the Kearneysville Tree Fruit Research and Education Center in West Virginia, for example, have excellent online guides, many with detailed color photographs showing pest and disease symptoms.

ORCHARD SCOUTING FORM

<table>
<tr><td>Date:</td><td colspan="4">Days from: □ first bloom □ full bloom □ shuck split</td><td colspan="2">Growing degree days:</td></tr>
<tr><td colspan="2">Current weather</td><td>Min temp:</td><td>Max temp:</td><td>Cloud cover:</td><td>Humidity:</td><td>Precipitation:</td></tr>
<tr><td colspan="7">Describe weather:</td></tr>
<tr><td colspan="7">Fruit variety:</td></tr>
<tr><td>Pest:</td><td colspan="4">Sampling method:
□ sticky trap □ pheromone trap □ visual □ beater tray</td><td colspan="2">Number of pests per
□ leaf □ branch □ trap</td></tr>
<tr><td colspan="7">Damage: □ none □ slight □ moderate □ severe</td></tr>
<tr><td colspan="7">Describe damage:</td></tr>
<tr><td>Pest:</td><td colspan="4">Sampling method
□ sticky trap □ pheromone trap □ visual □ beater tray</td><td colspan="2">Number of pests per
□ leaf □ branch □ trap</td></tr>
<tr><td colspan="7">Damage: □ none □ slight □ moderate □ severe</td></tr>
<tr><td colspan="7">Describe damage:</td></tr>
<tr><td>Pest:</td><td colspan="4">Sampling method
□ sticky trap □ pheromone trap □ visual □ beater tray</td><td colspan="2">Number of pests per
□ leaf □ branch □ trap</td></tr>
<tr><td colspan="7">Damage: □ none □ slight □ moderate □ severe</td></tr>
<tr><td colspan="7">Describe damage:</td></tr>
<tr><td>Disease:</td><td colspan="3">Severity: □ mild □ moderate □ severe</td><td colspan="3">Percentage of tree affected:</td></tr>
<tr><td colspan="7">Symptoms:</td></tr>
<tr><td>Disease:</td><td colspan="3">Severity: □ mild □ moderate □ severe</td><td colspan="3">Percentage of tree affected:</td></tr>
<tr><td colspan="7">Symptoms:</td></tr>
<tr><td colspan="7">Weeds present:</td></tr>
<tr><td colspan="4">Location of weeds:</td><td colspan="3">Severity of infestation
□ mild □ moderate □ severe</td></tr>
<tr><td colspan="2">Alley crop</td><td colspan="5">Condition of alley crop: □ poor □ fair □ good □ excellent</td></tr>
<tr><td colspan="7">Describe</td></tr>
<tr><td colspan="2">In-row living mulch</td><td colspan="5">Condition of in-row living mulch: □ poor □ fair □ good
□ excellent</td></tr>
<tr><td colspan="7">Describe</td></tr>
<tr><td colspan="2">Evidence of rodents: □ yes □ no</td><td colspan="2">Species:</td><td colspan="3">Severity of infestation:
□ mild □ moderate □ severe</td></tr>
<tr><td colspan="7">Evidence of rodent damage to trees: □ none □ slight □ moderate □ severe</td></tr>
<tr><td colspan="7">Describe damage:</td></tr>
</table>

Figure 10.1 · Sample orchard scouting sheet

Get help when you need it! Diagnosing plant diseases can be very difficult, even for specialists. While apple scab and codling moth are easy to identify, many other problems are not. Nutritional disorders and environmental damage (for example, frost or sunburn) often mimic disease symptoms, and many different diseases share common symptoms. Fruit specialists in provincial, state, and university programs can be invaluable aides in identifying and managing pest and disease problems.

Sanitation and Other Cultural Practices

I have had the great pleasure of visiting many high-quality commercial, home, and research orchards. One thing that has stood out is that these orchards are almost always clean and organized to the point of being immaculate. The cleanliness and order are not usually for appearances; rather they are essential strategies in maintaining a healthy and productive orchard.

Keep the orchard floor clean. While it may seem obvious, poor sanitation creates or aggravates many orchard pest and disease problems. Leaving diseased fruit or prunings in the orchard creates a reservoir of inoculum that can infect or reinfect trees. Rank weeds or unmanaged cover crops, unused equipment, boards, and other debris provide excellent habitats for voles, gophers, and other rodent pests.

When you are finished pruning, regardless of the time of year, remove the prunings from the orchard and burn, bury, or otherwise dispose of them. Chipping prunings can kill or reduce some pests, but it is best not to return the chips to the orchard. Either compost the chips or apply them as mulch well away from your orchard. Prunings from trees infected with fire blight or other pathogens that attack wood or buds should be burned or buried and not composted.

A top organic orchardist in Michigan described a situation where he deliberately pruned off some relatively large branches and fruiting wood throughout his orchard in late fall. The object was to provide voles and other rodents with enough food that they would not attack the tree trunks during the winter. Unfortunately, the Bud9 rootstocks that he used in his newer plantings proved more attractive to the rodents than did the prunings. Unless you have a very specific reason for leaving prunings on the orchard floor, remove large branch prunings from the orchard.

It is generally not feasible or necessary to remove thousands of thinned or June drop apples or other fruit, as these small fruits are not typically diseased and decompose quickly. Fallen mature fruits, however, should be removed from the orchard quickly, as they often harbor pests and diseases. As with prunings, fallen fruits are best burned or buried, especially if pests or disease symptoms are present. Healthy fruit can be composted, but it should be shredded first, if possible, to destroy pest larvae that might be present in the fruits. Be sure to turn the compost frequently to build enough heat to kill pests and disease spores.

Be careful not to move diseases, pests, and weeds from one area of your orchard to another. Use compressed air or a leaf blower to remove weed seeds from mowers and tractors before moving between fields. Low- or high-pressure water washes are valuable for cleaning tractors and tillage equipment to remove soil that can carry pests and diseases. If you know or suspect that a field is contaminated with soilborne diseases, such as Phytophthora root rot, steam cleaning or spraying tractors and tillage equipment with a 20 percent household bleach solution after water washing may help slow the spread of the pathogen to uninfected fields. Disinfecting boots, shovels, and other tools that contact the soil can also help slow the spread of soilborne diseases.

Manage the orchard floor. In chapter 9, we discussed the importance of carefully managing alley and in-row cover crops. Rank growth on the orchard floor is particularly a problem when it comes to rodent pests, but it can also weaken fruit trees and make them more susceptible to diseases and environmental damage. In humid areas, you may need to keep the alley and in-row cover crops mowed short to reduce humidity in the tree canopy. Going into winter, vegetation in the alleys and crop rows should be mowed very short.

If you use living mulches or leave organic mulch in the tree rows, also implement an aggressive rodent control program (see page 372). Keep mulch pulled at least several inches away from the tree trunks to reduce problems with collar and root rots. Avoid placing mulch on cold, wet soils.

Remove alternate hosts. In the site preparation chapter, we discussed removing plants from on and around your property that can harbor pests and diseases. It is best to remove wild and escaped fruit trees and bushes from fencerows and other areas on your property. Avoid planting eastern red cedar and junipers in your landscape because they are alternate hosts for rust diseases. Likewise, keep your property free of wild or unmanaged hawthorn, serviceberry (saskatoon or Juneberry), crab apple, and other fruit crops.

In fruit-growing regions especially, abandoned orchards often serve as reservoirs of pests and diseases. Encourage your neighbors to manage fruit crop pests and diseases in their landscapes and orchards. Offer to help them, if necessary.

Manage fruit crop nutrition. In home and market orchards, overfertilization is more common than underfertilization. In larger orchards and in orchards transitioning to organic, inadequate available nitrogen is common. Fruit crops are best managed with moderate levels of fertility. Excessive nitrogen can force dense, lush growth that is unproductive and highly susceptible to diseases. Excessive phosphorus fertilization can trigger iron chlorosis and other nutrient imbalances. Too little nitrogen, phosphorus, and other nutrients results in weak trees that bear little fruit and are susceptible to diseases and environmental stresses.

Use a soil test to guide you in adjusting soil pH and nutrients (see page 259). Repeat the soil pH tests for the centers of crop rows at least every second or third year. Regularly make visual observations of your tree growth and productivity (see page 254). Even better, have foliar samples tested annually to monitor fruit crop nutrition and make adjustments before problems develop. Strive for moderate levels of fertility and vigor.

Keep the canopy open. Be vigilant in your pruning and training. Of all of the cultural practices that we have discussed, maintaining an open canopy follows only choosing the right crops and varieties and maintaining good soil drainage in its importance to disease management. An open canopy greatly increases air movement throughout the trees and bushes, reducing humidity and keeping the foliage and fruit as dry as possible.

Many bacterial and fungal diseases require moisture for their spores to germinate and infect plants. The longer the plant tissues are wet, the greater the chance of infection. The situation becomes even more critical as temperatures rise. Open canopies allow more sunlight to penetrate, which helps the branches, foliage, and fruit to dry, and also dramatically increases the formation of flower buds on the interior of the trees.

When maintaining an open canopy, avoid the temptation to prune too often and too heavily. Excessive pruning, like excessive nitrogen fertilization, results in lush, upright, disease-susceptible growth and poor fruit crops. As we will discuss in chapter 12, combine your pruning with branch-spreading practices. Branches can be tied to trellis wires, spread with wooden sticks, or weighed down with water bottles to develop wide angles. Spreading the branches shifts the plant's resources from vegetative growth to reproduction and greatly increases the formation of fruit buds and spurs. Maintaining a sustainable fruit crop is the best means of managing vegetative growth.

Organic Fungicides and Bactericides

Organic fruit growers have relatively few fungicides and bactericides available to manage fungal and bacterial diseases. At the time of writing, the Organic Materials Research Institute listed only 29 fungicidal products.

Sulfur and copper have long been the staple materials for organic disease control, used alone or blended together. They attack pathogens at multiple sites on the fungal spores, which accounts for their effectiveness against a wide range of diseases.

Sulfur

Sulfur is among the oldest fungicides and has been used since antiquity. It remains the most important organic fungicide, particularly for such diseases as apple scab. Sulfur fungicides can be applied as dusts or sprays and are used to control a wide variety of diseases, primarily those caused by fungi.

Plant tissues can be damaged or killed by sulfur, and phytotoxicity is common on sensitive tissues, increasing at higher temperatures. Foliar applications of sulfur-containing products at temperatures above about 80°F (27°C) can cause phytotoxicity. Note that sulfur can also harm beneficial organisms. When used excessively or at the wrong time of the growing season, sulfur can be particularly harmful to predatory mites that help manage plant parasitic mites. Sulfur is a valuable component to plant protection, but it is not a silver bullet. Use it cautiously and no more than absolutely needed.

> **Caution!**
>
> Sulfur is toxic to apricots. Apply fixed coppers instead.

Copper

Copper acts against both fungal and bacterial pathogens and is particularly effective against bacteria. Different formulations are available. Copper sulfate is blended with sulfur in Bordeaux mix to provide a highly effective fungicidal and bactericidal treatment.

Fixed copper compounds include tribasic copper sulfate, copper oxychloride sulfate, and cupric hydroxide. These materials are effective in disease control and are easier to prepare and safer to apply during the growing season than Bordeaux mix. Unlike Bordeaux mix, they do not stain plant surfaces, which can be important in ornamental landscapes and orchards. Take care when using copper-containing materials in your orchard. Copper is a heavy metal and persists in soils. Confine your spray materials only to the aboveground portions of plants and follow label directions carefully.

Bordeaux Mix

Bordeaux mix is one of the most valuable sprays for organic orchards, combining copper and sulfur. The downsides are that it is sticky, is corrosive to metal, stains plant surfaces and clothing, and stinks. On the plus side, it is highly effective in controlling many trunk, cane, foliar, and fruit diseases. Bordeaux holds up well on the trunks and canes during winter rain and snow, providing long-term protection. It is usually applied between leaf fall and new spring growth. The following information was derived, in part, from information published by the University of California Integrated Pest Management program.

Many different Bordeaux mixture formulations are available. One of the easiest and best all-around formulations uses a 10:10:100 blend of copper sulfate (not fixed copper), dry hydrated lime, and water. Hydrated lime (a powdered material used to make plaster) and copper sulfate are readily available at garden centers. Use powdered copper sulfate and keep the copper sulfate and hydrated lime dry during storage. When they absorb moisture and become lumpy, they are hard to work with. Buy fresh hydrated lime every year or prepare a stock solution that can be kept indefinitely in a tightly closed container. While slaked lime can be used instead of hydrated lime, it is more difficult and dangerous to use and is best avoided.

Bordeaux Formulations

Table 10.1

Volume	Copper Sulfate	Dry Hydrated Lime	Water
Small	1.6 ounces (weight)	1.6 ounces (weight)	1 gallon
	3 tablespoons	10 tablespoons	1 gallon
	45 grams	45 grams	3.8 liters
Large	10 pounds	10 pounds	100 gallons
	4.5 kg	4.5 kg	380 liters

How to Prepare Bordeaux Spray Stock Solutions

To prepare enough stock solutions for 10 gallons of spray, follow these directions.

1. Dissolve 1 pound of copper sulfate in 1 gallon of warm water. Stir constantly while mixing.
2. Mix 1 pound of fresh, dry hydrated lime in a separate container holding 1 gallon of room temperature water. Prepare the lime suspension at least 2 hours before you need to make the Bordeaux mix. Stir constantly while mixing.
3. Once the solution and suspension are mixed, cap the containers tightly, label them, and store them out of the reach of children. The copper sulfate, in particular, is toxic.

Wear eye protection and a dust mask when preparing and applying Bordeaux mixes, and wear protective clothing during spray applications because the material stains badly. Stock solutions of copper sulfate and hydrated lime are toxic and must be kept away from children. See table 10.1 for the amounts to use. Prepared Bordeaux mixes are available from agricultural suppliers and garden centers. While some fruit specialists consider them inferior to preparations made from scratch, they can be easier and safer to use, particularly for home and market orchards.

Bordeaux sprays, whether they are made from dry ingredients or stock solutions, deteriorate rapidly, so make up only enough to use within that day. The materials in the mix settle quickly and require constant, vigorous agitation in the spray tank. For commercial sprayers, equipment that uses bypass agitation (pesticide solution recirculating into and out of the spray tank via a pump) will probably not be sufficient and paddle mixers are recommended. For handheld sprayers, vigorously shake the tank every few minutes during applications.

When mixing large quantities of Bordeaux mix for mechanical sprayers, fill the spray tank about one-third full of water, then begin adding the needed amount of copper sulfate by washing it through a screen into the tank with more water. Allow time for the copper sulfate to dissolve. If possible, constantly agitate the solution in the spray tank. Mix the required amount of lime in a plastic or stainless steel bucket and allow it to set for 2 hours with frequent stirring. Slowly pour the lime suspension into the spray tank while constantly agitating the mix. Fill the spray tank to the desired level with water. After spraying, thoroughly wash your spray equipment inside and out at least three times to remove the corrosive mixture. Add a small amount of vinegar to the rinse water to neutralize the lime.

How to Prepare Bordeaux Mix Using Stock Solutions

1. Add 2 gallons of room temperature water to a 3-gallon or larger container.
2. Shake the copper sulfate solution and hydrated lime stock suspension vigorously.
3. Add 1 quart of copper sulfate stock solution to the 2 gallons of water and mix thoroughly.
4. After the copper solution has been added and mixed with the water, add 1 quart of lime suspension to the container (total 2.5 gallons of spray mix) while constantly stirring the mix.

How to Prepare Bordeaux Mix Without Stock Solutions

If you do not wish to store stock solutions and need only 1 gallon or so of Bordeaux mix at one time, follow these directions.

1. Add 10 tablespoons of dry hydrated lime to ½ gallon of water and agitate the mix.
2. Allow the suspension to set for 2 hours.
3. Add 3⅓ tablespoons of copper sulfate to ½ gallon of water and agitate the mix.
4. Mix the hydrated lime suspension and copper sulfate solution together.

Lime Sulfur

Also known as calcium polysulfide, lime sulfur combines lime and sulfur to make an effective fungicide that also has some effectiveness against bacteria and insects. Lime sulfur ranks with Bordeaux mix in its value for organic orchards and is especially effective in controlling peach leaf curl. Commercial lime sulfur products are readily available. Certified organic growers must be sure the products they use are allowed by their certifying organization.

Sulfur and Crop Oils

Crop (horticultural) oils, which we will discuss in chapter 11 (see page 380), are valuable in controlling a wide range of pests. Some crop oils are also effective as fungicides. Stylet oil, for example, proved effective against powdery mildew on currants. OMRI presently lists one fungicidal oil product and many horticultural oils for insect and mite control.

While sulfur and crop oils are valuable components of an organic orchard pest and disease control program, they should not be used together. In general, you should wait at least 2, and preferably 4, weeks after applying a sulfur-containing material before applying horticultural oils.

Dormant Season Disease and Pest Control

A keystone of a successful orchard is an effective program of controlling diseases and pests during the dormant season. This particular program controls a wide range of fungal and bacterial diseases and is valuable in managing mite, aphid, scale, and other insect pests. The approach involves making a total of three spray applications from late fall through green tip.

Apply Bordeaux mix in late fall after the leaves have dropped. As we will discuss on page 362, some orchardists apply two additional Bordeaux sprays, the first immediately after harvest but before the leaves drop. For the program we are considering here, we are referring to a single Bordeaux mix application in late fall. Apply sufficient spray to cover the trunks, branches, and canes, but not to the point that the spray drips off. You want to minimize the amount of copper getting into the soil. The sulfur and copper act as fungicides and bactericides.

In late winter but at least 30 days before anticipated bud swell, apply a second Bordeaux spray or a spray of lime sulfur. Follow the directions on the label for the amount of lime sulfur to apply. Just as the buds are swelling in spring, up to the time a few green shoot tips are visible, apply a spray of dormant crop oil. Again, follow the directions on the label for application rates. The oil smothers overwintering adult pests and egg masses.

Neem Oil

Neem oil is derived from pressing the fruits and seeds of the neem tree (*Azadirachta indica*), which is native to India and is cultivated widely in tropical regions. While neem products are best known to organic farmers as insecticides, they also have some effectiveness against powdery mildew, anthracnose, rusts, and some bacterial spot diseases. Organic neem-based fungicides are approved by OMRI.

Plant-Derived Oils

Jojoba oil, rosemary oil, and other plant-derived oils are sometimes touted as organic fungicides, but they have yet to establish themselves as being valuable for crop production. Few, if any, commercial products that contain these materials are approved for organic use.

Cinnamaldehyde

Derived from the bark of trees in the genus *Cinnamomum*, cinnamaldehyde has proven useful as a fungicide in organic crop production. Its primary use is as an application to root systems, and its value in organic orchard production is questionable at this time.

Bicarbonates

Bicarbonates, both potassium and sodium, have shown some effectiveness in controlling certain plant diseases, most particularly powdery mildew. Commercial, organically acceptable products are available.

Hydrogen Peroxide

Sometimes described as an organic fungicide, hydrogen peroxide is allowed for use under the U.S. National Organic Program. At present, two commercially available products are approved by OMRI for plant disease control. In general, hydrogen peroxide is chemically active and quite effective in killing microbes under some conditions, making it a useful disinfectant for food processing and agricultural equipment. It is also used in disinfecting livestock.

Little is known, as yet, how hydrogen peroxide can be used to control plant diseases and how effective it is. Controlled tests on treating lettuce leaf diseases produced poor to mixed results. Hydrogen peroxide is a wide-spectrum disinfectant and does not distinguish between beneficial and pathogenic microbes. Proceed cautiously when testing hydrogen peroxide–based products, and ensure that any commercial products you apply are approved for organic food production. Some are reported to contain amendments that are not approved for certified organic production.

Tetracycline and Streptomycin

Tetracycline and streptomycin are bactericides that are approved under the U.S. National Organic Plan for plant disease control. Their primary use is preventing fire blight in pome fruits, but they are also used for a few other bacterial diseases. The move to include these antibiotics in the allowed list of synthetic materials was controversial and succeeded largely due to the importance of apple production and the difficulty of controlling fire blight. While they have proven valuable for controlling fire blight, there is evidence that bacteria are becoming resistant to the antibiotics. Use these products as parts of an integrated program to control fire blight, not as silver bullets.

Microbial Fungicides

OMRI-approved products containing *Trichoderma* or more generic microbes are available. *Trichoderma* is a genus of fungi that includes many naturally occurring strains found in soil and wood. These fungi are valuable biocontrol agents that help manage plant pathogenic fungi. Part of the value of *Trichoderma* is that it acts in many different ways to greatly reduce the risk of pathogens developing resistance to it. The *Trichoderma* can simply compete with pathogens for space and nutrients, and it can also alter the environment available to pathogens and stimulate plant defense mechanisms.

Surround

The product Surround is one of a handful of near-miracles for organic fruit production. This material is made of edible clay (kaolin) that is used in toothpastes, food additives, and in the treatment of diarrhea. Researchers found that applying a thin film of the clay to fruit trees helped repel and confuse insect pests. As the material became widely adopted by the orchard community, growers found that the clay layer covering the stems, leaves, and fruits also helped control powdery mildew and brown rot. Because Surround is primarily used in pest management, we will discuss it in detail in the following chapter.

Pome Fruit Diseases

Parts of the following discussion on pome and stone fruit diseases are adapted from the Pacific Northwest Plant Disease Handbook and materials published online by Ohio State University and the West Virginia University Kearneysville Tree Fruit Research and Education Center (KTFREC).

Scab Diseases

Several scab diseases affect pome and stone fruits. The best-known and most serious of these is apple scab. Pear scab and peach scab can also be occasional problems for fruit growers, particularly under humid conditions.

Apple Scab

Apple scab is one of the most serious apple diseases in North America and has severely limited commercial production in eastern North America. Crop losses of 70 percent or greater are common following cool, wet springs if the disease is not controlled. The disease is caused by a fungus (*Venturia inaequalis*) that infects leaves, blossoms, fruit, and sometimes new shoots and bud scales. Severe infections can defoliate trees and, if repeated, weaken the trees and decrease yields. Although infected fruit is edible, the scabby surfaces and deformed shapes can render it unmarketable.

Symptoms. Symptoms often occur first on the undersides of young leaves as velvety brown to olive-green spots with indistinct edges. As the leaves unfold, the spots appear on the upper leaf surfaces and leaf stems (petioles). The spots gradually grow and often merge, forming rough, brown scabs that can cover entire leaves. Heavily infected leaves, or those whose petioles have been infected, often fall from the trees. Symptoms on the fruit are similar and develop into rough, corky scabs.

When the flower and fruit stems (pedicels) are infected, the flowers and fruit can fall prematurely. The fruit often becomes deformed as growth stops in infected areas but continues in uninfected parts of the fruit. Large cracks can develop on the fruits, making them susceptible to infection by opportunistic pathogens and saprophytes (microorganisms that live on dead tissues). Early infections often appear first on the blossom end of the fruit. Late-season infections may not appear until the fruits are in storage.

Causes and timing. The apple scab pathogen can overwinter on the bud scales, but more commonly it survives on fallen leaves. Primary infections usually come from ascospores that develop in fruiting bodies on fallen leaves and fruit. Ascospores are dispersed from the fruiting bodies about the time of bud break and peak during bloom. Young tissues become infected more easily and quickly than older leaves and fruits. Once lesions form on the leaves, flowers, and fruits, conidial spores form in those lesions and can cause late-season, secondary infections.

For a crop to be infected by apple scab, temperatures must be above freezing and there must be drops or films of liquid water (free moisture) on susceptible tissues. Under continually dry conditions, apple scab creates few or no problems, even on susceptible varieties. As temperatures rise, the amount of wetting time required for infections to develop decreases. Table 10.2 allows you to predict when apple scab infections might develop and when to apply fungicides to control the disease.

Secondary infections caused by conidia from established lesions happen quickly, generally in 6 to 8 hours from the beginning of wetting when temperatures are above about 50°F (10°C). Unfortunately, prolonged rains make fungicide applications difficult. Also, organic growers do not enjoy the luxury of being able to use fungicides that are effective against pathogens once they are established. For organic growers, prevention is the key to managing apple scab.

Management. Many cultural practices reduce scab problems. Selecting a site with low humidity and good air movement helps minimize wetting periods and primary infections. An open orchard plan that emphasizes air movement and training systems that create open canopies are also important, as is regular and effective pruning. Remove cull and dropped fruit as soon after harvest as possible. Removing leaves from the orchard floor in autumn and composting, burning, or burying them helps reduce inoculum in the orchard. Collecting leaves is usually not practical in large orchards, however. An alternative is to shred the leaves with a mower and apply nitrogen to speed up decay of the leaves. In conventional orchards, urea is often used as a nitrogen source. In organic orchards, any liquid fertilizer such as fish emulsion or compost tea can help speed up decomposition.

Hours of Wetting for Primary Apple Scab Infection

Table 10.2

Number of hours at different air temperatures are approximate.

Avg. Temp. (°F/°C)	Hours of wetting required for infection from primary inoculum		
	Light Infection	*Mod. Infection*	*Heavy Infection*
78°/25.5°	13	17	26
77°/25°	11	14	21
76°/24.5°	9.5	12	19
63–75°/ 17–24°	9	12	18
62°/17°	9	12	19
61°/16°	9	13	20
60°/15.5°	9.5	13	20
59°/15°	10	13	21
58°/14.5°	10	14	21
57°/14°	10	14	22
56°/13.5°	11	15	22
55°/13°	11	16	24
54°/12°	11.5	16	24
53°/11.5°	11.5	17	25
52°/11°	12	18	26
51°/10.5°	13	18	27
50°/10°	14	19	29
49°/9.5°	14.5	20	30
48°/9°	15	20	30
47°/8.5°	15	23	35
46°/8°	16	24	37
45°/7°	17	26	40
44°/6.5°	19	28	43
43°/6°	21	30	47
42°/5.5°	23	33	50
41°/5°	26	37	53
40°/4.5°	29	41	56
39°/4°	33	45	60
38°/3.5°	37	50	64
37°/3°	41	55	68
33–36°/ 0.5–2°	48	72	96

Adapted from North Carolina State University and Michigan State University fruit publications and based on the original "Mills" chart developed by W. O. Mills of Cornell University and modified by A. L. Jones. The infection period starts with the beginning of rain.

Tests have shown that compost has the ability to help manage apple scab. Consider making fall applications of compost within tree rows to help break down infected leaves and destroy the overwintering fungus. Applying compost tea to the foliage, flowers, and fruit may help establish populations of innocuous or beneficial microorganisms that can help reduce apple scab infections by competing for space with the pathogen. Be sure when using compost tea that you have met the National Organic Program standards for composting that we discussed earlier (see page 270). Doing so is especially critical if animal manures are used in the compost.

By far the most important apple scab management practice for organic growers is to use scab-resistant varieties such as 'Freedom', 'Goldrush',

'Jonafree', 'Liberty', 'Prima', 'Priscilla', 'Pristine', and 'Redfree'. Table 5.2 lists scab resistance for popular apple varieties. For commercial growers, there is often a trade-off when it comes to selecting apple varieties. Many varieties that are highly resistant to apple scab presently are not popular with sellers. Some organic growers, particularly those in arid regions where apple scab is less serious, choose varieties that have some scab resistance but that are also profitable to market.

Fungicides serve as the last defense against apple scab. Sulfur, lime sulfur, and Bordeaux mix are the standard organic choices. All can be phytotoxic, so follow label directions carefully. Use as needed, but do not use excessively because sulfur can damage or repel predatory mites and other beneficial organisms. These fungicides have little or no eradicant activity, and you must apply them before infection takes place. This means you need to apply the dusts or sprays at the beginning of, during, or immediately after rain or other wetting event.

Table 10.2 will guide you on how much time is available to apply the fungicide. Once lesions develop on the leaves and fruit, you will need to continue applying the fungicide through harvest to prevent secondary infections by conidia. Particularly in long-season areas, consider applying post-harvest fungicides to reduce the amount of pathogens that overwinter. Leaf damage at this point is less serious than it would be earlier in the season.

Pear Scab

Pear scab is caused by *Venturia pirina*, a pathogen closely related to the one that causes apple scab. Although pear scab can create serious damage, it is more of a problem in Europe and Japan than in North America. Pear scab is sometimes called black spot.

Symptoms. Pear scab symptoms resemble those of apple scab. Unlike apple scab, pear scab attacks small stems, creating lesions that produce more spores and provide overwintering sites for the pathogen. Spores from the stem lesions and last year's fallen leaves infect developing fruits and are released over a 2-month period in spring and early summer. Infections on young fruits usually begin on the blossom ends and move up along the sides of the fruits. Small, velvety lesions darken to black points and can merge to form extensive scabs. Infected fruits are often deformed. Leaf infections are less common than for apple scab and appear as small, brown lesions, usually on the undersides of the leaves.

Management. Management methods are similar to those used for apple scab. While some pear varieties are resistant to pear scab in one area of the country, they may not be so in other areas due to different strains of the pathogen. Keep a record of pear scab problems to help plan management for the coming year. The same materials used to manage apple scab can be employed against pear scab, but often with fewer sprays. Use table 10.2 to time your sprays in relation to wetting events.

Fire Blight

Fire blight is caused by a bacterium that attacks many pome fruits and is a serious challenge to orchardists. The bacterium is particularly serious with pears and quince, and it can also be a serious problem for apples, mayhaw, medlar, and saskatoon. Fire blight also infects apricot, cherry, plum, crab apple, raspberry, blackberry, strawberry, chokeberry, cotoneaster, pyracantha, mountain ash, photinia, potentilla, roses, and other members of the rose family. The pathogen attacks the flowers, leaves, fruits, stems, and trunks. Infected trees, particularly young trees, can be killed. The bacterium overwinters in cankers on infected stems and trunks.

Fire blight is easily spread in splashing rain or irrigation water, on bees and other insects, and on pruning tools. Infection usually occurs through young flowers, but the pathogen can also infect young shoot tips and leaves. Wounds offer excellent entry sites for the bacterium. Infection only occurs when temperatures are 65°F (18°C) or greater and the humidity is at least 65 percent or the plant tissues are wet. In western North America, fire blight is not usually a serious problem except during unusually wet growing seasons. In eastern North America, fire blight is a common problem.

The increased use of fire blight–susceptible varieties in commercial orchards has increased disease problems. Most commercially important pear varieties are susceptible to fire blight, especially 'Bosc'. Apple varieties resistant to fire blight are described in table 5.2 (see page 152) and include 'Enterprise', 'Goldrush', 'Jonafree', 'Liberty', 'Melrose', 'Northwestern Greening', 'Nova Easygro', 'Prima', 'Priscilla', 'Pristine', 'Quinte', 'Red Delicious', 'Red Free', 'Sir Prize', 'Sundance', and 'William's Pride'. Moderately resistant pear varieties include 'Kieffer', 'Magness', 'Moonglow', 'Harrow Delight', 'Honeysweet', and 'Blake's Pride'. Fire blight resistance in other fruit crops is not well documented.

Symptoms. Infected tissues become dark, take on a water-soaked appearance, and may exude a yellowish-brown bacterial slime that looks like sugar water droplets. Sunken areas appear on infected spurs, shoots, and branches. Infected shoots and leaves turn dark brown to black and remain attached to the trees, creating conspicuous brown flags. Infected fruits shrivel into black mummies. Cankers girdle and kill infected stems, branches, and trunks. Be careful in diagnosing fire blight because the symptoms are similar to those of Pseudomonas blossom blast, Nectria twig blight, Phomopsis pear dieback, and twig borer damage.

Management. Grow fire blight–resistant varieties if available. In any case, you need a multipronged approach to managing fire blight.

Remove as many fire blight hosts as possible from your orchard, landscape, fencerows, and nearby areas. Remove blighted flowers, fruits, and stems as soon as you notice them, regardless of the time of year. Prune only when the trees are dry, however, to avoid spreading the pathogen even more. Carry a hand sprayer of Lysol (or similar disinfectant), 70 percent ethyl or isopropyl alcohol, or household bleach solution containing 9 parts water to 1 part bleach. The disinfectants and alcohol are less corrosive to tools than bleach.

Disinfect your tools after every cut. On cuts larger than 1 inch in diameter, also disinfect the cut surfaces. Be sure to cut at least 12 to 15 inches below sunken or discolored areas on the stem or other visible symptoms. The bacteria travel downward in the stems through the trees' vascular systems. Before starting your regular pruning, go through the orchard at least twice, removing blighted prunings. Bury or burn all prunings that might be infected with fire blight.

Maintain moderate growth in your trees. Avoid excess nitrogen, and keep pruning to a minimum to avoid forcing lush, blight-susceptible foliage. Instead, use spreading and other training methods.

When thinning fruit, remove late blooms. Remove and destroy severely infected trees if the main trunks show signs of infection; this is especially important in young trees. Avoid overhead irrigation, and use the practices on page 351 to reduce humidity in the orchard.

Control aphids, leafhoppers, and other insect pests.

Chemical and biological control. Applying Bordeaux mix, fixed copper, or fixed copper plus horticultural oil sprays at or slightly before bud break can reduce or delay inoculum development in cankers that you missed during spring pruning.

Tetracycline and streptomycin applications (be sure to use organically acceptable products) during mid- and late-bloom can protect blossoms and developing fruits. These products do not appear to protect against infections to shoot tips and leaves.

Several biological products can help reduce fire blight infections. They do not give complete control, but they can be part of an integrated program. Do not use biological and copper products together, nor apply biological products within 7 days of applying an antibiotic. Two types of biological, blight-control products are presently listed with OMRI. Blight Ban A506 (*Pseudomonas fluorescens* strain A506) is best applied during early bloom. Serenade (*Bacillus subtilis* strain QST 713) products act rather like antibiotics and should be applied with other biological products in mid- to late-bloom, rather than during early bloom.

Rust Diseases

Many related rust pathogens attack pome fruits, including cedar-apple rust (*Gymnosporangium juniperi-virginianae*), Pacific Coast pear rust (*Gymnosporangium libocedri*), quince rust (*Gymnosporangium clavipes*), and American hawthorn rust (*Gymnosporangium globosum*), among others. Together, these rusts have a wide range of hosts, including hawthorn, saskatoon, apple, crab apple, mayhaw, medlar, and quince. Unlike most fungal pathogens that can survive on a single host, these rust fungi require at least two different hosts in order to complete their life cycles. Plants in the Cypress family serve as the alternate hosts for pome fruit rusts. Particularly common and serious hosts are eastern red cedar, juniper, and incense cedar.

In arid and semiarid regions, rusts are seldom more than an annoyance and are easily controlled, due to low humidity and a lack of alternate hosts. In humid areas where the alternate hosts are found wild and in landscapes, rusts can cause severe damage to susceptible fruit crops.

Symptoms. Symptoms on fruit crops include scabbed and distorted fruits, often with bright yellow or orange pustules. Leaves become deformed and develop brown, necrotic spots on the upper surfaces and powdery yellow growths beneath. Infected twigs also become deformed. On juniper and cedar hosts, large galls develop during wet weather. The galls are usually orange and slimy, often with gelatinous horns or arms. Spores produced on the junipers travel up to about 3 miles to infect fruit crops. Spores produced on the fruit hosts likewise infect the cedar and juniper hosts.

Management. The best management strategy is to grow resistant varieties. Remove alternate hosts on and around your property. Also remove visibly infected fruit, leaves, and twigs from fruit crops, preferably before they discharge spores in midsummer. For infected cedar or juniper hosts that cannot be removed, prune out visible galls whenever you see them. When the galls on the juniper and cedar hosts enlarge and become slimy, they are discharging spores. At that time, apply sulfur fungicide sprays to your fruit crops.

Pear Decline

Pear decline is caused by the phytoplasma *Candidatus Phytoplasma pyri*. Phytoplasmas are organisms that have characteristics somewhere between bacteria and viruses. In this case, the pear decline phytoplasma is transmitted by pear psylla, infected rootstocks, or infected scions. The pathogen causes the cambium at the graft union to die, preventing new xylem and phloem from forming. Without these vascular tissues, sugars and other biochemicals produced in the leaves cannot reach the roots. As the roots die, the aboveground portion of the tree also declines and dies. The decline can be very rapid (weeks or months) or may take several years. In sudden decline, a secondary stress such as Phytophthora or Pythium root rot is typically involved.

Symptoms. With slow decline, a tree gradually loses vigor over several years. Terminal growth decreases. Leaves become smaller and may turn prematurely red in late summer. In rapid decline, the trees wilt, the leaves take on a scorched appearance, and the trees die. In both cases, a brown line of dead cambium cells will be visible below the bark at the graft union. Use a knife to carefully remove the outer layers of bark to reveal the cambium.

Management. Controlling pear decline involves five integrated strategies:

1. Avoid using *Pyrus ussuriensis* or *P. pyrifolia* rootstocks, if at all possible. Trees planted on *P. communis* rootstocks are much less likely to develop the decline. The Old Home × Farmingdale root-

stocks recommended in this book are generally resistant to pear decline.

2. Start with uninfected planting stock. Nursery stock can easily be tested for the phytoplasma, and you should ensure that the nursery you deal with tests its material for pear decline.

3. Keep trees vigorous by planting on well-drained soils, providing optimum nutrition, and following the other practices we have discussed to maintain healthy trees.

4. Control pear psylla. We will discuss this topic in chapter 11.

5. When budding European pear varieties onto Asian pear trees, insert your buds below the original graft union between the rootstock and Asian pear scion. Grafting a European pear scion above that original union will produce a tree that is highly susceptible to decline.

Fabraea Leaf Spot and Pear Leaf Spot

Fabraea leaf spot is caused by the fungus *Entomosporium maculatum* (*Fabraea maculata*) that infects apple leaves and the leaves, fruit, and shoots of pear and quince. A similar but less serious disease, pear leaf spot (Mycosphaerella leaf spot or *Mycosphaerella pyri*), infects pears, quince, and sometimes apples. Most European pears are susceptible to Fabraea leaf spot. These diseases are not much of a problem in arid and semiarid regions but can be very serious under humid and rainy conditions. Fabraea leaf spot is widespread in eastern North America.

Symptoms. Fabraea leaf spot symptoms begin as small purple dots on the lowest leaves. These dots gradually enlarge to form brown to black spots that are ⅛ to ¼ inch in diameter with purplish-black centers. Raised, blister-like fruiting bodies (acervuli) in the centers of the spots ooze a mucous-like material when wet, appearing as creamy or glistening areas on the leaves. In severe infections, defoliation can occur quickly and is common in humid areas. Fruits on infected trees develop similar spots and are usually small, misshapen, and cracked. Fewer fruit buds form, decreasing the following year's crop. Fall blooming may also occur. Superficial cankers form on the twigs but usually disappear by the following year. Trees can become stunted.

The pathogen overwinters mostly on fallen leaves and can also overwinter in twig lesions. New infections may not develop until late spring or early summer, and secondary infections begin developing about 1 month later during wet weather. Wind spreads late-summer spores. Infection can occur during wet periods lasting no more than 8 to 12 hours, so it is important to apply protective sprays before infections occur. Monitor the lowest 20 leaves on a test tree. From 1 to 10 spots indicate a moderate risk of serious damage. More than 10 spots, total, on the 20 leaves indicate a high risk.

Management. The same strategies are used to control Fabraea leaf spot and pear leaf spot. Remove or shred fallen leaves to reduce spore formation, as you would when managing apple scab. Early-season sprays used to control scab usually control leaf spot as well. Because older leaves and shoots are susceptible to infection, you must continue sprays later into the year than for scab. In Pennsylvania, one fungicide spray applied in June and another in July are usually sufficient. In more southerly locations, fungicide sprays during mid-August and September are also recommended, particularly during wet weather and on late-ripening varieties such as 'Bosc' pear.

Sooty Blotch and Flyspeck

These are common problems on apples, pears, and quinces in eastern North America. The diseases are caused by different pathogens, but they are usually discussed together because they behave similarly and are controlled in the same ways. These pathogens create superficial symptoms on the fruit surfaces. Although not damaging to the trees, cosmetic damage reduces the marketability of the fruit.

Symptoms. Sooty blotch symptoms include from one to many greenish or soot-colored colonies on the fruit surfaces. These symptoms may develop in as little as 3 to 4 weeks after petal fall but more commonly appear in mid- to late summer. Symptoms may not appear until the fruits are in cold storage. Flyspeck symptoms appear as small, circular patches of up to about 50 black spots on the fruit skin. Infected fruits have one to many such patches. The timing of infection and appearance of disease symptoms are similar to those for sooty blotch. Primary infections usually occur 2 to 3 weeks after petal fall, and secondary infections can develop in mid- to late summer during wet years.

The pathogens overwinter on infected twigs on the fruit trees and many other species of woody plants. Especially common and serious reservoirs are raspberries and blackberries, which should be removed from hedgerows. High humidity (greater than 90 percent) greatly increases the occurrence and severity of these diseases. Below 90 percent relative humidity, the pathogens do not grow.

Management. The cultural practices we have discussed to maintain open canopies, good airflow, and low humidity are critically important in managing these diseases. Diligently thin apples and pears to reduce humidity around the fruit clusters and improve spray coverage. Do not use overhead irrigation. In humid regions, keeping alley and in-row cover cops mowed short can help reduce humidity in the canopy.

Fungicides are an important step in managing these diseases in humid regions. As humidity and precipitation levels increase, so do the number of sprays needed to control sooty blotch and flyspeck.

Black Rot, Frogeye Leaf Spot, and White Rot

Black rot (on fruit) and frogeye leaf spot (on leaves) are common and sometimes serious diseases on apples. Both are caused by the same pathogen, *Botryosphaeria obtusa*. Girdling of the twigs and branches causes stunting, defoliation, and dieback. Infected fruits are small, may drop early, and can rot on the trees or in storage. Yields are poor. Unfortunately, no apple varieties appear to be resistant to the disease. 'Jonathan' and 'Winesap' are especially susceptible. White rot (*Botryosphaeria dothidea*) is a closely related disease and produces similar symptoms.

Symptoms. The first symptoms on the fruit usually appear on the blossom ends. The pathogens enter through natural openings or through wounds caused by pests, hail, or mechanical damage. One light brown spot usually develops on each infected fruit. These spots gradually darken and expand to form firm, leathery, bull's-eye-shaped rotted areas that can cover the entire fruit. Infected fruits shrivel into black mummies that may remain on the trees for more than a year. Fruits infected with white rot can take on a bleached appearance, especially in red-fruited varieties. The rotted areas may remain firm during cool weather or become soft and watery as temperatures rise.

Infected leaves develop spots up to ¼ inch in diameter with dark outer rings and lighter-colored centers that create a frog eye appearance.

Sunken lesions develop on twigs and form cankers up to 3 feet long or more on branches. The cankers may appear to be no more than rough areas on the bark, or they can be open wounds with raised edges.

Management. Good sanitation is the key for controlling these diseases. Prune out cankered wood whenever you find it because it is an important source of inoculum. To control both black rot and white rot, it is very important to carefully remove fire-blighted wood, as the pathogens easily enter blighted wood. Remove all prunings from the orchard and burn them. If you cannot remove the prunings, use a flail mower to chop them into small pieces that will decompose quickly.

Follow good nutrition, pruning, and pest and disease management programs, as we have already discussed, to maintain healthy trees that are resistant to the development of cankers. Plant rootstocks and varieties that are reliably winter-hardy in your area because the cankers develop more easily in winter-damaged wood.

Fungicides are an important part of an integrated program to control black rot and frogeye spot. Begin applying fungicides during bloom. If lesions develop on the fruits or leaves, repeat applications before or immediately after spring and summer rains. In the Mid-Atlantic and southeastern states, you will need to apply fungicides from bloom through harvest.

Bitter Rot of Apple and Pear

Also known as anthracnose on peaches and nectarines, bitter rot of apple and pear is widespread and can cause severe damage in warm, moist conditions. Apple and pear are most often affected, and infections are more common on fruit than on leaves and shoots.

Symptoms. Bitter rot symptoms caused by *Colletotrichum* species are easy to recognize. The first symptoms are one to many circular, light brown spots on the fruit skins. The spots gradually enlarge to 1 or more inches in diameter, with light brown outer rings and sunken, dark brown to black, saucer-shaped centers. The centers of the spots become cream-colored with pink spores, often arranged in concentric circles. Underneath the spots, the flesh rots and becomes watery. Infected fruits can shrivel and mummify.

Under optimal conditions (80°F [27°C] and 80 to 100 percent humidity), infections can occur in 5 hours. Primary infections usually occur during warm, wet conditions in early spring, and secondary infections can occur any time during the growing season. Common sources of inoculum are prunings left in the orchard, mummified fruits, and fire blight cankers.

Management. Plant apple varieties that have some resistance to bitter rot (see chapter 5 for some examples). Avoid planting 'Arkansas Black', 'Empire', 'Freedom', 'Fuji', 'Golden Delicious', 'Granny Smith', and 'Nittany' in areas with serious bitter rot problems, as these are especially susceptible.

Applying calcium as a nutritional supplement may reduce the incidence and severity of bitter rot in some years. Practice good orchard management and sanitation: Remove all prunings and fruit mummies from the orchard and burn them. Keep your tree canopies open, and maintain good air movement. In warm, wet areas, plan on applying fungicides from bloom through harvest.

Entomosporium Leaf Spot

The most serious disease of saskatoons, Entomosporium leaf spot, infects the leaves and fruit.

Symptoms. The first symptoms are small, brown spots on the leaves, which become yellow and may drop early. Small, gray lesions form on the fruits, which may become disfigured and crack. The disease can become serious under wet or humid conditions, and infections often develop following spring and early summer rains. The pathogen overwinters on infected twigs and fallen leaves.

Management. Some of the varieties described in chapter 5 demonstrate at least partial resistance to the fungus. Careful site selection and preparation, resistant varieties, good sanitation, proper plant spacing, and pruning to maintain good air movement go a long way toward managing Entomosporium leaf spot. The management practices described for Fabraea leaf spot will also control this disease (see page 350). Avoid sprinkler irrigation.

Bitter Pit

Bitter pit is a poorly understood physiological disorder (no pathogen is involved) that produces disease-like symptoms on apple fruits. It is caused by imbalances in mineral nutrients in the fruit. Although it is sometimes blamed on calcium deficiency, the absolute amount of calcium in the soil is not usually the cause. Excessive amounts of potassium and magnesium contribute to bitter pit, as does low soil pH, excessive nitrogen fertilization, excessive pruning, and excessive irrigation. Practices that result in very large fruit, including thinning fruit too early and overthinning, contribute to bitter pit.

Listed in the box below are highly susceptible varieties. Apple varieties that are less susceptible to bitter pit include 'Fuji', 'Gala', 'Golden Delicious', 'Haralson', 'Lobo', 'McIntosh', 'Red Gravenstein', 'Rome Beauty', 'Spartan', 'Stonetosh', and 'Winesap'.

Symptoms. Affected fruits develop dark, corky spots in the flesh just underneath the skin. The spots can grow to ½ inch in diameter. At about the time of harvest or perhaps not until the fruits are in storage, sunken spots develop on the fruit skins over the lesions, sometimes starting near the petal ends of the fruits. Fruits from young trees or from vigorous, upright shoots are most susceptible to bitter pit. With the exception of highly susceptible varieties, older, well-managed trees bearing high and sustainable yields are quite resistant to the disorder.

Management. The best control starts with good site preparation and crop management practices. Strive for moderate vegetative growth and high, sustainable yields of moderate-sized fruit. Calcium foliar sprays can help prevent bitter pit. Calcium nitrate is a synthetic product made

Apple Varieties That Are Highly Susceptible to Bitter Pit

- Baldwin
- Boskoop
- Bramley's Seedling
- Cleopatra
- Cox's Orange Pippin
- Golden Delicious
- Granny Smith
- Gravenstein
- Grimes Golden
- Jonathan
- Marigold
- Merton
- Northern Spy
- Prima Starking
- Red Delicious
- Rhode Island Greening
- Starkrimson
- Stayman
- Sturmer
- White Winter Pearmain
- Worcester
- Yellow Newtown
- York Imperial

by treating limestone with nitric acid. While it is used in conventional orchards as a fertilizer and for bitter pit control, its use in certified organic production appears to be prohibited.

For California apple growers, the University of California Extension recommends applying three to six sprays of calcium chloride at two-week intervals beginning after bloom. Apply 12 pounds of calcium chloride in 400 gallons of water per acre, and ensure that the fruit is thoroughly covered. Similar control programs should work in other growing areas. Be sure to apply the calcium sprays during dry weather and under conditions when the fruit will dry rapidly, to reduce russeting. While calcium chloride is a naturally occurring substance, it is prohibited by the National Organic Program except when derived from a brine process and used to treat physiological disorders associated with calcium uptake (e. g., bitter pit).

Dipping fruits into a 2 to 3 percent solution of calcium chloride can help prevent bitter pit from developing in storage. Dipping the fruit can damage it, however, and may not be acceptable to your organic certification organization. Certified organic growers should consult with their certification agencies before using postharvest calcium dips.

Commercial growers who have bitter pit problems in their orchards may choose to delay harvesting as long as possible, refrigerate the fruit, and delay packing for at least 4 weeks. By that time, surface symptoms will be visible, and affected fruits can be removed from the packing line.

Stone Fruit Diseases

CHERRIES, PEACHES, PLUMS, and other stone fruits are often called soft fruits. They typically ripen earlier in the season than most pome fruits and have short shelf lives. Stone fruits can be challenging to grow organically in humid areas due to brown blight and other diseases.

Brown Blight

Also called brown rot, blossom blight, and fruit rot, brown blight is widespread and among the most serious diseases of stone fruits. The disease attacks the flowers, fruits, twigs, and branches of many different crops, including apricots, cherries, nectarines, peaches, and plums. High humidity greatly increases brown blight problems, and the disease is especially common and severe in eastern North America.

Causes. Two different fungi, *Monilinia fructicola* and *Monilinia laxa*, cause this disease or complex of diseases. The pathogens survive on infected twigs and branches and in mummified flowers and fruits on the trees or on the ground. Air currents and splashing water spread the spores, which are released at temperatures above 40°F (4°C). Infections can occur at any above-freezing temperature when the trees remain wet for 24 hours or more. Following initial infection, secondary infections develop during the remainder of the season from spores that form on infected twigs, blossoms, and fruits.

Symptoms. Flowers can become infected at any time, but they appear to be most susceptible at full bloom. The flowers may become water-soaked and look as if they have been damaged by frost. They may also mummify and remain on the tree into the following year, serving as a source of additional infections. Small spots develop on the fruits and quickly grow in size. The infected fruits remain firm and dry. They may shrivel and mummify on the tree, drop to the ground, or remain symptomless until they are in storage. On peach trees, the disease can enter the twigs through infected flowers and fruits, developing cankers that girdle and kill branches. All infected tissues develop buff-colored or gray spores, depending on which pathogen is involved, sometimes in concentric rings on the fruits.

Management. Good crop management will help reduce disease problems by maintaining healthy trees and open canopies. Careful nutrition management is important because excessive nitrogen increases brown blight problems. Whenever possible, select fruit varieties that are resistant to brown blight.

Use aggressive sanitation practices. Prune out infected twigs and branches as soon as they appear, and remove all prunings from the orchard. Remove all mummified fruits from the trees and ground, and burn or bury them. Pick up and bury or destroy fruit that drops during thinning or naturally due to poor pollination. Remove susceptible hosts from the landscape and hedgerows. Manage insect and mite pests to reduce wounding the flowers and fruits. Harvest fruit carefully to avoid wounding it, and cool fruit as quickly as possible after harvest. Oregon State University suggests dipping harvested peaches and nectarines for 2.5 minutes in hot water (122°F [50°C]) to reduce postharvest rots.

Even with the best cultural practices, many growers will find that fungicides are necessary to control brown blight. Apply your sprays during bloom at early pink, full bloom, and petal fall. In California and other arid regions, one or two sprays should be enough. In areas with greater precipitation and humidity, you may need to apply fungicide sprays throughout the growing season. Do not apply sulfur to apricots. Use fixed coppers instead. As a biological control, *Bacillus subtilis* can be applied as a protectant. At the time of writing, OMRI listed Serenade MAX (*Bacillus subtilis* strain QST 713) for use as a biological control agent.

Bacterial Canker

This fungal disease is found across North America. The pathogen attacks sweet cherries in the Pacific Northwest and sweet and tart cherries in Michigan and Ontario and is a problem on peaches in the southeastern United States. Sweet cherries are most susceptible, but bacterial canker also attacks apricots, tart cherries, nectarines, peaches, and plums. The disease is caused by two closely related strains of the bacterium *Pseudomonas syringae*. Consult with fruit specialists in your region to determine if this disease is likely to be a serious problem for you.

Damage occurs not only directly to the plant tissues, but the bacterium also increases the risk of frost damage. It does this by producing a protein that causes ice crystals to form on plant tissues at higher subfreezing temperatures than normal. In a vicious cycle, the wounds formed by frost damage are susceptible to invasion by the bacterium. Anything that weakens the trees, such as incorrect pruning, poor mineral nutrition, improper soil pH, and damage from other diseases, increases the incidence and severity of bacterial canker. Trees on sites susceptible to damaging spring frosts are most at risk. Sites infected with the soilborne ring nematode are also at greater risk of bacterial canker.

Sweet cherry varieties 'Bing', 'Lambert', 'Royal Ann', 'Napoleon', 'Sweetheart', and 'Van' are very susceptible to bacterial canker, while 'Corum', 'Sam', and 'Sue' appear to be fairly resistant. In areas with a history of bacterial canker problems, clonal Mazzard F12-1 or Mazzard seedling rootstocks are recommended.

Symptoms. Signs of the disease include cankers on the trunks, limbs, and twigs; dead limbs; dead buds; and brownish, gummy sap being discharged from the bark and wounds. Girdled limbs may appear to grow normally during spring, then die suddenly at the onset of warm weather.

Management. Control involves using good management practices to maintain healthy trees with moderate growth. As with many other diseases, avoid using overhead irrigation. During dry weather in summer, prune to remove all cankers, disinfecting your tools between cuts using 70 percent alcohol or a solution of 9 parts water and 1 part household bleach. Delay your summer pruning until after harvest, when the weather is dry.

Completely remove limbs below cankers. Some fruit specialists recommend cutting away cankered areas to leave healthy tissue, but doing so may increase the risk of future infections. Completely remove and destroy badly infected trees, and do not allow the roots to regrow.

Copper bactericides are not always effective against bacterial canker, and resistance to copper has been observed. Continue your routine Bordeaux and copper treatments, but do not count on them to provide complete control of bacterial canker. Do not apply Bordeaux mix or other sulfur products to apricots. The best control is to use resistant varieties.

Bacterial Spot

A widespread and serious disease of stone fruits in eastern North America, bacterial spot affects apricot, nectarine, peach, and plum. The disease is primarily a concern in areas that receive more than 20 inches of precipitation per year. The bacterium *Xanthomonas pruni* attacks the twigs, leaves, and fruits of susceptible varieties.

Symptoms. Spring infections first appear as dark flecks on peach skins or water-soaked spots on smooth-skinned stone fruits. These spots gradually become sunken and can form very deep lesions on the fruit. Summer infections start as yellowish spots on the fruit. Angular lesions form between the veins on leaves and create a tattered appearance as the centers of the lesions fall out. Infected leaves turn yellow and drop early. On the wood, cankers develop near twig tips during spring, damaging or killing the tips. Summer infections appear as dark, irregular, sunken lesions on current-season shoots.

Management. Good orchard management practices to maintain healthy trees and open canopies help reduce the incidence of the disease. Avoid using overhead watering. Because this disease is caused by a bacterium, sulfur is not particularly effective in controlling it. Fixed coppers and Bordeaux mix can help control the disease, but they are phytotoxic to many stone fruits. In particular, do not use sulfur products on apricots. One strategy is to begin with a copper spray just before bud break followed by either weekly antibiotic treatments or alternating treatments of copper and antibiotic beginning at bloom. Note that copper damage on stone fruits can mimic symptoms of bacterial blight.

By far the most important strategy is to select varieties that are resistant to the disease. Also avoid planting near orchards containing susceptible varieties. High levels of inoculum can overwhelm even resistant varieties.

Black Knot

Caused by the fungus *Apiosporina morbosa*, black knot is widespread in North America and was once a serious orchard disease. Diligently following the management practices we have discussed should reduce the disease to a minor annoyance. Black knot is most often found on wild stone fruits, such as chokecherry, and in abandoned or poorly managed orchards. In orchards, black knot is most often a problem on plums and prunes, but it can infect apricot, peach, sweet cherry, and tart cherry.

Symptoms. The disease attacks only the wood and appears most commonly on twigs and small branches, but it can infect larger branches and trunks. The fungus forms very distinctive greenish to black corky swellings on the infected limbs, resembling a hot dog that has been skewered lengthwise. The fungus extends at least several inches underneath the bark from the ends of the visible swelling.

Management. Good cultural control practices are critically important for managing this disease. Follow the general cultural programs we have discussed to maintain healthy trees. Remove all wild and escaped stone fruit plants from hedgerows, woodlots, and other non-crop areas of your orchard. Before bud break in the spring and at any other time you find them, prune out the knots, cutting at least 4 inches below the knots themselves. Burn or bury the prunings.

Your normal fungicide program, including Bordeaux and lime sulfur sprays, will help manage black knot, provided you have followed the cultural steps above. Do not apply sulfur products to apricots. 'Bluefre', 'Damson', 'Shropshire', and 'Stanley' plums are highly susceptible;

'Bradshaw', 'Early Italian', 'Fellenburg', 'Methley', and 'Milton' plums are somewhat less susceptible. Resistant to highly resistant plum varieties include 'Formosa', 'President', Santa Rosa', and 'Shiro'.

Peach Leaf Curl and Plum Pocket

Peach leaf curl (*Taphrina deformans*) and plum pocket (*Taphrina communis*) are peach and plum diseases. Peach leaf curl is widespread and common. Plum pocket does not infect commercial varieties of European and Asian plums, but it does infect some species of native plums in North America.

Symptoms. Peach leaf curl symptoms first appear on expanding leaves in early spring. The leaves become thick and distorted and may become reddish or purplish. Infected leaves eventually become covered with silvery-gray spores, then drop early. New leaves may form in their place. Blossoms, young fruit, and shoots can also become infected and young shoots killed. Defoliation weakens the trees, making them more susceptible to winter injury and other diseases.

Plum pocket symptoms appear on the fruits as small, whitish spots. The spots enlarge, eventually covering the fruit. The plums become hollow and enlarge many times their normal size.

The pathogens overwinter as spores in the bud scales and in cracks in the bark. The spores germinate during cool, wet weather in early spring to infect the newly emerging leaves, shoots, and blossoms. Spores from these infected plant parts later cause secondary infections.

Management. These diseases are easily controlled with a fall spray of Bordeaux mix, followed by a spring dormant spray of Bordeaux mix or lime sulfur. If the diseases do develop, thin the fruits heavily and follow normal practices to maintain healthy trees. Avoid excessive irrigation and fertilization that might force late growth that would be susceptible to winter injury. Care for the trees, and prepare for next year's crop.

Peach Scab

Peach scab resembles apple scab but is caused by a different fungus (*Cladosporium carpophilum*). Rare in arid and semiarid regions, peach scab can become serious in humid climates, particularly in the Mid-Atlantic and southeastern states. Warm, wet days during midseason greatly increase the risk of the disease, which also infects apricots and nectarines.

Symptoms. Although the fungus also attacks the stems, symptoms are most visible on the fruits. Small greenish spots gradually enlarge, merge, and become blackened. The lesions often crack open, exposing the interior of the fruits to rot organisms. The disease overwinters in lesions on the most recently developed twigs and stems. By early the following season, these lesions appear grayish in color and slightly sunken.

Spores from the stem lesions develop in early spring and are splashed by rain onto young fruits and shoots. Water is required for infection. Fortunately, an incubation period of 40 to 70 days is required for symptoms to develop. Except in very late-season varieties, only infections that occur prior to pit hardening are likely to appear before harvest.

Management. Use all of the practices we have discussed to avoid humid conditions and create an open, well-pruned canopy with good air movement. Closely monitor your trees for signs of peach scab. If a tree shows infections one year, expect more serious problems the next year. Fungicides effectively control peach scab. Begin sulfur applications at petal fall, and continue every 10 to 14 days until at least 40 days before harvest. Do not apply sulfur products to apricots.

Plum Leaf Scald

A bacterial disease common in the southeastern United States, plum leaf scald is caused by the bacterium *Xylella fastidiosa*, the same pathogen that causes phony peach disease. The bacterium infects many other woody plants and is spread by glassy-winged sharpshooter and other small, wedge-shaped insects called leafhoppers that feed on sap in the plants' xylem. These insects attack an extremely broad range of host plants across the southeastern United States and into Texas. Infected trees can also infect other trees through naturally occurring root grafts.

Symptoms. Infected trees are smaller and appear leafier than normal due to abnormally short internodes on current-season shoots. The trees bloom earlier in the spring and drop their leaves later in fall than healthy trees. Infected plum leaves look scorched or scalded, although these symptoms do not appear in peach trees. Although the disease does not kill the trees, it makes them more susceptible to other diseases and shortens their life spans. Fruit size, quality, and yields are reduced.

Management. Avoid planting your trees next to an infected orchard. Infected trees cannot be cured, and you should remove them as quickly as possible to reduce the likelihood of infecting nearby trees. Scout your orchard and remove infected trees before starting summer pruning. Keep nitrogen fertilization moderate and avoid heavy pruning that stimulates the lush foliage that attracts leafhoppers. Remove all wild and escaped plums, cherries, and peaches in and around your orchard.

By far the most effective means of controlling plum leaf scald is to plant resistant varieties in the southeastern United States and other areas where the insect vector and disease occur. The Auburn University AU series of plum varieties are resistant to this disease.

Shothole Blight

Shothole or Coryneum blight is a fungal disease that is especially serious on apricot, nectarine, and peach. It also infects cherries and plums during long wet periods in spring and early summer. The pathogen *Wilsonomyces carpophilus* was formerly called *Coryneum beijerinckii*. Varietal resistance is rare, although peach varieties 'Lovell' and 'Muir' appear to tolerate the disease.

Symptoms. This fungus attacks the buds, twigs, stems, leaves, and fruits. The first symptoms are reddish, slightly sunken spots that appear on current-season and 1-year-old wood. The buds may be killed and the cankers can extend into and girdle 2- to 4-year-old wood. Gummy sap around wounds is common. Leaf symptoms begin as small tan or purplish spots that become scabby and fall out, producing the characteristic shothole effect. On infected fruits, similar spots develop and can become sunken with gummy exudates. On cherry and plum, usually only the leaves and fruit are infected, and then only during long wet periods.

Management. Maintain open canopies and good air movement in your orchard. Avoid overhead irrigation, and follow practices that minimize the humidity in the canopy. Prune out cankered wood and twigs, disinfecting tools between cuts and burning or burying the prunings.

Apply Bordeaux mix when half of the leaves have dropped in fall, and repeat when the trees are dormant. Follow up with a Bordeaux mix or lime sulfur spray as buds are swelling in spring. Do not apply Bordeaux or other sulfur products to apricots. Fixed coppers have also proven effective against the disease.

Diseases of Both Pome and Stone Fruits

SOME DISEASES HAVE WIDE HOST RANGES and attack both pome and stone fruits. Powdery mildew and root diseases are especially common, and some viruses can be troublesome for orchardists.

Powdery Mildew

Powdery mildew is caused by many different fungi and attacks the shoots, leaves, and fruits of many plants. In stone fruits, *Podosphaera pannosa* and *Podosphaera clandestine* cause powdery mildew. In pome fruits, *Podosphaera leucotricha* causes powdery mildew and appears to cause rusty spot disease on peaches and nectarines.

Symptoms. Signs of infection are similar in all crops. Leaves and new shoots become covered with powdery, grayish-white spores, often becoming distorted and developing brown necrotic areas. Infected new shoots often die back, and new twigs can grow deformed.

Infected fruits first become covered with white mold and gradually turn russetted and scabby. Rusty spot symptoms appear as small, round orange to tan spots that enlarge as the fruits grow. On pome fruits, infected fruits develop a network of roughened lines on the skins. Peach and nectarine fruits are susceptible to infection until the pits harden. Powdery mildew weakens the trees and reduces the fruit's marketability.

Management. Remove and destroy diseased wood and fruits while pruning and thinning. Unlike many fungal diseases, powdery mildew does not need free moisture (water droplets or films) in order to infect plants. Reducing humidity levels, however, can help combat the disease, so maintain open canopies and good air movement. Shading increases powdery mildew problems, so keep your fruit crops in full sun. Remove roses and wild and escaped stone and pome fruits from in and around your orchard.

Many fruit varieties are resistant to powdery mildew, and planting resistant varieties in areas where powdery mildew is a problem is a key to controlling the disease. In chapters 5 and 6, we describe disease-resistant varieties. Fungicides effectively control powdery mildew. Bicarbonate fungicides can give some control of early infections, and highly refined horticultural oil has proven surprisingly effective. Sulfur fungicides may also be used, but not within 10 to 14 days of applying horticultural oil. Sulfur can also russet sensitive fruits. Do not apply sulfur products to apricots.

Viruses and Phytoplasmas

Many kinds of viruses and phytoplasmas attack stone and pome fruits. A few include apple mosaic, apple proliferation, flat apple disease, rubbery wood, flat limb, cherry mottle leaf, cherry twisted leaf, leafroll, little cherry, and Prunus necrotic ringspot. In some cases, the viruses are carried by aphids, leafhoppers, mites, or nematodes. Most viruses can be transmitted during grafting or budding if using infected rootstocks or scions.

Symptoms. Signs of a problem include distorted limbs and fruits, leaf spots, distorted leaves, and even decline and death of trees. In some cases, the viruses kill the cambium at the graft union, causing the trees to die. Quite often, trees will harbor latent viruses that produce no symptoms.

Management. Start with virus-free rootstock and scion wood. Deal with reputable nurseries that test their planting stock for viruses. Once a tree is infected, it can never be cured. Where the virus is a threat to other trees in the orchard, rogue out infected trees as soon as you find them. Grind the stumps to speed decay of the roots and reduce the likelihood of the virus spreading through root grafts to other trees. An exception might be X-disease in cherries grown on Mazzard rootstocks where many trees in an orchard are infected. Infected cherries on this rootstock seldom die and can produce marketable crops for some years.

Test your soil before planting to determine if nematodes that serve as virus vectors are present. Control aphid, leafhopper, and mite infestations.

Root Diseases

Root diseases can be serious in orchards and, once established, are virtually impossible to remove. Conventional fruit growers have several fungicides available that suppress root rots, but none eradicate the pathogens. The soilborne pathogens can survive for many years in the soil, either in resting stages in the soil or on the roots of alternate host plants. The keys to avoiding them are to carefully select the site, prepare and maintain it well (ensuring good water drainage), and use healthy planting stock.

Phytophthora

A serious disease, Phytophthora attacks the roots and collars of many tree fruit crops. Several closely related water molds (*Phytophthora* species) can cause root rots in fruit trees. Infections can take place on fine feeder roots or through wounds on larger roots.

Symptoms. Infected roots die, and the pathogen moves upward in the tree to the collar where the trunk meets the root system. Most trees show gradual or rapid decline, with scorching and possibly wilting leaves, and thinning foliage on the upper branches.

Prevention. One of the most important factors in preventing Phytophthora root rot is well-drained soil. The pathogens require free water to move in and are seldom a problem in orchards grown on sandy or otherwise well-drained soil. Try to correct drainage problems before planting.

Start with healthy nursery stock from a reputable nursery. Avoid excessive irrigation. Do not mulch naturally cool, wet soils. Keep your soil well-drained and biologically diverse.

Most stone fruits are particularly susceptible to Phytophthora and should be planted on the best-drained soil available in your orchard. Of the pome fruits, mayhaw and pear are probably the most resistant to the disease, but they are still best planted on well-drained soil. For apples, avoid M26 rootstocks on poorly drained sites. For cherries, Mahaleb rootstocks are particularly susceptible to the disease.

Verticillium

Caused by a soilborne fungus (*Verticillium dahliae*), Verticillium can survive in the soil for many years and has an extremely wide range of host plants. Infection comes through the roots, but the pathogen moves upward in the trees through the xylem. Damage comes from the partial blocking of the xylem vessels responsible for transporting water and also from toxins produced by the pathogen that cause wilting.

Symptoms. Although younger trees are most easily attacked, the disease can infect older trees. Lower leaves may show discoloration and wilting, and those symptoms move upward. The foliage thins, and the tree takes on a bare look. The decline may occur over several years or shoots and twigs can die quickly, leaving flags of dead leaves attached to the tree. It is not unusual for a single branch or side of the tree to show symptoms, creating a prominent flag in the tree. Cutting into the xylem of infected wood with a knife often shows brown discoloration in the wood.

Management. If possible, do not plant where tomatoes, potatoes, peppers, eggplant, raspberries, blackberries, or strawberries have grown. They are common hosts and can facilitate the buildup of Verticillium in the soil. Phlox, geranium, shepherd's purse, lamb's quarters, and nightshade are other hosts and should not be allowed in the understory of an orchard. Where Verticillium is a problem, test your soil for fungal propagules before planting. If microsclerotia are present, you have a risk of the disease.

Do not interplant your orchard crops with Verticillium-susceptible crops. Control weeds, and avoid excessive nitrogen fertilization, irrigation, and severe pruning. The goal should be moderate vegetative growth and sustainable, high yields. Remove and destroy dead and diseased branches, preferably before the leaves fall and add more inoculum to the soil.

Armillaria Fungus

Armillaria or shoestring fungus is a fungal pathogen (*Armillaria mellea*). Although sometimes called oak root fungus, it attacks the roots of more than 500 species of woody plants, including orchard crops. Armillaria is widespread in North America and most often found where forests or woodlots have been cleared fairly recently. Honey-colored mushrooms that appear in fall or winter are the fruiting stage of the pathogen, but their spores seldom infect host plants directly.

Symptoms. The fungus grows underground, producing dark, root-like strands (rhizomorphs) in the soil. Infected plants often show characteristic white "fans" of fungal mycelium just beneath the bark on the upper roots and lower trunk, to about 12 inches below the soil surface. Use a knife to scrape away the outer layers of bark to expose the fan.

Aboveground symptoms can be very hard to detect. Shoots and leaves are small, but infected trees can appear reasonably healthy until suddenly dying, usually at the onset of hot weather.

Management. Simply removing an infected tree or shrub does not eliminate the disease, which can live for years on dead and decaying roots. Healthy plants become infected when their roots contact live or dead infected roots of another plant. Rhizomorphs that grow out into the soil from infected roots also carry the fungus to new plants. Control involves prevention. Even soil fumigation has not been very successful.

If you are clearing forest or woodlots for an orchard, take at least several years to prepare the site. Carefully examine the woodlot trees, and mark the locations of infected trees or clusters of trees. Girdle infected trees, and leave them standing for a year or two to speed the decay of infected roots. Remove all stumps, then deep-rip the soil, first in one direction, then at right angles. This practice helps drag roots from the soil. Remove and burn all roots larger than about 1 inch in diameter.

Fallow your site for at least 2 years using green manure crops of biofumigant varieties of rape, canola, or mustard. Alternate the green manure crops with disking or other cultivation methods to maintain bare soil. You can also grow cereal grains, which do not support Armillaria.

Before planting your fruit trees, isolate infected trees or clusters of trees by digging a trench 2 to 3 feet deep around them. Line the trench with plastic film, and backfill it to help prevent rhizomorphs from entering your planting area. Remove infected trees and as much of their roots as possible, as quickly as you find them. Do not replant in problem sites. Instead, grow nonhost plants and use them as insectaries for beneficial organisms.

Frost Damage

The pathogenic bacterium *Pseudomonas syringae,* which causes blossom blast or blight, increases frost damage by increasing ice formation on leaves and blossoms at higher temperatures than would be true if the bacteria were absent.

Where spring frosts are common, applying copper bactericides or *Pseudomonas fluorescens* in a product called Blight Ban A506 can reduce damage. This bacterium does not cause disease and helps protect fruit crops from frost and disease by competing with *P. syringae* for sites on the blossoms. At the time of this writing, Blight Ban A506 and many copper products were approved for organic fruit growers.

CHAPTER 11

PEST MANAGEMENT

- Survey of Pests | 369
- Pest Management Strategies | 375
- Insect and Mite Pests | 386

As organic fruit growers dealing with a whole-systems (holistic or ecosystem) approach, we need to integrate many pest management strategies into our overall orchard management plan. We place the emphasis of our prevention and pest control programs on beneficial organisms and cultural practices. Organic pesticides are also key elements in the program, after we have done everything else we can.

Survey of Pests

FRUITS AND FRUIT TREES ARE VALUABLE SOURCES of food and/or habitat for many different pests. Mammals, large and small, feed on the fruits, twigs, buds, bark, and roots of the trees. Many bird species feed on fruits, and woodpeckers and sapsuckers damage the trunks. Insect and mite pests attack fruits, leaves, stems, and trunks, and certain nematodes can damage roots. Fortunately, careful planning and management can control most of these pests.

Large Mammals

Moose are common and sometimes devastating pests for orchardists in northern latitudes across much of North America. From personal experience, I know that a single moose can destroy a remarkable number of fruit trees in a very short time. Deer are also serious orchard pests in many areas of North America.

Fruit growers have tried using repellant sprays made from rotten eggs or pepper, soap bars, bags of human hair, skunk and coyote urine, and many other repellant materials. The fact is that none of the materials works particularly well for more than a few days, and none will protect an orchard from hungry moose or deer. Noisemakers and visual scare devices have fared no better. Moose and whitetail deer have proven remarkably adaptable and quickly lose their fear of such things.

Large herbivores are best managed using exclusion fences. Many different types of fences are used for protecting orchards, and most work fairly well. At the high end, a New Zealand–style electrified fence made up of strands of high-tensile wire spaced about 10 inches apart and 8 to 12 feet tall works well for moose but is not reliable for deer. Single-strand fences, in general, are not effective against deer, which push through or under fences readily, even if the strands are electrified or made of barbed wire.

Some growers have used a double-row fence made up of two short (5 to 6 feet high) fences spaced about 4 feet apart. The theory is that deer are good at jumping heights but not particularly good at jumping long distances and hesitate to jump into the narrow space between the two fences. One creative Pennsylvania orchardist installed a single short fence, then placed short-legged dogs bred to herd reindeer in the orchard. The dogs could not run fast enough to catch or hurt the deer, but they were still effective at keeping the deer out.

The most effective means of excluding large herbivores is to use a fence that has about a 4-by-4-inch or smaller mesh along the bottom 5 feet, with wider mesh or single wires spaced 12 inches apart composing the rest of the fence to a height of at least 8 feet.

Figure 11.1 · **Typical Orchard Fences**

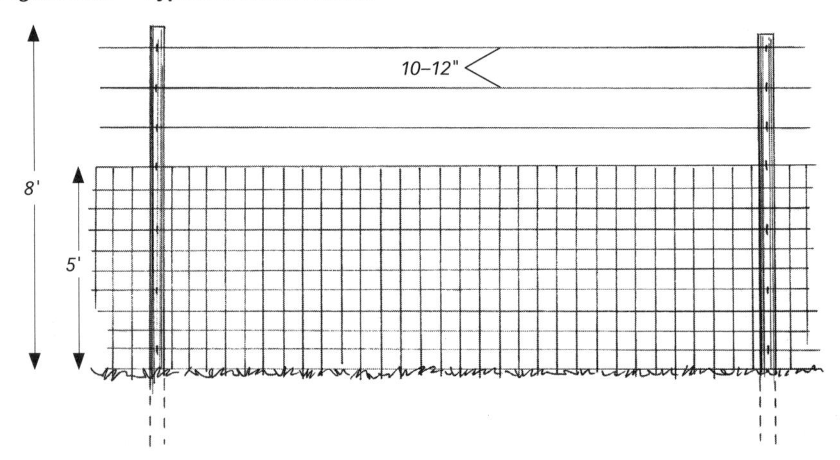

A) Mesh and single wire fence. The 4-inch-square mesh is 5 feet high with single-strand wires above spaced about 10 to 12 inches apart to a height of 8 feet.

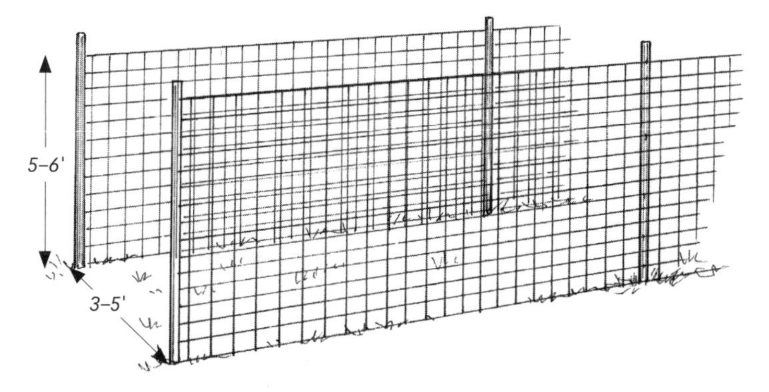

B) Double fence. The mesh fences are 5 to 6 feet high, spaced about 3 to 5 feet apart.

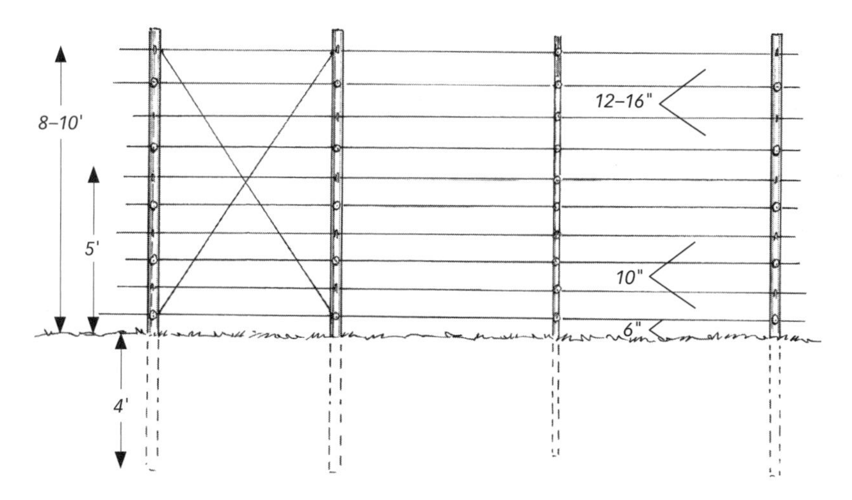

C) New Zealand electrified fence. The fence is 8 to 10 feet tall with every other wire electrified, starting with the bottom wire.

Keeping the mesh portion of the fence firmly on the ground provides the best deer control, but it also excludes foxes, coyotes, and other predators that feed on mice and voles that can attack your trees. An alternative is to keep the fence at ground level around the perimeter but leave a few small (1 foot square) openings in the fence for the predators to pass through and install a second barrier fence just inside or outside the opening to block the taller and longer deer. A similar design is often used for schoolyard fences to allow people to pass through but block bicycles and other vehicles. Keep the opening just large enough for a coyote or fox to squeeze through. Deer fawns are small and remarkably adept at crawling through or under fences. Figure 11.1 shows typical orchard fence designs.

Inexpensive nylon mesh fences resembling tall plastic chicken wire are marketed as deer fencing for orchards and gardens. While they are not as durable as metal fencing, they are less expensive and for several years proved surprisingly effective in protecting fruit plots from whitetail deer on one of my research farms. How well they work against moose is questionable. These fences are probably best used for home orchards or to provide temporary protection in larger orchards.

If raccoons or porcupines are a problem in your orchard, use a fence with small mesh and add electrified wires at the top, angled outward. Electricity can be produced using solar panels, although this strategy is not always effective during winter at high latitudes with very short days.

Regardless of the fence you choose, make sure the posts and corners are strong enough to support it. Keep the fence tight, especially at the bottom. Watch out for low spots, particularly where the fence crosses over gullies or water. An Alaska fruit grower that I worked with suffered damage to his crops when a moose calf fell into a bog at the edge of his nursery and swam under the fence where it crossed the bog. Once inside the fence, the calf was unable to find its way out again and survived by eating the fruit trees and bushes. The fence did keep out the cow moose, however.

Small Mammals

Rodents and rabbits can cause serious damage to an orchard by girdling trunks or feeding on roots. Because they are small, exclusion is generally not an option. Even if you use a small mesh fence to the ground, rabbits and other pests can tunnel under the fence. These pests can be managed in several ways.

Traps. Hand trapping is effective against pocket gophers and other larger, burrowing rodents. Trapping has generally proven more successful than using poisoned baits, as the pests learn to recognize the baits and avoid them. Anchor the traps firmly to the ground to prevent dogs and other predators from carrying away the trap after a gopher has been caught. Check the traps at least three times a week, from spring through fall.

Bare ground. Maintain bare ground within the tree rows using weed barrier fabrics, organic mulches, thermal weeding, organic herbicides, mechanical cultivation, and hand weeding. If this strategy does not fit into your orchard management plan, you can keep in-row cover crops and alley crops mowed short, particularly going into and through the winter. During the growing season, mow alternate rows every two weeks to reduce rodent habitat. If you blow mulch into the rows or grow in-row cover crops, plan an aggressive rodent control program.

Mice guards. Install wire mesh mice guards around fruit tree trunks to a height of at least 3 feet (these pests can use snow as an elevator to reach quite high up on trunks). Use galvanized hardware cloth with about ¼-inch mesh. Bury the bottom of the guard several inches in the ground and make it large enough in diameter so as not to girdle the trees. Use easy-to-remove fasteners to hold the edges of the hardware cloth together to allow easy removal of the tree guard for pruning and other trunk maintenance. Figure 11.2 shows such guards.

Good sanitation. Remove all plant debris in and around the orchard and compost, burn, or bury it. Don't keep brush piles or piles of wood or other materials. Harvest ripe fruits promptly. In barns and sheds, keep seeds and other plant materials that are attractive to rodents in sealed, plastic containers.

Cats and dogs. Cats and terrier-breed dogs have well-earned reputations as mousers and can also aid in managing gopher populations in small orchards. Consider keeping several neutered or spayed cats and dogs in your orchard.

Raptors and owls. Install raptor and owl perches and owl nesting boxes in and around your orchard. Providing habitat for these predators will help manage mammal and bird pests.

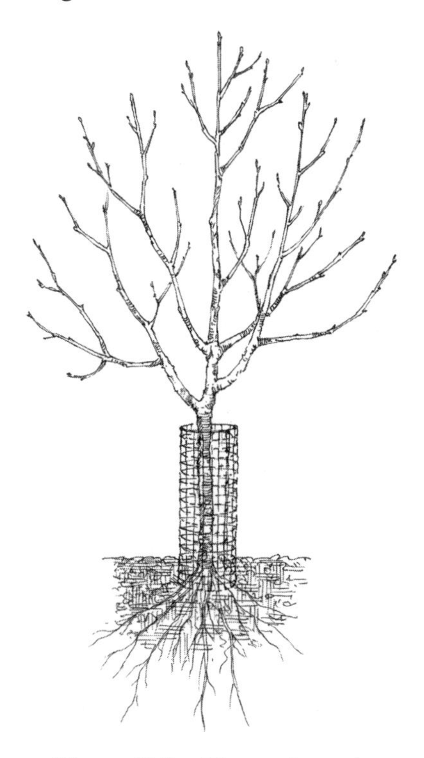

Figure 11.2 · Wire mesh guards prevent rodents and rabbits from damaging tree trunks.

Birds

Insect-eating birds are great beneficials. Some birds, however, can damage large amounts of fruit, often taking only a few pecks at a fruit but rendering it unfit for consumption. Bird damage is usually most serious in small orchards, where a few birds can destroy a relatively large percentage of the fruit. In large orchards, there is simply too much fruit for the birds to damage more than a very small percentage. As with deer, visual scare devices and noisemakers (including bird distress calls) work only for a very short time.

Site selection. If possible, avoid planting your orchard near woodlots that provide nesting and perching areas for birds.

Bird netting. For small orchards, install fine mesh bird netting over the trees and bushes. Put out the netting just before the fruits ripen, and keep the netting taut to prevent birds from becoming entangled. Mount the net so that it can easily be moved aside during harvest. Remove the netting immediately after the harvest is done.

Perches and nest boxes. Install raptor and owl perches in and around the orchard. To reduce predation on beneficial birds, keep the raptor and owl perches away from nesting boxes and other nesting areas. Install owl nesting boxes around the perimeter of the orchard near the owl perches. Face the boxes away from the nesting areas of beneficial birds, as we will discuss later in this chapter.

Insects, Mites, and Nematodes

Many insects, mites, and nematodes feed on fruit crops, and several cause severe and widespread problems in North America. Plum curculio, apple maggot, and codling moth are the three most serious pests.

In natural ecosystems, we usually see balances of plant pests and predators that feed on those pests. Many serious pest outbreaks can be traced to the introduction of exotic insects, mites, or nematodes that have no predators or diseases in the new areas to control their numbers. There are exceptions, of course. Plum curculio, long a limiting factor in organic fruit production, is native to North America. Growing crops in large monocultures can also upset the balance in favor of the pests.

Conventional fruit growers rely heavily on a wide variety of insecticides, miticides, and nematicides to manage pests in their crops.

Although organic growers have gained a few powerful pest control products in recent years, using cultural practices to develop an orchard ecosystem with a balance of pests and predators is still the best pest control strategy. Unlike with diseases, we have few fruit varieties that are resistant to pests.

The orchard floor management practices we discussed in chapter 9 play important roles in managing insects, mites, and nematodes, mostly by providing habitat for beneficial organisms. Because pest and disease pressures vary from one growing region to another, you will need to tailor those practices to your specific location.

Pest Management Strategies

PEST MANAGEMENT IS HIGHLY COMPLEX in all but the smallest orchards. Cultural practices must be integrated to reduce pest populations while protecting and enhancing beneficial populations. Tree nutrition, orchard floor management, pruning and training, use of insectary crops, pheromones to monitor (and sometimes reduce) pest populations, pest repellants, and pesticides all play roles in an effective orchard pest management program.

Scouting

As with disease management programs, scouting is extremely important in any orchard pest management program. During the growing season, spend at least 3 days per week in your orchard looking for signs of pests or the damage they leave behind. Useful scouting tools include a jeweler's magnifying loupe, a white sheet of poster board or a commercial beater tray, and a short stick. The loupe is helpful in locating and identifying mites, thrips, and other small organisms. By holding the poster board or beater tray under a branch and tapping on the branch with a stick, you can collect many pests that would otherwise be hard to spot.

A very helpful way to monitor apple maggot populations is to hang red, apple-shaped spheres smeared with petroleum jelly or other sticky material from the trees. Both yellow and white sticky cards placed throughout the orchard are valuable for monitoring pest and beneficial populations.

Commercially manufactured cards are available from orchard and nursery supply companies, or you can easily make your own from poster board. Cut the board into 4- to 6-inch square cards. Attach a string for hanging the cards from the trees, and coat the cards with petroleum jelly, or spray them with vegetable cooking oil. Likewise, pheromone-baited sticky traps can be used to monitor populations of codling moth, Oriental fruit moth, and other insects. We'll discuss these steps as they are used with specific pests. Figure 11.3 shows typical traps for monitoring insect pests.

Figure 11.3 · **Insect Monitoring Devices**

A) Red sticky spheres are commercially available and are useful in monitoring for apple maggot.

B) White, blue, and yellow sticky cards can easily be made and are also commercially available. They attract a wide assortment of insects.

C) Pheromone traps come in various sizes and shapes. Tent traps are very common and are used to monitor codling moth, Oriental fruit moth, and other pests. Pheromone baits attract insects to the traps.

Sanitation

As for the rodent and disease control programs, maintain a clean farmscape to manage pests. Remove and destroy or compost crop wastes promptly. Burn heavily pest-infested plant residues. If prunings cannot be removed from the orchard, use a flail mower to chop them into small pieces as soon after pruning as possible.

Beneficial Organisms

Beneficial insects, mites, nematodes, birds, and bats play vital roles in organic pest control programs. If beneficials are wiped out, a pest population can explode in a matter of days or weeks.

Managing natives. While you can purchase and release beneficial insects, mites, and nematodes, the best strategy is to manage those beneficials that occur naturally in your area. The huge advantage is that they are adapted to the climate and can provide abundant, long-lived populations.

Begin by learning to identify beneficial organisms that are native to your area. Integrated pest management specialists and entomologists at universities and state and provincial agricultural organizations can supply this information, much of which is readily available online. Particularly good guides and programs are available from Cornell University in New York, Pennsylvania State University, Washington State University, and the University of Kentucky. Common beneficials include green lacewings, ladybird beetles (ladybugs), big-eyed bugs, damsel bugs, parasitic wasps, and predatory mites.

Using the resources mentioned above, learn to identify beneficials in your area, and conduct a survey on your site to determine which beneficials are present and how large the populations are. Repeat this survey each year at the same time to monitor your pest management program.

Releasing beneficials. If you plan to release beneficials in your orchard, the most effective will be beneficials that are native to your area but not to your site. Nonnative beneficials might be useful in certain circumstances, but they seldom produce long-lived populations and must be released on a regular basis. Be very cautious in releasing nonnative organisms, as they often disperse out of the orchard and you do not want to accidentally introduce another exotic pest into the ecosystem.

Growing insectary crops. In chapter 9 (see page 306), we discussed the use of insectary crops. Insectary crops provide food, shelter, and egg-laying sites for beneficial insect and mite adults and/or juveniles. Pollen, nectar from flowers, and exudates from the glands on plant stems and leaves are important food sources for certain beneficial insects during at least part of their life cycles.

In many cases, insectary crops support prey for the beneficials. The prey might be the same pests that attack your fruit crops, related species, or an unrelated and innocuous insect or mite. The important point to remember is that without prey to survive on, predators and parasites will either die or move out of the orchard to areas where prey are available. In conventional orchards, fruit growers often try to eradicate a pest using insecticides or miticides. In organic systems, the goal is to maintain pest populations that are too low to cause significant damage to your crops but large enough to support healthy populations of predators and parasites.

Common insectary crops include dill, chamomile, hairy vetch, spearmint, Queen Anne's lace, buckwheat, yarrow, white clover, cowpea, cosmos, and other species. Some grasses, particularly those that have glandular leaves, can also provide habitat for beneficials. In some cases, you might be able to include herbal or ornamental crops in the insectary planting for cash flow or personal use. Using a mix of plants that flower throughout the summer can help support bees that you can use to pollinate your orchard in spring.

One strategy for an organic orchard is to maintain a planting of insectary crops within 50 feet of all trees. This might be accomplished by including insectary crops in alleyways, in in-row plantings using the "sandwich system" described in chapter 9, or in non-tree rows throughout the planting.

Noninsect and Non-mite Predators

Don't forget that many animals feed on insects, chief among them being birds and bats. Start by finding out which species of birds and bats in your area feed on insects that might become pests in your orchard. Learn about the nesting habits and living requirements for these beneficial birds and bats, and design nesting facilities and perching sites to support them in and around your orchard.

Wild Birds

If you live in an area where bluebirds live, for example, consider installing nesting boxes on the fence line around your orchard. About 80 percent of a bluebird's diet is insects. Be aware, however, that the other 20 percent comes from berries and fruits from small trees and bushes. You want to maintain enough of a bird population to help control insects without seriously damaging your fruit.

Swallows feed almost exclusively on flying insects, often over fields and water. They construct nests under building eaves, in barn lofts, in culverts, and in cliff pockets. Chickadees eat seeds and berries, but much of their diet, even in winter, consists of insect adults, larvae, and egg cases. Wrens are aggressive birds that subsist almost entirely on insects. While wrens can be valuable assets in an orchard, however, they also attack the eggs of other cavity-nesting birds, such as bluebirds. Give some thought as to how well the different species of birds in your area will work together in your orchard before building nesting boxes and other facilities to support them. Remember to supply sources of clean water for your bird and insect beneficials. Birdhouse plans for many species of North American birds are available online.

Chickens

Chickens have long been used in orchards and gardens to control insect pests. They eat adults and larvae and are especially valuable in scratching up and eating pupae in the soil, including apple maggot. Unfortunately, organic certification programs generally do not allow livestock, including chickens, in a fruit planting within a certain timeframe before harvest. The reason is to reduce the likelihood of human pathogens contaminating the fruits. Organically certified tree fruit growers will need to remove the chickens from the orchard at least 90 days before harvest, and bush fruit growers will need to do so 120 days before harvest. In such situations, consider placing chickens in the orchard immediately after harvest to help reduce the number of overwintering pests. Use pens and fences to manage the chickens and protect them from predators.

Bats

Many bat species live on insects. The little brown bat, for example, is native to the northern United States and Canada as far north as the

Yukon. Bats can consume half their body weight each day in insects, and lactating females can eat more than their body weight in insects daily. Little brown bats are often found near streams, ponds, and lakes because part of their diet comes from aquatic insects. Forty-four other species of bats are native to North America and are extremely important to agriculture, partially by pollinating certain crops but also by consuming night-flying insects.

Find out which bats are found in your area and consider constructing shelters for them around the orchard but away from storage buildings and other areas where people work. This separation will reduce human/bat interactions and keep the bats' strong-smelling feces away from areas where people congregate. Note that bat populations in North America are presently threatened by a serious disease called white nose syndrome (WNS) that was first observed here in about 2006.

Insecticides

Even with the best cultural practices, most commercial organic fruit growers need insecticides to produce marketable crops. While relatively few organic insecticides are registered, several new materials have become available in recent years. The following materials are allowed in certified organic orchards under the U.S. National Organic Plan and are generally allowed under other certification programs. If you are a certified organic grower, be sure that the products you use are approved by your certification organization.

Use even these organic insecticides with great care and follow label directions carefully to minimize damage to beneficial organisms. Organic certification programs usually require that you use all available cultural practices to control pests before applying a pesticide, and that you document this in your orchard records.

Horticultural Oils

Horticultural oils have been used to protect crops for more than 150 years, and many different types are available. They may be produced from petroleum, fish, or plants. Heavier oils are applied to dormant trees and bushes while lighter-weight oils (summer oils or superior oils) can be applied to foliage and fruit during the growing season.

In general, oils smother pests and eggs and have been especially valuable in controlling mites and scales. Oils have also proven effective in controlling aphids, caterpillars, leafminers, and psyllids. Highly refined petroleum oil has proven very effective in controlling powdery mildew on cherries and peaches. Be sure that the product you use is certified for organic production and labeled for your crop.

Oils can be phytotoxic, particularly when applied with sulfur. Avoid applying oils when the humidity is high and temperatures are above 80°F (27°C). Use large volumes of water (200 gallons per acre or more), and set your sprayer equipment to produce fine droplets. Make sure the oil has an emulsifier to allow it to mix with water, use constant agitation when spraying, and follow label directions very carefully.

Horticultural oils are generally safe for mammals and birds when used carefully and according to label directions. They can be toxic to beneficial mites and result in pest mite outbreaks when used incorrectly. In an orchard, an application of dormant crop oil just as buds swell in spring is very effective in controlling mites and scales when the beneficial organisms are not yet in the trees. To avoid phytotoxicity, be careful not to apply too soon before or after sulfur sprays.

Insecticidal Soaps

Insecticidal soaps are made from potassium or ammonium salts of fatty acids and are applied to plants to control insect and mite pests. The soaps act by smothering pests or damaging their cuticles (outer layers) and allowing the pests to dry up and die.

Insecticidal soaps work best against small, soft-bodied insects like aphids, mealybugs, thrips, whiteflies, and mites. Soaps are also toxic to caterpillars and leafhoppers but generally do not provide an effective level of control against them, especially against older, larger larvae. These products have little impact on hard-bodied insects, including pest and beneficial beetles and syrphid flies, and do not usually harm bumblebees. They have a very low level of toxicity to mammals and birds, although the soaps are irritating to the skin and eyes. Ammonium-based soaps are sometimes touted as being repellent to deer and rabbits, but do not count on them to protect your trees or bushes against these pests.

Although pesticidal soaps are very useful, they have definite limitations and must be used as part of a system that integrates other pest controls. Their effectiveness is rather low, perhaps 40 to 50 percent, against small, soft-bodied pests. Soaps kill only pests that are contacted directly by the soap solution and in relatively large amounts. Once the soap solution dries on the plant, it has no effect on pests. You need to spray thoroughly to ensure the upper and lower surfaces of branches, leaves, and fruits are covered. This means using large amounts of soap solution, usually at about 2 percent concentrations.

Hard water makes soaps ineffective. If your water is hard, you may need to treat the water to soften it. Some products are available that can be added directly to the water to improve insecticidal soap effectiveness. Follow label directions carefully. Soap sprays work best when they dry slowly, such as on cool, cloudy days, or in the evening.

The soaps can be phytotoxic if used too frequently or at concentrations that are too great. Similar soap products have been developed as organic herbicides, as we discussed in chapter 9 (see page 297). Fruit crops generally tolerate pesticidal soap applications well, but fruit can be damaged if large amounts of spray collect at the bottoms of fruits. Soaps are also highly unpalatable and have offensive odors. Be sure your fruits are free of soap residue before using or marketing them.

Microbial Insecticides

Microbial insecticides include assorted strains of *Bacillus thuringiensis* and *Beauveria bassiana*.

Bacillus thuringiensis

Bacillus thuringiensis (Bt) is found naturally in soil and on plants, and many different subspecies or strains exist, some of which attack insects. It is very important to know that not all strains attack all insects. Some strains are effective against caterpillars (moth and butterfly), while others attack beetles or whiteflies. These products are generally nontoxic to mammals or birds, although a very small percentage of people are allergic to *B. thuringiensis* in large concentrations, and the bacteria can enter the human body through wounds or contact with mucous membranes.

Bacillus thuringiensis products work only after an insect ingests them. A crystalline protein produced by the bacteria attaches to the gut of the insect, perforating the gut wall. The insects often stop feeding rather quickly after ingesting the bacterium but may take several more days to die. *Bacillus thuringiensis* products have a short lifespan in orchards and typically do not remain effective for more than a few days after being applied to crops. Be sure that the product you use is certified for organic programs (there are some genetically engineered Bt products available), and be sure your product is labeled for the specific pest you are trying to control. Resistance to *B. thuringiensis* has been found in some populations of diamondback moth and Colorado potato beetle, probably due to overuse of these bacterial insecticides. Use Bt products in rotation with other pesticides in an integrated and diverse pest management program.

Beauveria bassiana

Beauveria bassiana is a soilborne fungus that is found worldwide and that attacks some insects. Two strains are commonly used to control pests and are made using fermentation to produce spores that are applied to plants in sprays.

Pests for which *B. bassiana* products are registered include ants, aphids, caterpillars, grasshoppers, mealybugs, thrips, weevils, and whiteflies. Infection usually occurs when the spray contacts an insect or the insect contacts the spores left on plant surfaces after the spray dries. The spores germinate on the host body and penetrate the cuticle, killing the pest in 3 to 5 days. The infected bodies produce more spores that can infect additional pests.

Beauveria bassiana products work best in cool, cloudy, moist conditions at temperatures below about 80°F (27°C). They are likely to be of more use to growers in moderate climates than in warmer regions. Apply the product only when the pest is present, and target young pests early in an infestation. The effectiveness of *B. bassiana* products has been variable, with more research studies showing poor pest control than showing fair or good control. *Beauveria bassiana* does not discriminate against pests and beneficials. Avoid spraying *B. bassiana* products when pollinators are active in your fruit crops.

Spinosad

Spinosad is a patented product that is derived from the fermentation of the very rare actinomycete bacterium *Saccharopolysora spinosa*. This product is a broad-spectrum pesticide that is useful against many insects. The material works mostly as an ingested poison, although insects directly contacted by the spray can also be affected. Essentially, the toxin causes the nervous system to maintain a high state of agitation until the insect dies from exhaustion.

Spinosad has very low levels of toxicity for mammals, is slightly toxic to birds, and is highly toxic to bees. Spinosad persists for several days on leaf and fruit surfaces and is most effective against caterpillars and beetles that feed on the leaves. Its control of aphids, whiteflies, and thrips has been mixed. Beneficials are generally not affected much by the material once the spray has dried.

Spinosad can persist for months in water and soil that is not exposed to sunlight. It is moderately toxic to fish and aquatic invertebrates and should not be allowed to contaminate surface waters. While spinosad is a breakthrough for controlling leaf rollers and other orchard caterpillars and plant-eating beetles, it is not a silver bullet. Resistance can build up when the material is overused, so make it part of an integrated program, and rotate it with other pesticides such as neem products and *Bacillus thuringiensis*. Several commercial products containing spinosad are available. Ensure the products you use are registered for your crops, pests, and area.

Neem Products

Neem products are derived from seeds of the neem tree, *Azadirachta indica*, as discussed in chapter 10. Neem products have been used as pesticides for centuries, although serious research on them only began in the 1920s in India and spread to other countries in about 1959. There are three basic types of neem products: azadirachtin-based, neem oils, and neem oil soaps.

Neem oil is produced by crushing the seeds and extracting active ingredients using water or alcohol. The extraction and processing methods that are used affect the oil's effectiveness as a pesticide, meaning not all neem products are created equal. Although azadirachtin was thought to be the primary pesticidal compound, it now appears likely that other chemicals in the neem seeds also help protect crops. These compounds act like hormones in insects, preventing juveniles from maturing into adults. They also reduce insect feeding and egg-laying on treated plants.

In fruit crops, neem products have effectively controlled aphids (including rosy apple and wooly apple aphids), leafhoppers, spotted tentiform leafminer, and tarnished plant bug. It has partially controlled white apple leafhopper, caterpillars, and mites. It has not performed well against beetles, psyllids, and scales. Mixing azadirachtin-based and neem oil products has sometimes produced better results than using the materials alone. Mixing neem oil and neem oil soap products can cause phytotoxicity.

Pyrethrum Products

Pyrethrum products are derived from the dried flower heads of several species of pyrethrum daisies. They are effective against a wide range of insects and paralyze the insects by attacking their nervous systems. Pyrethrums are especially known for knocking down insects quickly. Unfortunately, the insects are not killed quickly and can recover from the paralysis if the pyrethrum dose is too low.

Pyrethrums are also quickly broken down by sunlight, water, and soil, meaning they have very short life spans in an orchard. To be effective, the spray must actually contact the pest. Synthetic pyrethroids have a much longer active period and are more effective pesticides but cannot be used in organic programs. Many pyrethrum products are on the market. Be sure you purchase those approved for organic use. Pyrethrum has provided fair to good control against green peach aphid and good control of western grape leafhopper but poor control of other orchard pests in limited tests.

Surround

Surround represents a breakthrough approach to pest management called particle film technology. It is not, however, a silver bullet and must be used as part of an integrated system. Made from an edible clay (kaolin) that is used in processed foods, toothpaste, cosmetics, medications, porcelain, and paper, this material is also valuable in orchard pest and disease management. The product consists of very finely ground kaolin clay that is applied to orchard crops as a spray, with the wettable powder suspended in water. Once dry, the kaolin forms a thin, white film over the leaves, fruit, and wood. Although nontoxic, the film prevents some pests from recognizing the host plants. Other pests appear to be repelled by the film. Note that Surround is a special formulation of kaolin. Simply spraying raw kaolin on fruit trees can, reportedly, damage or kill them.

Surround has reportedly done a good job of controlling leaf rollers and leafhoppers and has partially controlled apple maggot, codling moth, plum curculio, thrips, and stinkbugs. Weekly sprays usually begin at petal fall and continue for about 8 weeks. Prebloom sprays are also useful in managing several pests, as discussed later in this chapter. The kaolin is not applied during bloom. If necessary, the sprays can be continued until harvest, although the fruits must then be washed to remove the white film. During the growing season, the film washes and rubs off the trees, making it necessary to apply it frequently in wet and windy climates. Fruit specialists in high-rainfall areas of the southeastern United States report problems keeping enough kaolin on the trees to be effective in controlling pests.

The added benefits of Surround include improved photosynthesis (probably by deflecting sunlight and cooling the leaves) and improved coloring on some apples. The kaolin film appears to reduce symptoms of fire blight, sooty blotch, and flyspeck. It is compatible with most organic pesticides, but do not tank-mix it with sulfur or Bordeaux mix.

Insect and Mite Pests

MANY INSECTS AND MITES FEED ON ORCHARD CROPS. Most cause minor damage and are easily controlled. In order to develop an effective pest management program, you need to know which pests are found in your area. As for diseases, Cooperative Extension and Ministry of Agriculture offices often publish regional guides for fruit growers. Many of these guides are now available on the Internet for regions across the United States and Canada. The following section includes the more serious and common orchard pests found in North America.

Sucking Pests

Sucking pests include insects and mites that feed directly on the liquid contents of leaf and fruit cells. Typical orchard sucking pests include aphids, leafhoppers, scales, and mites. Because they feed on sap from inside the cells, ingestion-type insecticides are generally ineffective in controlling these pests. Beneficial organisms play important roles in controlling many sucking pests.

Common Aphids

In North America, the most common and serious aphid pests include:

- Apple grain aphid
- Black cherry aphid
- Green apple aphid
- Green peach aphid
- Hop aphid
- Leaf curl plum aphid
- Mealy plum aphid
- Rosy apple aphid
- Rusty plum aphid
- Spirea aphid
- Thistle aphid
- Wooly apple aphid

Aphids

Aphids are among the most common horticultural pests and affect an enormous range of plants, including orchard crops. Of the more than 4,000 known aphid species, about 250 attack agricultural crops and at least 12 can cause moderate to serious damage to pome and stone fruits in North America. All are soft-bodied insects that feed by sucking the phloem contents out of leaves, young shoots, developing fruits, and other tender tissues. Some aphids also feed on plant roots. Feeding causes leaves, young fruit, and sometimes shoots to become deformed. Leaves curl up around the aphids, making them hard to reach with pesticides.

The aphids excrete honeydew that covers leaves and fruits with a sticky coat and often becomes black as sooty mold grows on the honeydew. The aphids' feeding weakens the trees and makes the crops more susceptible to other pests and diseases. Aphids also serve as disease vectors and can infect orchard crops with viruses.

Being soft-bodied, aphids are relatively easy to control, but they need to be managed with diligence and accurate timing to prevent large populations from building up. The key is to manage the aphids from the start of the growing season, before the crop is damaged. Start by applying horticultural oils (usually heavier, dormant oil formulations) at the beginning of bud swell through the emergence of green tips on the buds. Follow up, if necessary, with a pesticide as new leaves and flowers emerge. Neem products, insecticidal soaps, and Surround are often recommended to control aphids. Surround is applied after petal fall to control aphids. The timing of the sprays will depend on the pest and the product used. Follow label directions carefully. Aphids have many natural enemies, including ladybird beetles and green lacewings. Protect these beneficials and provide habitats for them in your orchard by planting insectary crops.

Mealybugs

Apple mealybug, Comstock mealybug, and grape mealybug somewhat resemble aphids in size and habit. They have sucking mouthparts and feed on juices from the phloem, causing stippling of the leaves. Although heavy infestations can weaken trees, the economic damage comes from the mealybug's excretions, which drip onto the fruits, creating russetted patches. The pests can also enter into the calyx ends of fruits, infecting the fruits with rot pathogens.

The pests overwinter as eggs or larvae on host plants or litter on the orchard floor. All pome and stone fruits can be attacked, as well as many other species of plants. You can use aphid control programs to effectively combat mealybugs. Parasites and predators are important parts of the management program.

Leafhoppers

Several species of leafhoppers, including potato leafhopper, rose leafhopper, and white apple leafhopper, attack pome and stone fruit crops. Leafhoppers are small, wedge-shaped insects about ⅛ inch long with sweptback wings. They have sucking mouthparts like aphids but are much more mobile. Although they feed only on the leaves, they damage fruit by depositing their feces on them. These deposits create spots on the fruit resembling tobacco juice and support the growth of sooty mold. A far more serious impact is that some leafhoppers vector plant diseases such as fire blight, plum leaf scald, and assorted viruses. Large infestations of leafhoppers create clouds of insects in orchards and are irritating for orchard workers. Leafhoppers are found across North America.

Programs that control aphids also work well for leafhoppers. It is very important to implement control programs early, before bloom, to prevent populations from building. Eggs are deposited on the host plant in cracks in the bark. The first larvae begin emerging at about the time of bloom. Applying dormant-type horticultural oil sprays at the beginning of bud swell through green tip may help smother eggs. Because these are sucking insects, ingestion-type pesticides have little effect on them. Azadirachtin products, insecticidal soap, and Surround appear to offer the best controls during the growing season.

Mites

Many mite pests infest pome and stone fruits in North America (see box on page 390). Spider mites feed on phloem juices and cause leaf stippling and defoliation that weaken the trees. Eriophyid mites are very small and cause similar symptoms to those of spider mites, but some can also cause leaves to become deformed and develop blisters, galls, and velvety or discolored areas. Feeding on flower clusters early in the season can also cause fruit deformities. Otherwise, healthy trees can support large mite populations without serious harm.

In well-managed organic orchards, predators usually control pest mite populations effectively. Serious mite outbreaks are often caused by pesticide programs that kill off predators, including predatory mites. It is important to make dormant applications of lime sulfur 30 days or more before bud swell and applications of horticultural oil during bud swell through green tip. At those times, predatory mites are generally not in the trees and are not harmed by the pesticides. During the growing season, sulfur fungicides will reduce mite populations. Here, however, is one of the serious weaknesses in organic orchard pest and disease management. The sulfur applications needed to control serious fungal diseases are also toxic to beneficial predatory mites. The overuse of sulfur fungicides during the growing season can lead to outbreaks of pest mites. As you develop a pest and disease management plan for your orchard, use as many strategies as you can to reduce the number of sulfur sprays applied to the trees during the growing season.

Pear psylla. This serious pest was introduced into North America. The psylla resembles a small cicada about 1⁄10 inch long or less and ranging from greenish-brown to very dark, depending on the time of year. Adults overwinter in bark crevices or on fallen leaves and emerge before bud break to lay eggs at the base of buds.

The larvae hatch in 2 to 5 weeks and begin feeding on leaves and buds, sucking the juices from the phloem. These larvae pass through a series of five stages (instars) in 30 to 50 days, and there may be three to five overlapping generations per year, depending on the length of the growing season. The psylla's feeding weakens the pear trees, but far more serious damage is done by the virus-like pathogen (phytoplasma) carried by the psylla. This pathogen causes pear decline (see page 348). The psylla also injects substances into the pear trees that cause shock and damage to the trees, and the honeydew the psylla excretes contaminates the fruits and supports the growth of sooty mold.

Fortunately, it is relatively easy to manage pear psylla. Many predators and parasites attack pear psylla, including adult and larvae ladybird beetles (ladybugs), green lacewings, plant bugs, minute pirate bugs, earwigs, and parasitic wasps. It is very important to maintain healthy and abundant populations of these beneficial insects in the orchard. However, predator and parasite populations typically build too slowly in the spring to provide the control that is necessary to prevent large summer populations of psylla. Instead, you can make dormant applications of lime sulfur 30 days or more before bud swell, followed by an application of horticultural oil during bud swell through green tip to very effectively reduce psylla populations. Applying Surround pre- and postbloom, as well as azadirachtin products during and after bloom, should provide good control of this pest, provided you have an effective population of predators and parasites. You may increase the effectiveness of azadirachtin by tank-mixing it with summer-weight horticultural oil. Be careful, however, because some pear cultivars can be damaged by azadirachtin products.

Types of Mites

Eriophyid mites include:

- Apple blister mite
- Apple rust mite
- Bigbeaked plum mite
- Peach silver mite
- Pearleaf blister mite
- Pear rust mite
- Plum rust mite
- *Prunus* rust mite

Spider mites include:

- Brown mite
- European red mite
- McDaniel mite
- Two-spotted spider mite
- Yellow spider mite

Moth Pests on Leaves and Fruit

Many moth (lepidopterous) pests attack pome and stone fruit crops, typically feeding on leaves, fruits, and/or twigs and trunks. Some, such as codling moth, can cause severe crop loss and require aggressive management.

Codling Moth

Codling moth is one of the most serious fruit tree pests across North America and has limited organic fruit production in many areas. The adults are brownish-gray moths with a wingspan of about ¾ inch and bronze bands on the wing tips. The juveniles are pinkish-white, worm-like larvae that tunnel in and around the fruit core. This pest typically produces two generations per year. Codling moth is the proverbial worm in the apple. Even a small percentage of moth-damaged fruit can render a crop unmarketable, and it is critical to aggressively manage this pest.

Organophosphate pesticides were long the primary means of controlling codling moth. Today we use pheromone-baited traps to monitor populations and pheromone-impregnated plastic strips tied in the trees to disrupt mating. These strategies are integrated with applications of insecticides and Surround to provide organic control.

To monitor flights, attach codling moth traps impregnated with pheromone baits to branches in the top third of your orchard canopy. If more than five moths are captured during the first generation, or two moths are captured per trap in the second generation, you should apply an insecticide.

By the time the first flower buds show white or pink color (first pink), you should have 200 to 400 mating disruption dispensers per acre in your trees. These dispensers (often resembling twist ties) are impregnated with a pheromone that mimics the one produced by female moths ready to mate. The large numbers of mating disruptor strips attract the male moths and make it difficult for the males to locate females. Bear in mind that these dispensers will attract moths to your orchard and can increase damage to your crop if you have a badly infested orchard nearby.

Beginning at first bloom, apply an approved organic product containing codling moth granulosis virus. This is a naturally occurring organism that is very specific for codling moth. Repeat sprays of the granulosis virus every 7 to 10 days as long as adult moths are present. The spray will kill the larvae, but not until some damage has occurred. Beginning at petal fall, begin applying Surround, neem products, and spinosad products according to label directions as long as the pheromone traps are capturing adult codling moths.

You can also use parasites to manage codling moth populations (see the Oriental fruit moth section below). Maintain insectary crops in your orchard. Purchase and release the parasites, if needed.

During the 1960s, fruit researchers in California tested ultraviolet (black light) traps to monitor codling moth flights. The light traps were hung in the orchards just before the codling moth flights were expected to begin and turned on at dusk. In addition to codling moths, the traps caught other orchard moth pests and other insects, averaging 80 other insects for each codling moth trapped. These traps can be used to supplement the management program described above. They will not be as accurate a monitoring device as pheromone traps, and it is unlikely that they will provide adequate control of codling moth populations by themselves. To reduce their impact on non-pest insects, use pheromone traps to monitor codling moth flights and turn on the black lights only when the moths are flying.

To target pupating larvae, loosely wrap strips of cardboard around the tree trunks beginning with the spring flight. Larvae that crawl down the trunks seeking a place to pupate hide under the cardboard and spin cocoons. Examine the cardboard strips weekly after you observe moth flights and replace as necessary to destroy the cocoons.

Oriental Fruit Moth

Oriental fruit moth is an introduced pest that can cause very serious damage in fruit crops and is considered more difficult to control than apple maggot. It was once considered to be primarily a problem on peaches, but it also attacks apricots, nectarines, almonds, apples, quince, pears, plums, and cherries, as well as many woody ornamental plants. The adults are brownish to greenish moths with a wingspan of about ½ inch. They lay eggs on tender, young shoot tips, and the larvae bore into and feed on the tips. The feeding kills the twigs and shoot tips, causing the fruit trees to have a bushy appearance. The larvae also feed on young fruit (young injury) and nearly ripe fruit (old injury). Damage from both young and old injuries is clearly visible. In "concealed injury," which is common in peaches, the outside of the fruit may appear perfect but conceal a larvae feeding within, usually near the pit.

The methods of control used for Oriental fruit moth are similar to those used for codling moth, with slight exceptions due to the fact that OFM produces three generations per year. Pheromones are available for monitoring populations, and studies of mating disruption are now being conducted. As with codling moth, Oriental fruit moth is susceptible to beneficial parasites. According to Michigan State University, *Trichogramma minutum* (an egg parasite) and *Macrocentrus ancylivorus* (a larval parasite) may parasitize 50 to 90 percent of Oriental fruit moth eggs or larvae in an orchard. The parasites alone, however, do not provide sufficient control for commercial crops and must be included in integrated pest management programs.

Cutworms

A number of pests fall under the category of cutworms, including Bertha armyworm, spotted cutworm, and variegated cutworm. These lepidopterous pests are night-flying moths that overwinter as pupae in the soil and emerge in early spring. The larvae range from ¼ to 1 inch in length and curl up when disturbed. They feed on foliage and fruit at night. It is important to use predators and parasites to control cutworm populations, but sporadic outbreaks of the pests can occur. Codling moth management programs should also control these pests. Insecticides of choice include neem products, *Bacillus thuringiensis*, and spinosad products.

Eyespotted Bud Moth

Eyespotted bud moth is a minor pest for apple, cherry, pear, and plum trees. The adults are small, brown moths, and the larvae are small, brown caterpillars that feed in buds, twigs, unfolding leaves, and young fruits. The leaves may be folded over. It is not usually necessary to control them, but this pest will be controlled by codling moth and Oriental fruit moth programs.

Fruitworms, Leaf Rollers, and Loopers

Many similar pests fall under the category of fruit worms (see box on page 394). Although they are different species, they are all moth (lepidopterous) pests that are biologically similar and can be controlled with the same practices. Some of these pests overwinter as eggs and others

Types of Fruitworms

- Filbert (European) leaf roller
- Fruit tree leaf roller
- Green fruitworm
- Lacanobia fruitworm
- Leaf crumpler
- Lesser apple worm
- Mineola moth
- Obliquebanded leaf roller
- Pandemic leaf roller
- Redbanded leaf roller
- Redhumped caterpillar
- Tufted apple bud moth
- Variegated leaf roller
- Winter moth

as mature larvae. Feeding begins as early as prebloom and can result in deeply scarred fruit. The most common symptoms include partially eaten and rolled leaves and larvae feeding on and in the fruits. Programs to control codling moth should also control most fruitworms and leaf rollers. *Bacillus thuringiensis*, spinosad products, and Surround are all important parts of a control program.

Gypsy Moths

Gypsy moths are introduced lepidopterous pests from Europe and Asia and are slowly moving across North America. Because they can defoliate large expanses of hardwood plants, they are closely monitored by state, provincial, and federal organizations. Aggressive community and regional eradication efforts are often used when new populations are detected. Practices that control codling moth, Oriental fruit moth, leaf rollers, and fruitworms will also control gypsy moths. Tanglefoot sticky wraps placed around the trunk can help prevent the night-feeding larvae from reentering the trees. During heavy infestations, however, the larvae remain in the trees feeding around the clock. Pheromone traps for monitoring gypsy moth and mating disruption materials are available but should not be needed in most areas.

Tent Caterpillars

Western tent caterpillars and forest tent caterpillars can become problems in orchards, and their combined ranges spread across the United

States and Canada. These pests form conspicuous, web-like nests in trees, and the larvae can quickly defoliate an orchard tree. Remove and destroy the tents whenever you find them. Programs to control other lepidopterous pests will also control tentworms.

Apple and Thorn Skeletonizer

The small, reddish-brown apple and thorn skeletonizer moth has a wingspan of less than ½ inch and is found from Virginia to California and north into Canada. The adults overwinter in bark cracks and crevices of the trees, emerging in spring and laying eggs on the undersides of leaves. Although its preferred host is the apple tree, this pest also attacks hawthorn, pear, cherry, and plum trees. The larvae are about ½ inch long, yellowish-green caterpillars with black spots. They can often be found dangling from silken threads from infested trees and blow readily from one tree to another. The larvae feed on the leaf tissues between the veins, creating a skeletonized or lace-like appearance. They roll the leaves and pupate within the shelter, emerging as adults to start a new generation.

Severe infestations of apple and thorn skeletonizer can virtually defoliate the trees, and infestations are most serious following mild winters. In most fruit-growing regions, expect at least two generations per year. Spinosad and *Bacillus thuringiensis* products are effective against apple and thorn skeletonizer, and rotating between these two products will help reduce resistance buildup. Neem products should provide partial control, and applications of Surround may reduce feeding and egg-laying.

Leafminers

At least three species of leafminers infest pome and stone fruit crops across the United States and Canada, including apple blotch leafminer, spotted leafminer, and western tentiform leafminer. These pests are small moths about ⅒ inch long. Adults have brownish wings with white bands that appear silvery as the moths fly.

The adults deposit eggs on the undersides of the leaves, and the larvae tunnel into the centers of the leaves and tunnel through the leaf between the upper and lower surfaces. The first three stages (instars) of the larvae have sucking mouthparts, and the fourth and fifth stages have chewing mouthparts and eat the leaf tissues. The later larval stages web together the sides of the tunnels, creating tent-like shapes. The larvae then pupate inside the fallen leaves, emerging as adults in early spring as new leaves begin unfolding.

Parasites normally control these pests, and serious outbreaks are infrequent. During mid-spring, examine 100 leaves per block of trees. If there are fewer than two or three tunnels per leaf, you probably do not need to control them. If there are more than two or three tunnels, you may want to begin a management program. Applying azadirachtin products, insecticidal soaps, and spinosad products just before bloom should help manage leafminers. Good sanitation practices that involve removing or destroying overwintering leaves on the orchard floor may also help reduce leafminer populations.

Flies, Sawflies, and Midges

Several fly, sawfly, and midge insects are serious pests on fruit crops. Among the most serious are apple maggot and various fruit flies. While damage to the fruit may often appear minor, even a small percentage of infested fruit can render a fruit shipment unmarketable.

Apple Maggots

One of the most widespread and serious fruit pests in North America, apple maggot has long limited organic production. This insect primarily attacks apple trees, but it also infests cherry, crab apple, hawthorn, pear, plum, and quince trees. Adults are black flies that are a bit smaller than houseflies and have black bands on their wings. They emerge from the soil in spring to lay eggs under the fruit skins. The eggs hatch into small, white larvae that tunnel throughout the fruits, eventually dropping to the ground and forming pupae that overwinter. There is only one generation per year. Maggot-damaged fruit is unmarketable and usually unsuitable for home use.

Control starts with early detection. Adult flights can be detected and monitored using yellow sticky traps or red sticky spheres hung in your trees at about eye level. Unfortunately, the traps do not capture enough flies to manage the population. In western North America, the adults start laying eggs about 10 days after they begin flying. Applying Surround, beginning at petal fall, will somewhat control apple maggot. Spinosad significantly reduced maggot damage on apple trees in Cornell tests. Begin these applications at petal fall.

Fruit Flies

The most serious fruit flies for stone fruit growers are cherry fruit flies, of which there are several species ranging across North America. Fruits damaged by maggots are unusable, and maggots in processed cherries are considered contaminants. The real problem is for commercial growers because there is a zero tolerance policy for maggots in packed fruit. A single larva found in a shipment will cause that shipment to be refused. In the Pacific Northwest, the western cherry fruit fly is the primary commercial cherry pest and requires perfect control. Spotted-wing drosophila was recently introduced into California and is considered an extremely serious emerging pest of cherries, raspberries, strawberries, blueberries, and perhaps of other stone fruits because of similar quarantine issues.

Adult cherry fruit flies resemble small houseflies, and the larvae are white maggots that feed inside the cherries. The pest overwinters as pupae in the top 4 inches of the soil. It emerges beginning about five weeks before harvest, and peak emergence is generally about harvest time. The adults spend 5 to 10 days feeding before laying eggs under the skins of the fruit. After feeding inside the fruits, the larvae drop to the ground, tunnel in, and pupate until the following year. The fruit flies usually remain very close to where they emerge. Using effective control measures within an area will generally reduce or eliminate the pest until it is reintroduced. An important control measure is to eliminate all wild and escaped cherries in and around your orchard.

To date, biological controls do not provide the perfect control needed for commercial fruit. Yellow sticky traps and sticky red spheres can help monitor the pests, but results are poor to variable. According to Washington State University, cherry fruit flies do not respond to attractant pheromones.

Early-maturing cherries are least susceptible to these pests. Pick all fruit from the trees, as even a few remaining fruits can support a new generation of flies. Harvest fruit as soon as it is ripe, and destroy dropped and culled fruits. Surround provides some protection, and spinosad-impregnated baits splattered in the trees have provided at least partial control of adults before they can lay eggs. Spinosad also appears to be effective against spotted-wing drosophila. At least two sprays are required, the first just as the fruits are turning pink and the second 7 to 10 days later.

European Apple Sawfly

The European apple sawfly is an introduced pest that is now found in the northeastern United States and Ontario. The adult flies are ¼ to ⅓ inch long with yellow heads, nearly black backs, and yellow to orange lower abdomens. They emerge from cocoons several inches below the soil surface at just about apple bloom time. The females deposit their eggs in apple blossoms, and the resulting larvae feed near the surface on developing fruits, causing prominent scars that make the fruit unmarketable. In later feeding, the larvae tunnel toward the core of the fruit. A single larva can damage all the fruits in a cluster.

Mature larvae grow to about ⅜ inch long and resemble caterpillars. Sawfly larvae can be distinguished from caterpillars by having seven pairs of prolegs (stumpy leg-like appendages) rather than five. Damaged fruits often drop about the time of normal June drop. The larvae emerge from the fruits, tunnel into the soil, and pupate until the following spring. Although two generations per year occur in Europe, normally only one generation per year occurs in North America. So far, parasites and predators have not effectively controlled European apple sawfly. You can monitor populations of this pest by hanging white sticky cards in your trees.

Although the larvae look like caterpillars, they are not affected by *Bacillus thuringiensis*. The best organic control appears to be two prebloom applications of Surround, followed by an insecticide spray at petal fall. Be very cautious. Pesticides that control these pests also kill bees. Programs that control spring-feeding moth larvae and tentiform leafminer should also help control European sawfly. There have been very few published recommendations on how to control European sawfly with organic methods. Neem and spinosad products appear to be the most likely candidates to treat the pests at petal fall.

Cherry Slug

Also called pear slug and pear sawfly, cherry slug is widespread across North America. It is easily controlled and usually causes only minor damage, but it can be serious in poorly managed orchards and on young trees. The adults are shiny black flies that are ¼ inch to nearly ½ inch long. They emerge from the soil in early spring and lay eggs in slits they make on the leaves of cherry and pear trees.

The eggs hatch into slime-covered, olive-green larvae that feed on the upper surfaces of the leaves, skeletonizing them. The larvae drop to the ground, pupate several inches below the soil surface, and emerge as new adults during July and August. Heavy feeding can weaken the trees; insecticidal soap, spinosad products, and Surround should provide good control.

Pear Leafcurling Midge

Pear leafcurling midge is an introduced European pest that attacks only pear trees. The adults are tiny flies that pupate underground and emerge at about bloom time and lay their eggs on new, folded leaves. The grub-like larvae hatch in 3 to 5 days and feed within the folded leaves, with as many as 30 grubs per leaf. The feeding causes the leaves to remain tightly folded, protecting the grubs from predators.

When they mature, the grubs chew out of the leaves, fall to the ground, and pupate. Depending on the growing region, you may experience four to five or more generations per year. Mature trees tolerate the damage, and control is seldom needed in established orchards. Newly planted pear trees in orchards and nursery trees can be severely stunted. In conventional orchards, organophosphate pesticides are used to control the pest, but recommendations for organic control in North America are lacking. In New Zealand, the closely related apple leafcurling midge is effectively controlled by parasitic wasps, pirate bugs, and earwigs. Where the midge is a problem, consider applications of neem and spinosad products at petal fall and focus on maintaining healthy populations of beneficial insects.

Boring Insects

Many insect pests damage fruit trees by boring into trunks, branches, and twigs, including ambrosia beetle, American plum borer, dogwood borer, lesser peachtree borer, Pacific flatheaded borer, peachtree borer, peach twig borer, and shothole borer.

Ambrosia Beetle and Shothole Borer

These are general terms that include species that range from the southeastern United States to the Pacific Northwest and into Canada. Other names include fruit-tree bark beetle, lesser shothole borer, European ambrosia beetle, and Asian ambrosia beetle. These pests were introduced from Europe and Asia and are serious pests of forest, fruit, nursery, and landscape trees. The adults are small, dark brown to black beetles with hard bodies. They bore into trunks and stems, creating multiple-chambered tunnels or galleries in the inner bark and outer sapwood.

The adults lay eggs in the galleries, and both adults and young inhabit the tunnels, feeding on ambrosia fungus that is cultured and tended by the adults. The larvae typically pupate in the tunnels and emerge as adults by boring a hole out through the bark. The many entry and exit holes produce the characteristic shothole effect. The tunnels damage the phloem and weaken the trees, and the ambrosia fungus can block the vessels that carry water and food. Fusarium and other pathogens can also be introduced into trees. There are no effective chemical controls for organic fruit growers. The best strategy appears to be keeping your trees healthy because these pests sometimes prefer to target weakened trees.

Pacific Flatheaded Borer

The larvae of this reddish-brown or coppery beetle bore beneath the bark and can girdle the trees. Apples, cherries, plums, and other stone and pome fruits are hosts of this borer. Young trees are at greatest risk of girdling, and the adults target sunburned or otherwise damaged areas of the trunks. There are no recommended organic chemical controls. Wrapping the tree trunks and shading them with boards may help prevent sunburn and make the trees less attractive to adults. Applying Surround should also help reduce sunburn.

Peachtree Borer and Lesser Peachtree Borer

Peachtree borer and lesser peachtree borer are common pests across North America, particularly east of the Rocky Mountains. The peachtree borer adult is a steel blue moth, and the females have a distinctive orange band around their abdomens. Besides peaches, these borers attack cherries, plums, prunes, nectarines, and apricots. They overwinter as mature larvae in tunnels near the soil surface. The larvae pupate on the soil surface and emerge as adults, typically from June to September. The adults lay eggs singly or in masses near the soil line, and the young tunnel into the trunk and large roots near the soil surface.

Young trees can be girdled or killed by a single larva, and older trees can be damaged. Mounds of sawdust-like frass and small holes, often with gummy exudates near the soil line, indicate that the pests are present. You can protect trunks by tightly wrapping them with plastic or metal that extends several inches into the soil and fits tightly against the trunk for several inches above the soil. Heavy-duty aluminum foil works well for this purpose. For small plantings, excavate about 3 inches into the soil around the trunks and kill individual larvae using a wire to probe into the feeding holes.

To reduce egg laying and feeding in the Pacific Northwest, horticulturists recommend applying Surround to the base of trees in early spring. Making weekly applications of *Bacillus thuringiensis* to the base of the trees from July through August can kill the larvae as they begin feeding on the trunks. The parasitic nematode *Steinernema carpocapsae* can help manage these borers when applied as a drench to the lower trunks and soil during warm weather in spring and fall. You can also use mating disruption twist ties (Isomate) to manage pest populations.

Peachtwig Borer

Peachtwig borer is an introduced moth pest that attacks twigs, shoot tips, and newly emerging shoots of peaches, plums, and apricots. The pest overwinters in limb crotches inside chimney-like structures called hibernacula, which are made up of sawdust-like frass. The larvae emerge early in spring and begin feeding on newly emerging shoots before pupating in protected areas on the trunk or inside fruits.

The adults emerge to lay eggs on new shoots, fruits, and young branches, and there are multiple generations each year. Later generations of larvae feed inside the fruits, usually entering near the stem. Applying dormant lime sulfur 30 days or more before bud swell and dormant horticultural oil sprays at bud swell through green tip are quite effective ways of managing this pest. Spinosad can be tank-mixed with the oil. Summer *Bacillus thuringiensis* sprays can help manage the larvae and mating disruption strips are available. According to the National Sustainable Agricultural Information Service, "The peach twig borer has many natural enemies and parasites, including the parasitic wasps *Paralitomastix varicornis*, *Macrocentrus ancylivorus*, *Euderus cushmani*, *Hyperteles lividus*, *Erynnia* species and *Bracon gelechiae*, as well as the grain mite *Pyemotes ventricosus*. The California gray ant, *Formica aerata*, can be beneficial when it preys on peach twig borer, but it unfortunately also protects aphids and scales. Other predators of the peach twig borer include lacewings, ladybugs, and minute pirate bugs. Beneficial, predatory insects can be attracted to the orchard by habitat plantings, cover crops, and hedgerows."

Beetles and Plant Bugs

Several beetle and plant bug species are pests of fruit tree crops in North America. The most serious is plum curculio, which attacks a wide range of stone and pome fruits and has severely limited organic tree fruit production in eastern North America.

Plum Curculio

Plum curculio is an extremely serious pest of pome and stone fruits in eastern North America, roughly east of the longitude line running through Manitoba and Texas. Adults are beetles with characteristic weevil snouts. They emerge from cocoons in the soil in early spring and fly to the trees, where they feed on buds, flowers, and developing fruits. They lay eggs in developing fruits, and the female cuts a crescent-shaped slit underneath the eggs, creating a flap that prevents the eggs and larvae from being crushed by the expanding fruit cells.

The larvae tunnel through the fruits, feeding near to but not on stone fruit seeds. The larvae will feed on pome fruit seeds, but they are crushed by the expanding cells inside apples that remain on the trees. In pome fruits, plum curculios only complete their life cycles in fruits that drop prematurely. When mature, the larvae leave the fruit and enter the soil, where they pupate 1 to 2 inches below the surface. The pests overwinter as adults in litter on the orchard floor or in other protected areas. Beside the adults that overwinter in your orchard, adult plum curculios are strong fliers and can enter your orchard from nearby orchards and hedgerows.

Damage from plum curculio includes early-season feeding by adults, oviposition injuries where eggs are laid, internal injury by larvae, and late-season feeding injury by adults. All make fruit unmarketable and, in many cases, unusable for home consumption.

It has historically been very difficult for organic fruit growers to control plum curculio. Pheromone traps and mating disrupters are not yet available. Routinely disking the entire orchard floor destroys many of the fragile pupal cases, but it severely damages the soil. Chickens and geese are effective in scratching up and eating the pupae, particularly when grain seed is mixed lightly into the soil on the orchard floor, but they are generally not allowed in certified organic fruit tree orchards later than 90 days before harvest. Using Surround to reduce plum curculio has proven quite effective and is probably the strongest weapon in your arsenal to manage this pest. You must apply it very thoroughly and begin applying just before bloom, and again at petal fall. You will need to repeat applications, particularly after rain.

The Integrated Pest Management group at Michigan State University has evaluated several organic approaches to control plum curculio. In addition to Surround, which they do not recommend for cherries, they suggest using Pyganic pyrethrum insecticide. Unfortunately, this material has an effective lifespan of only about 12 hours and requires frequent applications. It is also very toxic to predators and bees, so wait until petal fall to apply pyrethrums.

The MSU group has also suggested using trap crops, in which case the center of an orchard is thoroughly sprayed with Surround to protect the trees and a few border rows are left untreated but are heavily sprayed with pyrethrum. Reportedly, 'Liberty' apple is highly attractive to plum curculio and can be used as the trap crop. This push-pull approach is designed to push the adults out of the center of the orchard and pull them to the perimeter, where they can be more easily controlled and where there will be less damage to beneficial insects and mites in the main orchard. In a similar approach, called mass trapping, the orchard is sprayed with Surround. Areas outside the orchard are stocked with many plum curculio traps that collect the adults before they can reproduce.

The MSU group also reports that *Beauveria bassiana* fungi and *Steinernema* nematodes appear to be promising ways of controlling plum curculio between the spring and summer generations, when the pest is on the ground or in dropped fruit. Apply these organisms following irrigation or rain.

In home and smaller orchards, it is very important to remove dropped fruit throughout the season. In larger orchards, using hogs for controlled grazing reduced subsequent plum curculio generations three- to five-fold, according to MSU. Unfortunately, the National Organic Program requires removing livestock from the orchard at least 90 days before harvesting tree fruits. The grazing strategy might be effective with cherries, where the animals are placed in the orchard after harvest.

Japanese Beetle

Japanese beetle was introduced to the eastern United States in about 1908 and has been moving westward with outbreaks across the United States and Canada. The adults are bright, iridescent beetles with coppery wings and feed on the leaves and fruits of stone and pome fruits and on a large range of other plants. The white grubs feed on the roots of trees, shrubs, turfgrasses, and other plants. There is one generation per year, and damage from adults usually appears as irregular holes on or skeletonizing of the leaves. Fruits may also be attacked.

Although pheromone traps are available, studies have shown that they attract more beetles than they catch. If you use these traps to monitor Japanese beetle populations, place the traps well away from your orchard. Most Japanese beetle control programs target the grubs. Milky spore virus and special strains of *Bacillus thuringiensis* are available for this purpose. The parasitic nematodes *Steinernema glaseri* and *Heterorhabditis bacteriophora* have proven effective in controlling the grubs.

Two wasps that are native to North America parasitize Japanese beetles. *Tiphia vernalis* looks like a black, winged ant and is found throughout the northwestern United States and south to North Carolina. The female wasps enter the soil and lay eggs on the grubs. The wasp larvae then consume the beetle grubs. The adult wasps feed on honeydew excreted by aphids that live almost exclusively on maple, cherry, and elm trees and peonies. Tulip poplar nectar is another important food source for adult wasps.

Istocheta aldrichi adult wasps lay eggs on the throats of adult Japanese beetles. The wasp larvae bore into the body of the beetle, killing it before it can reproduce. Aphid honeydew deposited on Japanese knotweed (*Polygonum cuspida*) appears to be a food source for the adult wasps. Because Japanese knotweed is a serious invasive weed, consider including benign relatives, including rhubarb, baby's breath, carnations, and buckwheat, in your insectary crops and in-row cover crops.

Plant Bugs

These include a large number of pests such as mullein plant bug, lygus bug, western tarnished plant bug, consperse stinkbug, say stinkbug, and western boxelder bug. Although this is a large and varied group, they all damage fruit by feeding on the developing buds or young developing fruits, causing them to be deformed. Peaches may develop humpy shapes or catfacing. Other tree fruits can be deformed or have corky areas. The damage caused by plant bugs often leaves the fruits edible but unmarketable. Bugs collected with the harvested fruits can create disagreeable odors and render your fruit unusable.

Plant bugs are associated with a wide range of hosts, including blackberry, raspberry, mustard, and alfalfa. You can monitor pest populations by using an insect sweep net in your alley, in-row, and insectary cover crops. Also monitor inside your fruit trees using a beater tray.

Although no threshold numbers have been established, if you find the pests, you might want to take action, particularly early in the season between bud break and June drop. During this period, avoid harvesting or mowing alfalfa, rape, canola, or mustard crops in or around your orchard. Mowing these crops can drive the adult bugs into your trees to feed on the blossoms and fruits. During the growing season, mow alternate alleys and in-row crops at least 1 to 2 weeks apart to leave habitat on the orchard floor for the adults and reduce the numbers in your trees.

Remove wild and escaped brambles from your windbreaks and fencerows. You can make prebloom and postbloom applications of Surround to help prevent feeding damage and use azadirachtin sprays to kill the pests in the trees. Monitor pest populations in your trees during the growing season using beater trays. If the pests are present, you may want to apply control sprays 1 and 2 weeks before harvest to reduce the contamination of harvested fruits.

Campyloma Bug

The campyloma bug is rather a paradox. The young larvae resemble fast-moving, translucent, greenish-white aphids. They cause damage by feeding on developing apples less than about ½ inch in diameter. Yellow and green apples appear more susceptible to damage than red varieties. Damage appears as small corky spots on the young apples, and severe infestations can cause catfacing on the fruits.

Adults are reddish-brown, oval plant bugs with dark spots on their legs and antennae. Unlike the nymphs, the adults are beneficial insects, feeding on orchard pests. Monitor populations using a 20-inch-diameter round or 18 × 18-inch square white tray and padded stick to collect the juveniles. Hold the tray under the blooms, and strike the branches with the stick. Sample one branch per tree from at least 10 trees in each 5-acre block. If there is less than 0.1 larva per tree for yellow or green apples, and less than one insect per tree for red varieties, no treatment is needed.

If you must spray, treating at prebloom and bloom times is more effective than treating postbloom. Unfortunately, organic insecticides, other than rotenone, are not generally listed for control of this pest. Because of potential human health problems, rotenone is typically no longer available to organic fruit growers. Neem products, insecticidal soaps, and spinosad products are about the only materials available for this pest. Apply insecticides from prebloom until the first blossoms begin opening, and again at petal fall.

Scales

Scales are sucking insects that include San Jose scale, oystershell scale, and fruit lecanium scale. Unlike aphids, which are soft-bodied and easy to kill, adult scales cover themselves with hard shells that fit tightly against the bark, leaves, and fruits. Scales are common and serious pests of tree fruits across Canada and the United States. The adults' and nymphs' feeding weaken and can kill trees, and the scales on fruit make the fruit unmarketable.

These pests overwinter as eggs and larvae in fruit trees and many other woody hosts. It's important to scout carefully for them. During the dormant season, look for the flattened shells on twigs and limbs. In early spring, about the time codling moth emerges, use pheromone traps to monitor the flights of male San Jose scales. From spring through fall, attach black sticky tape, with the sticky side out, around twigs and small branches to detect crawlers. It is at the crawler stage that the scales are most susceptible to predators and pesticides, but with up to five generations per year, you can almost always find the crawlers during the growing season.

A very effective way of controlling scales is to apply a dormant lime sulfur spray about 30 days before bud swell, followed by a dormant horticultural oil spray during bud swell through green tip. Using summer-weight horticultural oils during May and June can also be effective, but they may cause phytotoxicity on your fruit crops. Azadirachtin can control crawlers and mobile adults during the growing season.

Thrips

These pests include pear thrip and western flower thrip. Despite their names, thrips attack a wide range of fruit hosts across North America. Thrips are tiny, about 0.05 inch long, and range from nearly black to brown to straw-colored. The adults and nymphs feed primarily on flowers and can damage and deform them. They also feed on developing fruits, causing cosmetic stippling and small scars. The late-season damage they do to cherries leaves silvery rings where the cherries touch each other.

Thrips live on flowering plants in and around the orchard. Even in a clean, well-managed orchard, these pests can fly in from surrounding areas. One effective strategy is to focus pesticide treatments on the perimeter tree fruit rows or trap crop rows. Avoid mowing or cultivating the orchard during bloom, so as not to drive more thrips into your trees. Apply azadirachtin and spinosad products beginning at petal fall. Later in the season, lacewings and pirate bugs should provide adequate control in most orchards. It may be helpful to make mid- and late-season applications of azadirachtin or spinosad products to late-maturing varieties of cherries such as 'Lapins' and 'Sweetheart'.

Earwigs are minor, but common, pests in orchards and are easily recognized by the prominent pincer-like structures at the end of their abdomens. They are opportunistic, usually night, feeders, and they prey on soft-bodied insects as well as decaying vegetation. They also feed on ripening and overripe fruit, leaving shallow, irregular holes. Earwigs emit a foul-smelling liquid when alarmed and can contaminate fruit during harvest.

Control earwigs by keeping vegetation and debris away from the base of the trees and generally following good sanitation practices. Harvest ripe fruit promptly, and remove overripe and cull fruit from the orchard. Applying a sticky wrap (Tanglefoot) around the trunks can prevent adults and juveniles from climbing into the trees at night. When earwigs become a problem, spinosad products are recommended for organic control. Because they help manage aphids, psylla, and leafcurling midge early in the season, delay controlling earwigs, if needed, until a few weeks before harvest.

PRUNING AND TRAINING

- Types of Pruning Cuts and Plant Responses to Them | 411
- Where Should You Make Pruning Cuts? | 413
- Pruning Tools | 418
- Training Systems | 421
- Training Trees, Crop by Crop | 425

To maintain healthy and productive orchards, it is critically important to prune and train fruit trees and bushes. Although people have pruned and trained plants for thousands of years, there are many approaches, and it is easy to obtain conflicting advice. The reality is that many of the different approaches work well, as long as they are based on a few simple concepts.

Pruning involves removing parts of a plant. Regardless of when and how it is done, it has a dwarfing effect; pruned trees are always smaller than equivalent trees that have not been pruned. We prune our orchards to:

- Develop tree and bush shapes and establish strong structures for freestanding trees
- Control plant size
- Remove diseased or damaged wood
- Admit light to maintain growth and fruitfulness
- Provide good air movement through the canopy
- Control fruit load
- Invigorate older plants
- Remove old or poorly located branches

Dormant pruning is carried out in late fall, winter, or early spring, when all plant growth has stopped. This makes up most of the pruning in orchards. Dormant pruning invigorates and stimulates new growth at and slightly below the cuts. When no leaves are present, it is easy to see where cuts must be made and to make them. It is easier to prune at this time when there are few other orchard tasks. On the downside, dormant pruning in larger orchards sometimes means working in cold, wet conditions. Dormant pruning can stimulate too much growth, and pruning in fall or winter can create open wounds that are exposed to pathogens for extended periods of time.

Summer pruning can be carried out at any time during the growing season and includes everything from rubbing off new shoots with your fingers to cutting deeply into 2-year-old or older wood. The plant's response depends on the timing and the amount and age of the wood that is removed. Summer pruning usually has more of a dwarfing effect than dormant pruning. Pruning before the terminal bud sets or before active growth ceases can stimulate new growth, while summer pruning after active growth stops causes little growth. Cutting into 2-year-old or older wood produces less growth than cutting into current-season or 1-year-old wood.

During the late 1900s, when large fruit trees were still common in orchards, summer pruning was sometimes used to open up apple tree canopies and to allow more light to reach the fruit and color them up. During July and August, workers would cut back into 2-year-old wood. The work was difficult and extremely labor intensive. With today's smaller trees and open training systems, this practice has largely fallen out of favor. Summer pruning remains important, however, especially for removing unwanted shoots before they are more than a few inches long.

Precautions to Take When Pruning

- Don't leave stubs that will rot later. The exception is for sweet cherries and plums, discussed later in this chapter.
- Make cuts cleanly using sharp, properly adjusted tools. If you are struggling to make a cut, either the tool you have is too small or you have the wrong tool. Use the proper tool, not brute force!
- Don't allow the bark to tear.
- Avoid pruning in the late fall or early winter whenever possible.
- When cutting out diseased wood, disinfect your tools between cuts.
- Make cuts at a slant so that water cannot accumulate on the cut surface.
- When heading, cut back to a bud that is pointed in the direction that you want the new growth to go.
- Make a pruning cut only when you have a specific reason to do so. Think before you cut.

The advantages of summer pruning are that you generally have good working conditions and, when it is properly done, the trees regrow relatively little. The disadvantages are that there are many other tasks that must be done during the growing season, and the leaves can make it difficult to see the overall tree or shrub shape.

Types of Pruning Cuts and Plant Responses to Them

DESPITE THE MANY DIFFERENT ORCHARD CROPS and training systems, there are only two types of pruning cuts: heading and thinning.

Heading. This involves cutting a stem between branching points and is often done on 1-year-old wood just above a strong bud. Heading shortens and stiffens the limbs. It eliminates apical dominance for a time and usually causes a strong growth response at buds several inches below the cut. Apical dominance is caused by hormone-like chemicals produced in the tip of a trunk leader or shoot and results in straight, unbranched growth. The more wood that is cut off, the greater is the growth response. You normally see less regrowth when horizontal limbs are cut than when upright branches or shoots are headed. Stiff branches are desirable for many training systems, especially for freestanding trees. If the branches become too stiff, however, they can become more susceptible to breaking due to heavy crop and snow loads.

Hedging is a type of heading cut that is sometimes used for fruit trees, grapes, and berries. Beginning in the late 1900s, hedging became a popular means of reducing labor costs in commercial orchards and vineyards. Tractor-mounted, sickle bar–type implements were used to cut the sides and tops of trees and vines to remove unwanted vegetative growth and maintain desired shapes. Hedging cuts are indiscriminate, however, and typically produce a thin layer of densely branched, twiggy growth immediately below the cuts, leaving a barren, woody interior. This dense outer layer can reduce the amount of light that penetrates to the center of an individual plant or canopy, and orchard hedging should be followed up with more detailed hand pruning. Hedging can be used with newer, high-density training systems, but it requires a substantial amount of skill and careful follow-up corrective pruning.

Thinning. In this type of pruning, cuts are made at the junctions of branches. These cuts result in less growth than with heading, and in a stronger, more natural structure. Thinning is important for selecting scaffold limbs and keeping a plant open, which allows light to penetrate throughout the canopy and ensures healthy and fruitful growth on the lower and internal parts of a plant.

In thinning, you should generally avoid making "bench" cuts (figure 12.1). Bench cuts happen when an upright branch is removed and an outer secondary branch is left. These cuts usually occur in training trees to open centers and can occur on any fruit tree. They are especially common on peaches and nectarines. The problem is that the benches are structurally weak and can leave the tree open to breaking due to crop load, snow load, and wind. A better strategy is to carefully select scaffold limbs and use spreaders and weights throughout the spring and summer to develop desirable shapes, rather than simply relying on dormant pruning.

Figure 12.1 · **Types of Cuts**

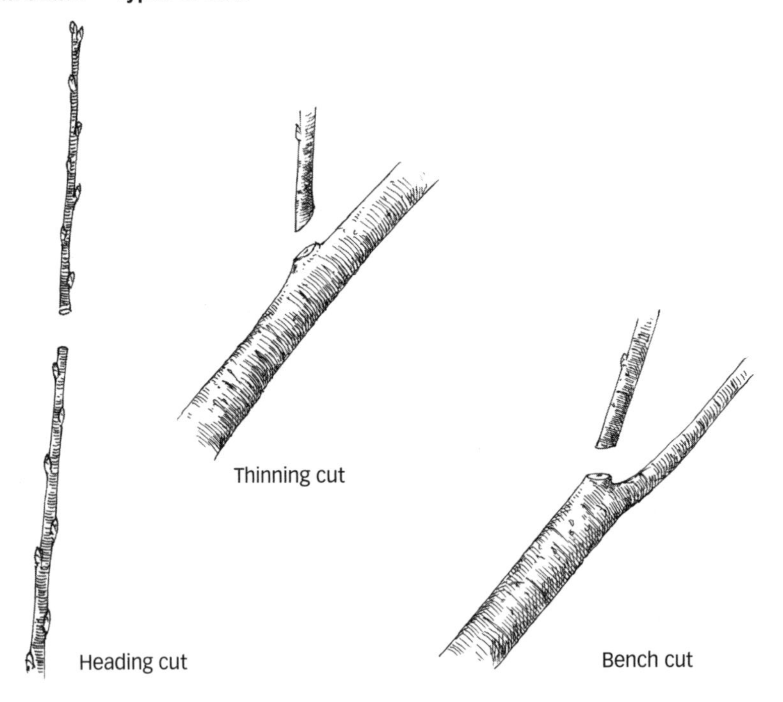

Where Should You Make Pruning Cuts?

DISEASED AND DAMAGED WOOD CAN, and should, be removed whenever you find it, regardless of the time of year. Where diseased wood is present, wait for dry weather to make the cuts and disinfect your pruner and saw blades between cuts with a 70 percent alcohol solution or a solution containing 9 parts water and 1 part household bleach. Thinning cuts are usually best for these situations.

When removing branches at the trunk, cut outside of the bark swelling or collar at the base of the branch (figure 12.2). Flush cuts callus over more slowly than those where the collar is left, and they leave the tree open to infection by wood rot organisms and cankers. Do not apply tree wound dressing to cuts. Such materials can actually slow the development of new bark across wounds. When you are removing large branches, use the three-cut method illustrated in figure 12.3 to prevent tearing the bark below the cut.

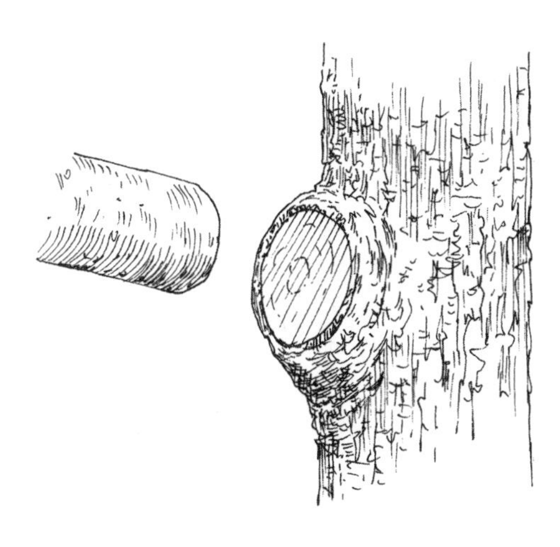

Figure 12.2 **· Preferred Method of Cutting Off a Branch at the Trunk**
Cut to the outside of the swelling at the base of the branch. Cutting flush with the trunk increases the risk of infection and tree death. Avoid leaving a stub, with the exceptions of certain pruning cuts on sweet cherry and plum and when heading back water sprouts to leave some shade on exposed limbs. Exceptions are described in the text.

Figure 12.3 · **How to Remove a Large Limb to Prevent Bark Tearing**

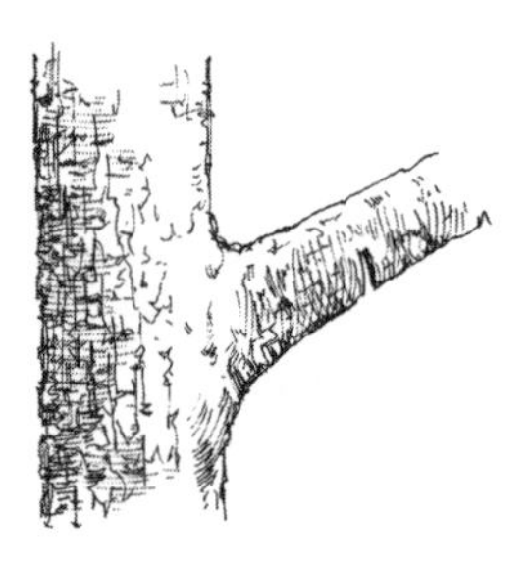

1) Make your first cut on the underside of the branch about 6 to 12 inches away from the trunk.

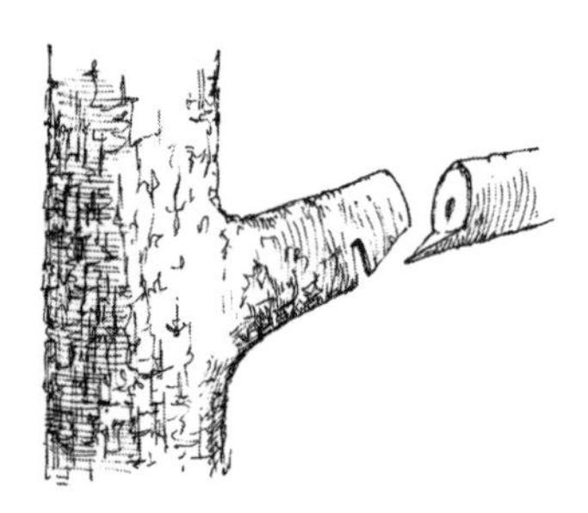

2) Cut the branch off outside of the first cut.

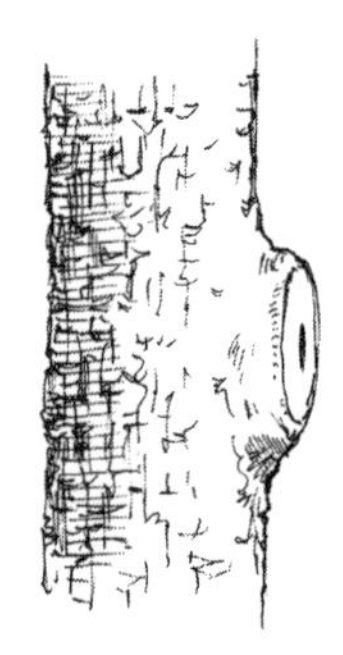

3) Cut off the branch stub just outside of the swelling at the base of the branch.

Figure 12.4 · **Inclusion**

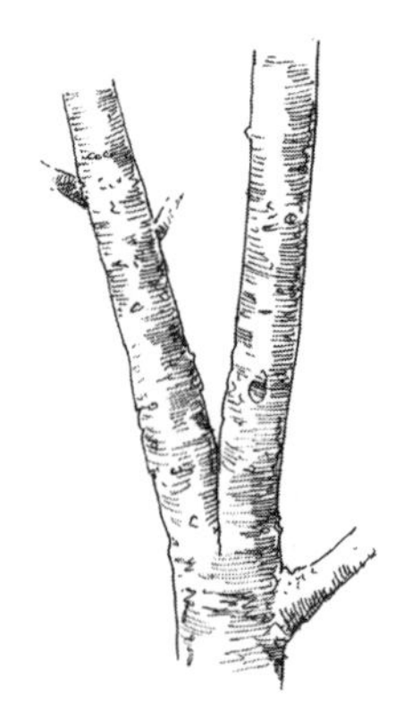

A) A bark inclusion resulting from a narrow branch angle

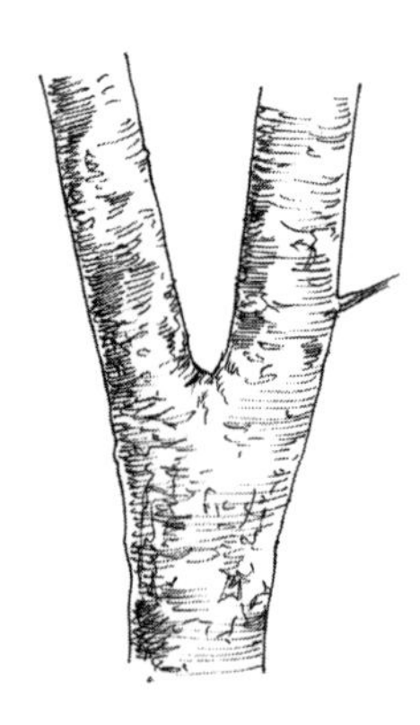

B) A ridge of bark between the branch and trunk shows no inclusion is present.

When two branches or a branch and trunk develop with a very narrow angle between them, problems result. Bark will continue to form between the two pieces and can form a bark inclusion, as shown in figure 12.4. Water and microorganisms become trapped in the inclusion and contribute to the development of wood rot. Insects also find narrow crotch angles to be a good entry site, and trapped water can freeze, causing branches to split apart. If you can see a ridge of bark between the branch and trunk, there is no bark inclusion, but you may still want to spread the branch to increase the branch angle.

With narrow branches, take action as early as possible, preferably when the branches are no more than green shoots a few inches long. Remove poorly located branches using thinning cuts. Use weights, ties, and spreaders to develop strong, wide branching angles. Figure 12.5 illustrates these techniques.

Water sprouts (vigorous, vertical shoots) should be removed, except where they are needed to replace leaders. Water sprouts are unfruitful, create shading problems, and interfere with the development of strong scaffolds. Completely cut off sprouts on interior parts of a tree. In apples and pears, some sprouts can be headed back to be several inches long for 2 or 3 years in a row to develop fruitful spurs.

Water sprouts often develop at the site of a pruning wound. Bending limbs down, such as during trellising, will also cause more water sprouts to form. Horizontal and down-pointing limbs produce more water sprouts than limbs that angle upwards. It is extremely important to manage water sprouts in all orchards.

When sprouts arise from exposed limbs in the top of a tree, removing them entirely can cause the main limbs to become sunburned. For such exposed limbs, remove about half of the sprouts entirely and head the others back to be 1 foot or less. The headed sprouts will need to be headed several years in a row, but the shade that they produce will help prevent sunburn (figure 12.6).

It is best to do summer pruning or pinching when the unwanted growth is just a few inches long. When done correctly, this may involve pinching off new shoots with your fingers or rubbing them off with your thumb (leather gloves are an asset here). Except when removing diseased or damaged older branches, you should not need more than a pair of hand pruners.

Figure 12.5 · **Methods of Positioning Branches**

Several methods can be used to position branches to prevent narrow branch angles and bark inclusions or for positioning branches to desired angles for various training systems.

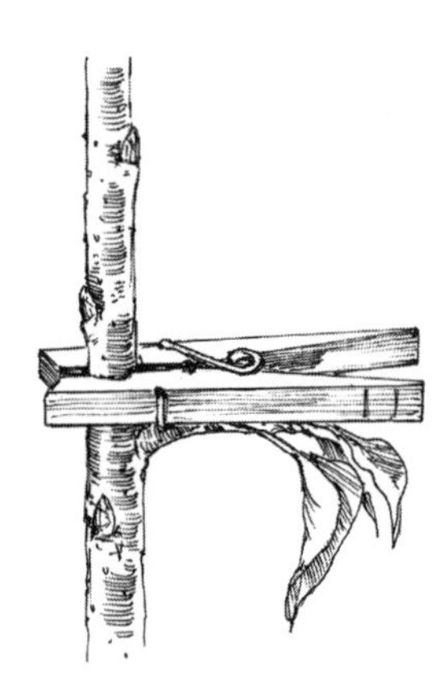

A) A spring-type clothespin is clamped around the trunk, and the tail of the clothespin holds the new shoot horizontally.

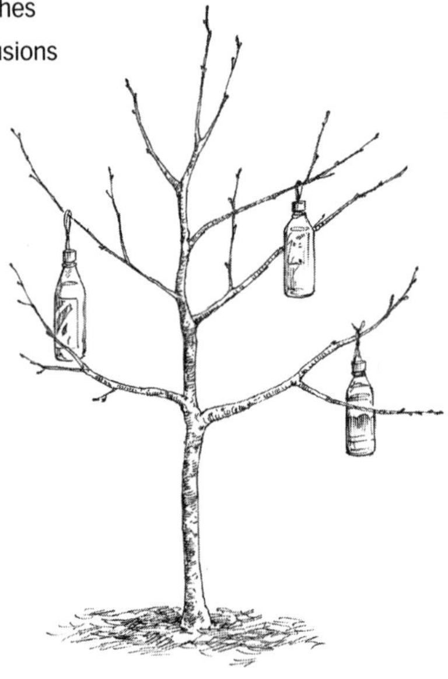

B) A weight tied to a branch. Plastic water bottles and milk jugs work well and allow you to adjust the weight by adding or removing water.

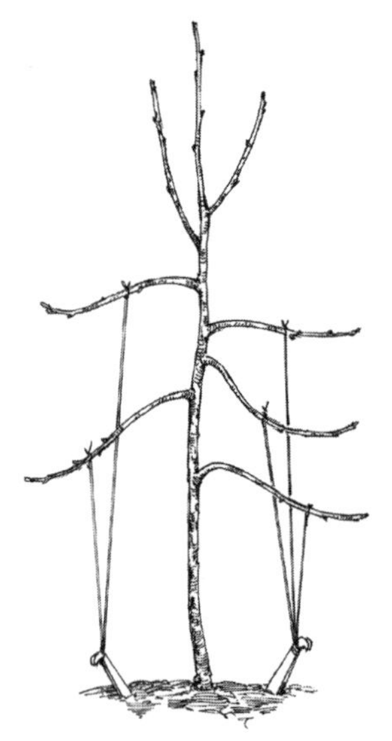

C) Tie-downs attached to stakes in the ground

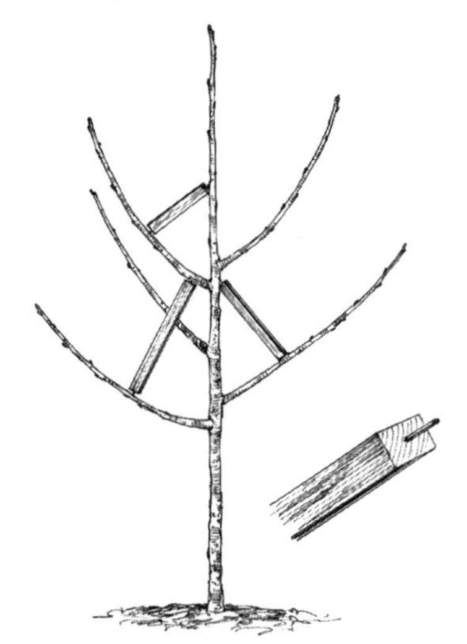

D) Wooden spreader bars with finishing nails inserted in the ends

Figure 12.6 · **Branches with and without Water Sprouts**

The tops of shaded branches can sunburn if all water sprouts are removed (left). Leaving some of the sprouts but heading them back repeatedly over several years (right) can provide partial shading to protect the main branch.

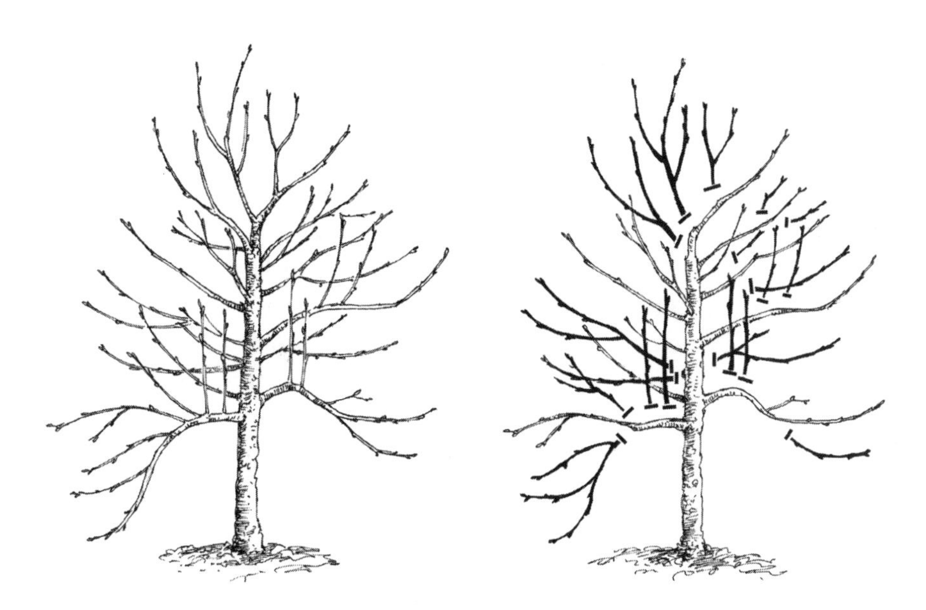

Figure 12.7 · **Fruit Tree before (left) and after (right) Dormant Pruning**

Note the removal of the large scaffold limb on the lower left. The shoot retained just above the removed limb will provide renewal fruiting wood.

How to Prune

1. Remove all diseased and dead wood.
2. Remove branches that rub against or cross each other.
3. Remove suckers (shoots growing from the ground or bottom of the trunk).
4. Remove water sprouts, but do not denude exposed limbs on older trees.
5. Remove multiple trunks unless they are part of the training system.
6. Now look at the tree. Decide if there are branches that need to be removed to:
   - Maintain a central or modified central leader
   - Form a vase or open center shape
   - Develop a particular shape
   - Correct narrow branch angles
   - Remove inwardly growing branches
   - Remove weak, dangling limbs

Pruning Tools

WHETHER YOU HAVE ONE TREE OR 10,000, pruning is an essential task. The right tools make the job go faster and produce the best results. Avoid the temptation to buy cheap pruners and saws; they often produce inferior results and make pruning more difficult. Tools that are well designed and constructed will last for many years.

Hand shears. Anvil-type pruners are suited for stems that are up to about ½ inch in diameter, but they tend to crush stems if they are not very sharp, and they require more force than do bypass or scissor types.

Bypass or scissor-type pruners are less likely to crush stems and usually make cleaner cuts than do anvil shears. These pruners allow you to make clean cuts very close to the trunk, speeding the healing process. They may or may not have a ratchet device to provide a mechanical advantage during especially thick or difficult cuts (up to ¾ inch in diameter), but they generally require less effort than anvil shears to make the same cut. Better-quality pruners often have replaceable blades and other parts. Given equal-quality construction, bypass designs are generally better for orchard work than anvil styles.

Loppers. These are long-handled pruners used for large cuts (up to 1½ inches in diameter). They come with either anvil or bypass cutting edges, and the blades are sometimes replaceable. Handles can be hollow metal, wood, or fiberglass. Wood and fiberglass handles are generally quite durable, and high-quality aluminum- and steel-handled pruners can be very durable. The handles on cheap pruners, however, often bend easily. Loppers come in lengths of about 12 to 48 inches, and short loppers can often be used in place of hand pruners. Their longer handles and two-handed grip reduce hand fatigue and allow you to make cuts with less effort.

Pruning saws. These saws have narrow blades with coarse teeth that are designed to cut on the "pull" stroke. Both straight and curved blades are available. Saws may be fixed into one piece or can fold, which makes them very convenient to carry when you are pruning many trees. Wooden-handled folding saws are generally more durable than those with plastic handles. Saws may also be fixed onto long poles for taller trees, often with a hook and pruner head that is operated by pulling a rope. It can be difficult to make clean, accurate cuts with pole saws and rope-operated pruners, so take great care.

Chain saws. Chain saws are used to remove trees entirely and to make large cuts. They are often used to head back large limbs in preparation for cleft or bark grafting in a process called topworking. Use topworking to replace the scions without replanting. While chain saws are used in orchards, particularly commercial orchards, they increase the risk of personal injury and damage to trees. Unless you are rehabilitating older trees, topworking, or removing trees entirely, chain saws are most often used only in larger commercial orchards. Chain saws may be gasoline, electric, pneumatic, or hydraulic.

Pneumatic, hydraulic, and electric pruning tools. These pruners use air pressure, hydraulic liquid pressure, or electrical solenoids to operate. They have two great advantages in that they require less hand pressure to make the cuts and they provide greater reach. Pruners range from about 8 inches to 12 feet in length and often consist of a cutting head mounted on interchangeable or adjustable extension rods. Chain saws, likewise, have handheld units that can be mounted on extension rods up to nearly 12 feet long. These powered tools are extremely valuable for commercial orchards because they increase speed and efficiency while reducing fatigue. Most cuts can be made from the ground, reducing the need for hauling and setting up ladders. On the downside, they are much more expensive than handheld pruning tools and require dragging hoses or cables about the orchard. They also require separate air compressors, hydraulic motors, or electrical generators.

Figure 12.8 · **Typical Orchard Pruning Tools**

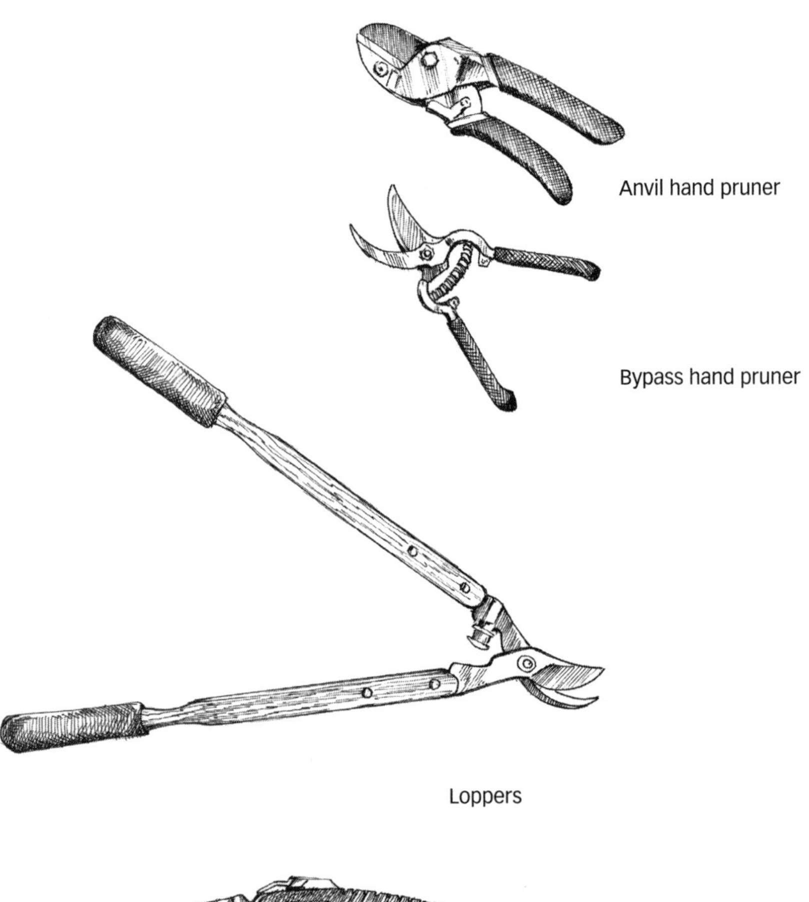

Training Systems

TRAINING IS USED TO SHAPE A TREE to a desired architecture, and many variations on training are available to fruit growers. The basic concepts, however, are quite simple. For freestanding trees, we use training to create strong structures that produce and bear large fruit crops well while maintaining good light penetration and air movement through the canopy. Some training systems support the trees on wires or poles, reducing the need for a strong structure and emphasizing fruit production. Other training systems are used for decorative effects in landscapes.

Freestanding Trees

For thousands of years, domestic fruits have been grown as freestanding trees. Freestanding trees are the least expensive to establish because there are fewer trees per acre than in high-density systems and no supports are needed. Depending on the crop and rootstocks, trees are allowed to grow to about 7 to 20 feet tall and are generally trained to a central leader, modified central leader, or vase shape. For most training systems, early training involves developing strong trunks and branches, rather than fruiting. Some of the newer freestanding systems place greater emphasis on early yields and frequent branch replacement. Apricot, bush cherry and plum, European and Japanese plum, tart cherry, loquat, medlar, quince, and saskatoon are most often grown commercially as freestanding trees or bushes.

Growing freestanding apple, pear, peach, nectarine, and sweet cherry trees is less technically challenging and requires less-intensive management practices than some of the modern higher-density systems. The trade-offs are that your early yields are lower, it can be more difficult to control pests and diseases, and the labor costs for pruning and harvesting are typically greater than for smaller trees on supports. Freestanding trees take longer to come into bearing and generate less early and overall profit per acre than higher-density systems.

For home and market fruit growers who do not need the highest yields nor want to pay the high establishment costs or perform the intensive management practices that go with high-density plantings, freestanding trees remain a good option.

Whether to use freestanding trees for a commercial orchard depends on your goals and philosophies. If economic returns over a 10- to 15-year period are your driving concern, you probably should not go with freestanding apple, pear, and sweet cherry trees. If you are planning to keep your orchard longer than 15 years, freestanding trees can provide sustainable yields and eliminate the need to install and maintain support systems. Economically, they are never likely to produce the profits available from early-producing orchards using the latest, high-value varieties. On the other hand, they will not require the same level of intensive management.

Trees on Supports

During the 1980s, trellis systems became popular, and many different designs emerged. Some were tremendously complex and expensive or difficult to work with, and they did not remain popular for long. Since the 1990s, several training and support systems have emerged that meet growers' needs more cost-effectively. The primary systems in commercial apple production today use high-density plantings, tree supports, and intensive management practices to produce marketable fruit within 2 or 3 years of planting. Pears can also be grown on temporary or permanent supports in high-density plantings. Quince and medlar can be adapted to these training systems, although they are more commonly grown as freestanding trees or large shrubs. High-density support systems are sometimes used for commercial peaches, nectarines, and sweet cherries, and they can be adapted to be used for apricots.

Tree support systems include vertical and split canopy designs that support the trees on poles and/or horizontal wires. Trees usually range from 6 to 14 feet tall, depending on the crop and system.

Support systems have the primary advantages of producing marketable yields quickly while reducing labor costs and better managing pests and diseases, compared with low-density, freestanding trees. High-density apple trees that use precocious rootstocks, for example, come into bearing as early as their second growing season (second leaf). Some designs create narrow, wall-like crop rows as little as 2 feet wide, although rows that are 3 to 4 feet wide are more common. Fruits on most support systems receive abundant light and usually color up well. Trees on supports can be spaced closer together than those in freestanding sys-

Vertical or Split Canopies?

Fruit trees can be trained to either vertical shapes or split canopy V or Y shapes. Split canopies in low and moderate tree densities increase yields and intercept more light. They are most often used for apricot, peach, nectarine, and pear and are sometimes used for sweet cherries. From a practical standpoint, split canopies can be more expensive and difficult to establish than vertical systems, and they can make it more difficult to scout for pests and diseases and apply spray materials. Older split canopy systems made getting tractors and other equipment through an orchard problematic and were generally miserable to work around. Some of the newer designs are more practical.

tems, which increases yields. You can also use relatively small tractors and other equipment, compared to those used with standard-sized, freestanding trees.

On the downside, support systems make it more costly to establish and maintain an orchard. Systems that use horizontal wires make it difficult for workers to move between rows. Depending on the system, supported trees can also require greater skill and time to manage than non-supported trees.

Trellis Systems

Trellis systems have traditionally used two to five horizontal wires to support trees that are up to about 6 feet tall. Trellises are used almost exclusively for apples and pears, and many variations exist, with trees trained vertically, diagonally, or as palmettes. The idea is to create a narrow wall or hedge of fruit, with the branches tied to and supported by the trellis wires. Trellised trees are easy to prune and harvest, and they make it easy to control pests and diseases. For an organic home landscape or orchard, trellising apple and pear trees can work very well. Taller trellises may also be suitable for new commercial orchards, and they are particularly valuable where space is limited. Trellises lend themselves well to U-pick orchards.

Because traditional trellises produce lower yields than spindle-trained trees, they have largely fallen out of favor for new commercial orchards, even though they are generally less expensive to install than the taller spindle systems. The lower yields are due to the fact that the trees are kept short and the crop rows are narrow. Researchers at Pennsylvania State University determined that, in terms of canopy height and density, trellis systems are about as productive as the taller vertical axis designs for the first 10 years after planting. They are now testing tall trellises (about 10 feet in height) for commercial orchards.

It's important to select a rootstock that will create enough vigor to fill the canopy but not so much as to require excessive pruning. Trellised trees are usually spaced about 3 to 6 feet apart in rows that are 10 to 12 feet apart, and they require 660 to 1,452 trees per acre. The more closely spaced the trees, the more expensive it is to establish the system and the more intensive the management becomes.

Spindle Support Systems

These became very popular for high-density apple orchards during the 1980s and 1990s. The slender spindle was one of the first designs and is simply a post supporting a tree, usually to a height of about 6 feet. Newer designs can be up to 10 feet tall. No horizontal wires are used and the trees are typically trained to be the shape of a Christmas tree or pyramid. Vigorous, upright leaders are cut off, and weak shoots are tied up in their places to form slow-growing leaders as part of an orchard-size control program.

Due to their relatively high cost and low yields, slender spindle systems have become less popular than newer training systems. The newer super spindle system is a viable design for commercial apple orchards. While it resembles the earlier slender spindle, the trees are supported on wires without the use of posts, and the trees are planted about 2 feet apart to quickly form a dense canopy. They are costly to establish due to the large numbers of trees needed: up to 2,178 per acre. Space rows about 10 feet apart.

Vertical Axis Designs

There are many vertical axis designs, and they are often used in new commercial apple orchards and are adaptable to pears and sweet cherries. Other names you may hear include the "axe," "V-axe," "French axe," and "slender axis." All are similar and differ mainly in pruning techniques and how the trees are managed. The V-axe uses more trees per acre and splits the canopy into a V shape.

Axis systems are typically 10 to 14 feet tall. Some designs use a metal electrical conduit, bamboo pole, or wooden stake placed next to each newly planted tree for support. Wires run the length of the tree row, supporting the tops of the poles and keeping them in line, either horizontally or at an angle for V systems. Additional horizontal support wires can be used, depending on the design. In some systems, no poles are used and the trees are supported entirely by the wires. The trees are tied loosely to the poles or wires so that as they grow they form straight trunks. Depending on the crop and design you choose, trees are spaced about 3 to 7 feet apart in rows that are 10 to 16 feet apart. About 389 to 1,452 trees are planted per acre. In cooler climates where trees are less vigorous, it is best to space trees closer together within rows.

Training Trees, Crop by Crop

Apples are remarkably adaptable trees that can be trained to large pots for patios or as freestanding giants. For organic fruit growers, freestanding or supported trees 6 to 12 feet tall work well.

Training Apple Trees

If you have an established orchard of freestanding trees, do not rush to remove them and replant. What you have may already be suitable, and replanting costs can be high. Begin using the pruning and training practices described in this chapter to improve fruit quality and pest and disease management. In particular, open up the trees to ensure good light penetration and air movement.

Freestanding, central leader design. To train an apple tree to a freestanding central leader design, you may start with either an unbranched whip or feathered tree (figure 7.1). Provided that the branches on the feathered trees are in suitable locations, you can save a year in training and come into production sooner. Poorly located branches, however, are no bargain.

Traditionally, whips were headed back to several inches above where you wanted the first scaffold branches to start. By heading back the leader, you eliminate apical dominance and stimulate the first three or four buds below the cut to grow. These new shoots form the scaffolds, and one shoot is trained up to form the modified central leader.

If you can avoid heading back the leader, your trees will develop faster. If you have a well-feathered tree, simply prune off the branches that you do not want. Central leader trees are traditionally started with a cluster of three or four branches about 3 feet above the ground. Space the branches in the whorl 3 to 6 inches apart vertically so that no two or more branches arise at the same height on the trunk, and evenly distribute the scaffold branches around the tree. Leave the trunk bare for about 18 inches above the topmost branch of the first whorl, and repeat the process to form a second whorl. Some central leader or modified central leader trees have three or four whorls.

An alternative to heading back the leader to stimulate branching is to remove some of the buds from the whips. Rub off all of the buds up to about 3 feet above the ground and rub off the first six buds directly below the terminal bud. Leave roughly every second or third bud for now, choosing those that point in the direction you want the branches to grow. This approach produces branches that are distributed evenly along the trunk, rather than in distinct whorls. Figure 12.9 shows the stages involved in developing a central leader or modified central leader tree. The main difference between the two approaches is that in a central leader tree, the leader is not headed to stimulate branching.

The final tree shape should be conic-shaped or pyramidal, with a relatively wide base and narrow top to allow light to penetrate to the lowest branches and the interior of the trees. In some orchards, the trees are allowed to become rounded with a wide top. The shading from the upper limbs, however, reduces the fruit quality and production on lower branches. Distribute your branches around the trunk so that one branch does not lie directly above another. Leave enough scaffold branches to fill in the canopy, but not so many as to create excessive shading.

Freestanding, open center design. To develop a freestanding, open center tree, you also begin with either whips or feathered trees. Develop a whorl of branches as you would for modified central leader trees, described above. The difference between mature open center and central leader trees is that the open center trees have only a single whorl of scaffold

Figure 12.9 · **How to Train a Tree to a Modified Central Leader**

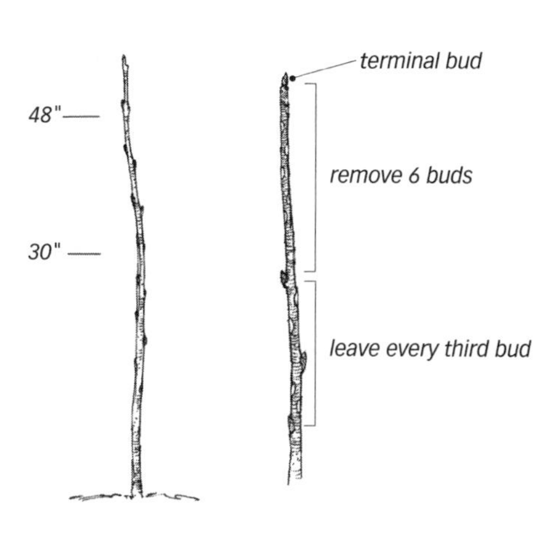

1) If starting with a whip, rub off the first six buds below the terminal and leave about every second or third bud after that between 30 and 48 inches above the ground. Keep three or four branches well distributed around and along the trunk on this section of the trunk.

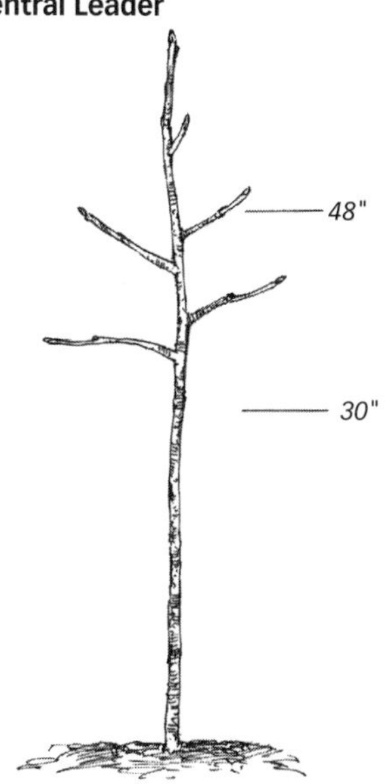

2) With a feathered tree, select three or four branches 30 to 48 inches above the ground.

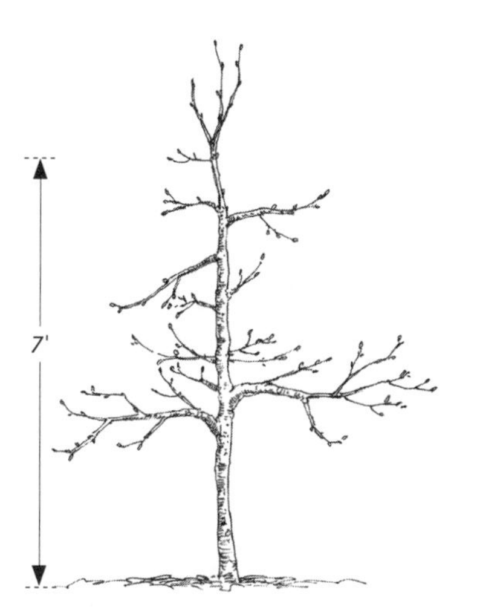

3) Try to avoid heading the leader. Remove buds and select branches as for steps 1 and 2 to establish three or four more branches to about 7 feet above the ground.

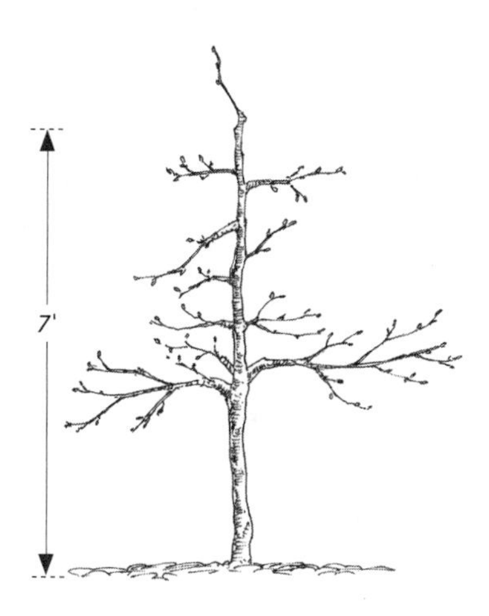

4) Thin out the leader to a weak lateral.

Figure 12.10 · **How to Train a Tree to an Open Center**

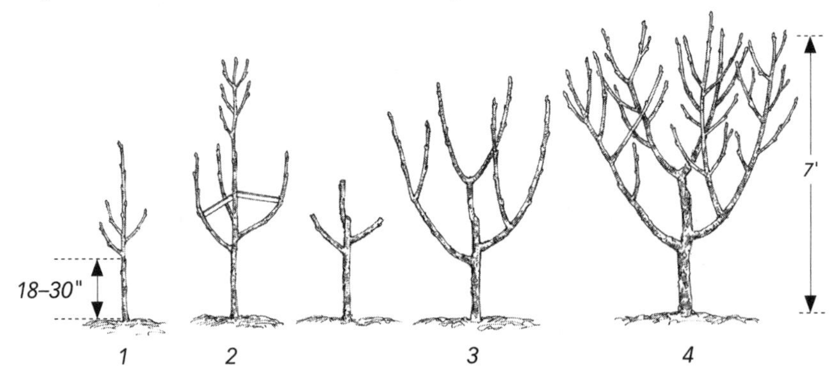

1) Start with a whip or feathered tree, as described for figure 12.9. Select three to four branches that are well distributed around the trunk and spaced 3 to 6 inches apart vertically, starting 18 to 30 inches above the ground. Keep the central leader for now, but remove all the branches above those you have selected. Spread the branches with clothespins, weights, spreader bars, or tie-downs.

2) Allow the main scaffolds to develop, keeping them spread to a uniform bowl or vase shape.

3) Remove the central leader, and head the main scaffold branches about 18 inches above their bases.

4) Prune to keep an open canopy. Head and thin as necessary to maintain the desired tree height.

branches set atop a short trunk. How high you should develop your scaffold depends on your preferences and your orchard equipment. Scaffolds are typically developed at a height of 18 to 36 inches above the ground.

Open center trees usually have three to five scaffold branches. It's easier to scout for pests and diseases and to spray trees with three or four main branches. Distribute the branches evenly around the trunk and space them vertically 3 to 6 inches apart on the leader. For now, leave the central leader in place.

As the scaffold branches grow, use spring-type clothespins, weights, tie-downs, and spreader bars to position them so that they form a uniform vase shape with wide branch angles at the trunk. When the main scaffold branches are 12 to 18 inches long, head them back or select vigorous side branches to form V-shaped pairs of branches. Remove the central leader. Depending on the desired height of the tree, each of these secondary scaffold branches may also be headed back or thinned to form more V shapes. If you can avoid heading the leaders, your tree will come into production sooner. Figure 12.10 shows the stages in training an open center tree.

Supported Apple Trees

Apples and pears have long been trained on posts and wires. One such supported design is the slender pyramid, which was mentioned in chapter 3. Although trees in this design produce greater yields than freestanding central leader and open center apple trees, they are slow to come into production and produce lower yields than trees trained with the other high-density designs discussed later in this chapter. For those reasons, I do not recommend it.

Trellises

Trellises are well suited to apples and pears. While they were probably originally intended for ornamental landscaping, the designs are also useful in high-density orchards. Because they are more costly to establish and produce lower yields than axis and tall spindle systems, trellises are not widely used in commercial apple production. (Taller trellises that produce yields comparable to other high-density systems are being tested in Pennsylvania at the time of publication.) Trellises can still be excellent choices for home and market apple orchards, however, and they make attractive fences for landscapes.

The objective is to develop more-or-less permanent scaffold limbs. The scaffold limbs are tied to horizontal trellis wires and can be horizontal or angled upward. The trunks can be vertical or slanted. The high-density double leader pear system is a modification of the trellis design.

For trellis systems, it's best to use the more-dwarfing rootstocks to prevent unwanted vigor, but avoid the extreme-dwarfing M27, P22, and V3 rootstocks. Apple rootstocks Bud9, M9, M26, Geneva G11, P22, Vineland V2 and V7, and Ranetka crab apple work well. Pears are more easily trained to trellises than are apples, and OHxF 40, 69, and 217 rootstocks are adequate. In warm, dry climates where fire blight and cold injury are not problems, quince rootstocks provide greater dwarfing for pears.

Installing trellises. Install the trellis posts before planting your trees; you can install the wires after the trees are planted. Use strong end posts, and space posts every 25 feet within rows. Brace the end posts to prevent them from tipping inward. For trellises longer than 50 feet, it is best to use high-tensile wire. Adjustable ratchets can be used to keep the wires tight. Remember that most of the weight of the fruit crop will be borne by the wires and posts. The same techniques are used for wire-supported axis and spindle systems. The heights of the wires can be adjusted to meet your needs, but it is common to have three to five wire designs that are 5 to 7 feet tall. Use taller trellises for higher yields. Trees are typically spaced 3 to 6 feet apart, and rows are typically spaced 10 to 11 feet apart.

Figure 12.11 · **Methods of Training an Apple or Pear Tree to a Trellis**

Trees are trained to a trellis using one of the following designs. Trellises range from about 3 to 10 feet tall.

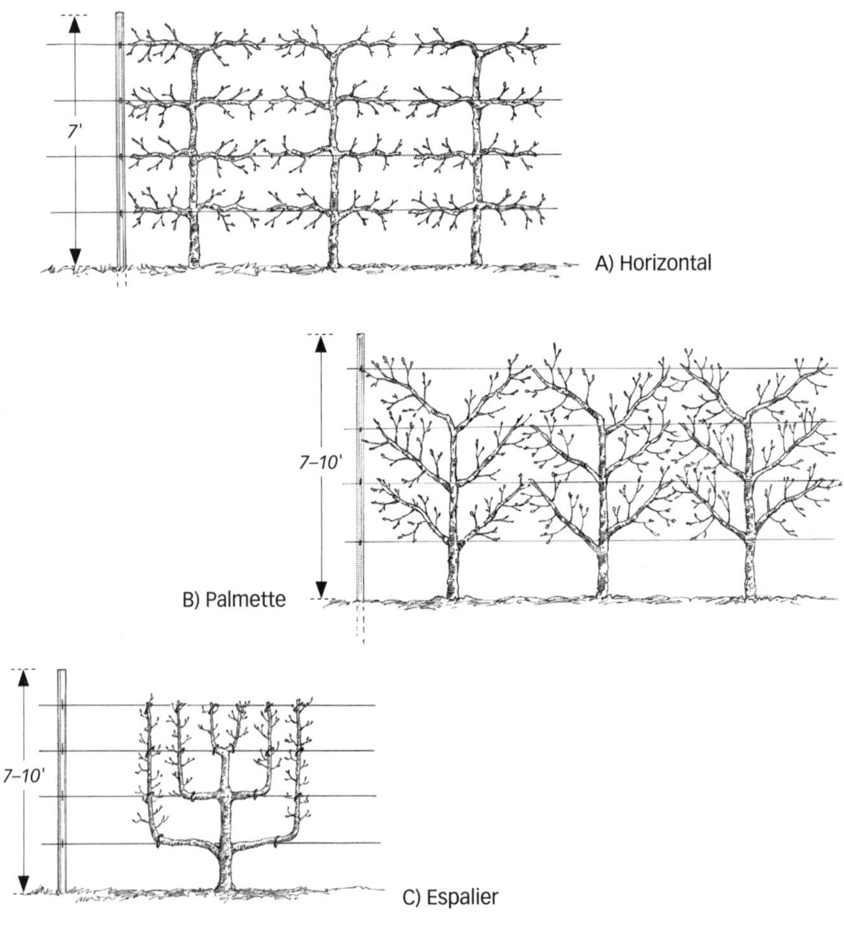

Training trees to trellises. You can use either whip or feathered trees, although feathered trees can save you training time and you will produce your first crop sooner than you would with whip trees. For feathered trees, simply select the branches you want and remove the others. Tie the selected branches loosely to the trellis wires. Check the ties frequently during the growing season, and replace them before the ties cut into or girdle the branches.

For whips, you can head the leader back to the first wire to develop lateral branches and repeat as needed for each additional wire or scaffold. This is a slow method and delays time to fruiting. A better method is to leave the terminal bud and rub off the next six buds. Leave several buds on the leader at each wire or desired branch location and rub off all of the other buds just as they are swelling in spring. As the new shoots form, keep those that fit the trellis well and remove the others.

You can train the branches horizontally, at about 45 degrees, or vertically. Horizontal branches typically produce the most undesirable water sprouts, while palmette designs can be easier to manage. Pruning established trellises mostly involves removing water sprouts, keeping the rows narrow, and occasionally renewing the fruiting wood beginning about 6 years after planting. Figure 12.11 illustrates different trellis designs.

Slender spindle design. One of the earlier innovations in high-density apple plantings was the slender spindle design. Apples grown on M9, Bud9, M26, and similarly vigorous rootstocks are planted next to strong, rigid posts, and the trunks are tied loosely to the posts. Allow the trees to grow about 6 to 8 feet tall, and train them to conic shapes. The trees are typically planted about 4 feet apart in rows that are spaced 10 feet apart, with 1,089 trees per acre.

Slender spindles work well when space is limited, but they do not provide the high yields needed for today's commercial orchards. It's also very expensive to install the many trees and posts that are needed. For these reasons, slender spindles are no longer popular commercially, but they still have their place in home and market orchards. A higher-density version called the super spindle uses horizontal wires to support the trees, rather than solid posts. In the super spindle system, trees are planted 2 to 3 feet apart.

An ideal method is to plant stock that has four to eight feathers that begin about 18 inches above the ground. The feathered trees save training time by eliminating the need to head the leader to force lateral branches.

Figure 12.12 · **How to Train an Apple Tree to a Slender Spindle Design**

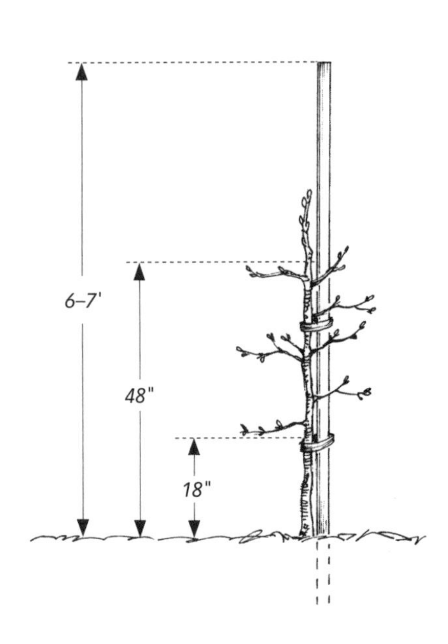

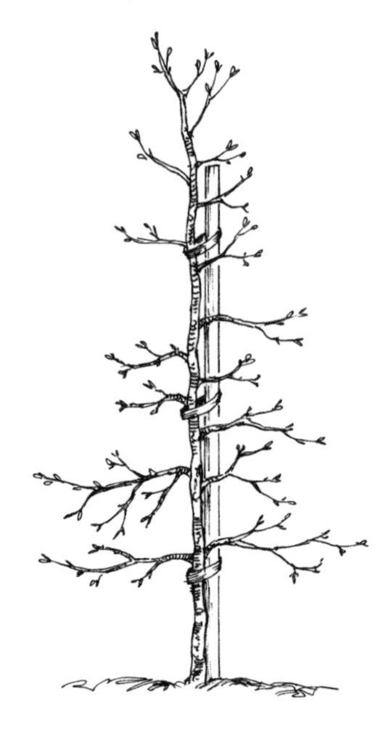

1) Use a whip or feathered tree as for figure 12.9 to establish four to eight branches between 18 and 48 inches above the ground.

2) Use weights or ties to train the branches nearly horizontal.

3) If possible, avoid heading the leader. Use the bud-thinning steps described in figure 12.9 to form more branches to a height of about 6 to 7 feet. Thin back the leader to a weak lateral, and tie up the lateral as a new leader.

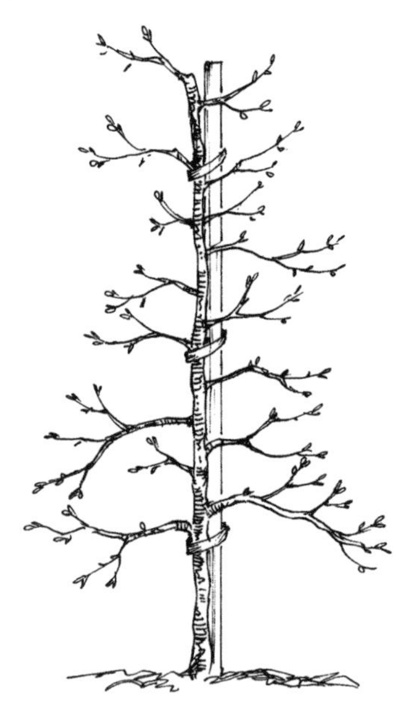

Distribute the branches around the trunk, several inches apart vertically. Use clothespins and weights, if necessary, to form angles that are nearly, but still above, horizontal.

As the trees mature, you will need to remove some of the scaffold limbs to prevent crowding and shading. By removing older branches, you also promote the formation of new fruiting wood and spurs. One strategy for maintaining mature trees is to thin out some of the largest branches each year so that all branches are replaced about once every 5 years. To keep the trees short, cut off vigorous leaders using thinning cuts, and tie up weak lateral branches near the tops of the trees. This leader renewal method produces a zigzag top and is a critical part of controlling tree vigor. Prune to maintain a conic shape with a narrow top and a wide bottom. Figure 12.12 shows the steps in training a tree to a slender spindle design.

Tall spindle design. The tall spindle appears to be one of the best high-density axis designs today for commercial apple plantings and comes largely from work done at Cornell University. This design is quite similar to the axis and other spindle training systems described earlier in this chapter. According to Michigan State University's Crop Advisory Team, the tall spindle system is excellent for eastern apple growers and has the following advantages:

- It is the most cost-effective, efficient apple training system.
- It produces significant yields of high-quality fruit in early and mature-bearing years.
- It is highly adaptable to machine-assisted practices (pruning and harvest).
- It is a simple system and easy to learn.
- It fits the natural growing characteristics of a high-density apple tree.
- It captures the maximum amount of available sunlight.
- It has little to no shaded (or wasted) space in the tree.
- It maximizes yield per acre due to its tall (10- to 11-foot) tree height.
- It is one of the best apple production systems to maintain a low carbon footprint.

In tall spindle designs, each tree may be supported on an individual pole or the trunks may be supported only on horizontal wires, with three to five wires forming the trellis. The bottom wire is usually about 2 feet above the ground and can support a drip irrigation line. The top wire is usually 9 to 12 feet above the ground, and you will need stout support poles and bracing at the ends of the tree rows. When poles are used, they are done so mostly to keep the trunks straight, not to support the weight of the crop. Wooden poles, metal electrical conduit, or 1-inch-diameter PVC pipe will all serve this purpose. Training without poles can be very successful and is less expensive than training with poles.

Space the trees 3 to 4 feet apart in rows that are 10 to 12 feet apart on level ground, and 12 to 13 feet apart on slopes. This will allow for between 837 and 1,452 trees per acre. Rootstocks that have performed especially well are M9 and Bud9; Geneva G41, G11, and G16 also work well. Tall spindles can be used for any apple variety.

The system begins with feathered trees with trunks about ½ to ⅝ inch in diameter. The first side branches should be about 3 to 4 feet above

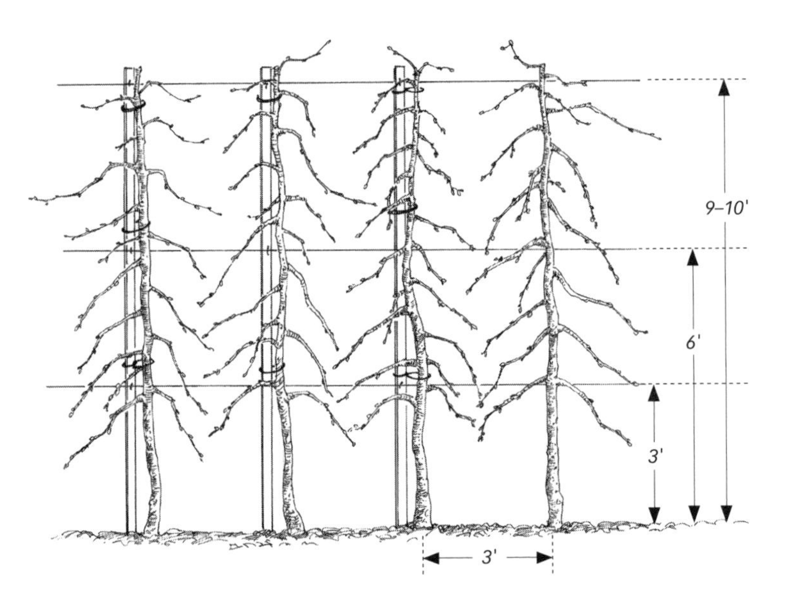

Figure 12.13 · **Apple Trees Trained to a Tall Spindle Design**

At this stage, the poles used to train and support the trunks remain in place. With support poles, the 3-foot and 6-foot wires are optional. The tree at right is shown without a support pole. The trunks can also be trained without poles, using only the three to five support wires and U-shaped wire clips to hold the tree trunks in position.

the ground, with at least four to eight branches between 3 and 4.5 feet above the ground. If you do not use support posts, you can place commercially available U-shaped wire clips around each trunk to hold it in place against the horizontal wires. The clips install quickly and last for the life of the orchard. Figure 12.13 shows a tree trained to a tall spindle.

You need to do little or no pruning during the planting year or the second year. During the planting year, or no later than the second year, tie the laterals down to below horizontal. Wires that are about 12 inches long and have open hooks at both ends work well for this procedure and are available commercially. Simply hook one end of the wire to the outer

How to Prune Established Tall Spindle Trees

Pruning established trees involves three steps for most apple varieties, plus a fourth step for 'Gala'.

1. Prune off the leader at the optimum height by making a thinning cut at a small side branch. The optimum height is 90 percent of the row width; for example, 9 feet tall for 10-foot-wide rows and 10 feet tall for 11-foot-wide rows.
2. Prune off the two largest branches along the trunk. These branches will typically be ¾ to 1 inch in diameter. It does not matter which branches you remove or where they are located. By removing these large branches, we remove unwanted secondary fruiting branches on the main laterals and keep the canopy young and productive.
3. Remove the side branches on main laterals. The goal is to produce long, columnar fruiting branches that have no side branches. This allows excellent light distribution, which is essential for developing color. Varieties that bear fruit at the tips, such as 'Fuji', weigh down the branches (which is desirable) and prevent you from having to mechanically bend down each branch.
4. On 'Gala', the long laterals tend to bend down too much, resulting in little shoot extension and thin wood bearing clusters of small apples at the tips. Head back laterals to a point where they are about the diameter of a pencil.

portion of the branch, pull it down, and hook the other end of the wire to the trunk. Weights also work well, but they take more time to fasten to the trees. Some growers prefer to tie the branches down to stakes set into the ground. By bending down the branches, you prevent them from developing into competing leaders and induce early fruiting, which is critical for maintaining desirable tree vigor. The goal is to produce fruit by the second year and production should become substantial in years 3 and 4. Do not, however, sacrifice early leader growth at the expense of fruit production. The trunk should reach the top wire by the third growing season. Filling in the side branches may take a year or two more. Continue developing new fruiting branches as the trees grow taller.

Training Pear Trees

Commercial pear production differs from apples in several ways. First, very few pear varieties make up the bulk of world production, and we do not see the rapid introduction of hot, new varieties as we do with apples. 'Bartlett', as it has for decades, still makes up the majority of pear production domestically and worldwide. There is much less need to force early production to capitalize on the short market life of new pear varieties, as is now done with apples. In addition to pear varieties being limited in number, so are pear rootstocks. Some of the dwarfing rootstocks needed for the high-density and very-high-density pear orchards are not adapted to many growing regions, and others have yet to be tested across North America.

This is not to say that new commercial pear orchards should not be designed using high-density systems. We do, however, have much to learn about high-density pear production systems, and there is an extremely limited amount of technical support for growers using them in North America. Growers in Australia, Italy, the Netherlands, and Japan are further ahead in developing high-density pear production. In short, do your research and exercise caution before investing heavily in a high-density pear orchard. In colder climates, you may find that medium-density orchards and moderately vigorous rootstocks will be more reliable than higher density systems using dwarfing rootstocks.

Especially for organic growers, be careful to design your orchard to allow sustainable yields of high-quality fruit to be produced at reasonable costs for the life of the orchard. Avoid the temptation to chase early, high yields with a design that may make long-term production difficult.

Freestanding and palmette-trained trees on simple trellises are still well suited to home and market orchard pear production. For a new commercial orchard, where the climate allows, several high-density systems will likely prove to be better investments than low-density plantings. Some high-density systems make it easy to pick without the need for ladders and they work well for home and market orchards, U-pick, and more intensive, grower-pick orchards.

Freestanding Pear Trees

Both central leader and open vase training systems have long been used and generally work well.

One difference between pears and apples is that pears have a greater tendency to produce very upright growth, multiple leaders, and narrow branch angles with their associated bark inclusions. It is especially critical to train pear trees early, and you must pay careful attention to removing unwanted leaders and removing or spreading upright branches.

Trellises

Pears are among the easiest trees to train and they can be given almost any shape you want. Trellises have long been used to support and train pears, two of the most popular systems being the palmette and espalier designs described earlier for apples and shown in figure 12.11. These designs use from three to five or more horizontal wires. Newly formed leaders and branches are tied to the wires to hold the desired shapes until enough wood and stiffness develop to hold the shapes. Depending on the size of the tree and the training system, the supporting wires may or may not be left in place permanently.

High-Density Pear Systems

In 2008, California researchers reported on 13 years of trials comparing freestanding pear trees to those in high-density Tatura and parallel hedgerow systems. They found that the Tatura system was the highest-yielding. The high-density systems came into bearing sooner than the freestanding trees and would have become profitable in the sixth growing season, compared with the ninth growing season for the standard planting. At full production, the high-density plantings produced about

50,000 pounds of fruit per acre, compared with 40,000 pounds for the freestanding trees. The researchers estimated that a grower would recover establishment costs in year 10 if using the high-density systems and year 21 if using the freestanding trees. These results are similar to those found in pear trials conducted in other countries.

Several divided canopy designs are available for pear growers. These systems train four leaders per tree, or train alternate trees to lean one way or the other in a V-shaped row. The four-leader systems are used because they reduce the number of trees needed per acre compared with single-trunk, single-leader trees.

V-Hedge

This design (figure 12.14) uses single rows of well-feathered trees spaced 4 feet apart in rows that are spaced 11 to 12 feet apart. Four leaders are developed at the top of a short trunk, about 18 inches above the ground. Wooden stakes or bamboo poles position the leaders and are supported by horizontal wires about 6 to 7 feet above the ground, creating a double V shape about 4 feet wide at the top. Four poles are used for each tree, with one leader being trained per pole. Keep the fruiting laterals short on each leader to maintain an open canopy.

Open Tatura

This high-density pear system is used in Europe. Early Tatura systems developed double leaders on short trunks, forming a Y shape perpendicular to the crop rows. This design resulted in tall trees that stuck out into the alleys and interfered with the movement of equipment and other orchard operations. Newer Tatura designs use shorter trees, more vertical trunks, and rows more closely spaced than did early designs.

The open Tatura design uses double rows of trees, with each member of a tree pair leaning in an opposite direction toward the alleys and creating V-shaped rows of trunks supported on poles and horizontal wires. The two rows of trees in each crop row are separated by a strip of open ground about 20 inches wide running along the center of the crop row. The row centers are 11 to 12 feet apart. The trees are planted in an offset pattern between rows to maximize the distance between the trees. Within each row, the trees are planted 3 to 6 feet apart (there is 1.5 to 3 feet between trees in the paired rows). The support posts for the trellis wires are similar to the one shown in figure 12.14b.

Figure 12.14 · **Training Pear Trees to a V-Hedge**

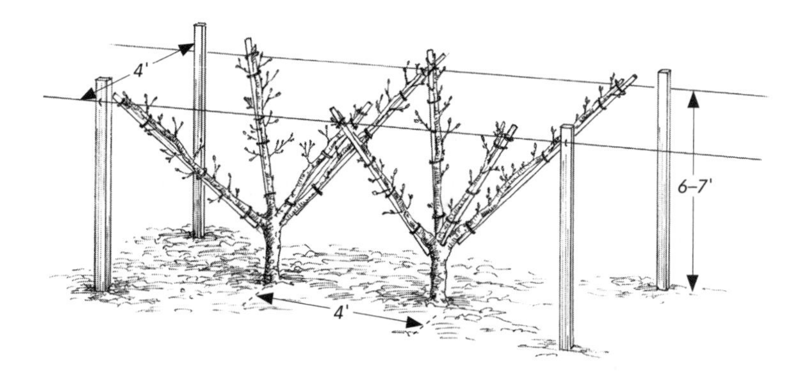

A) Pear trees trained to a V-hedge. A bamboo or other similar pole is used to support and train each of the four leaders on each tree. These poles attach to the support wires.

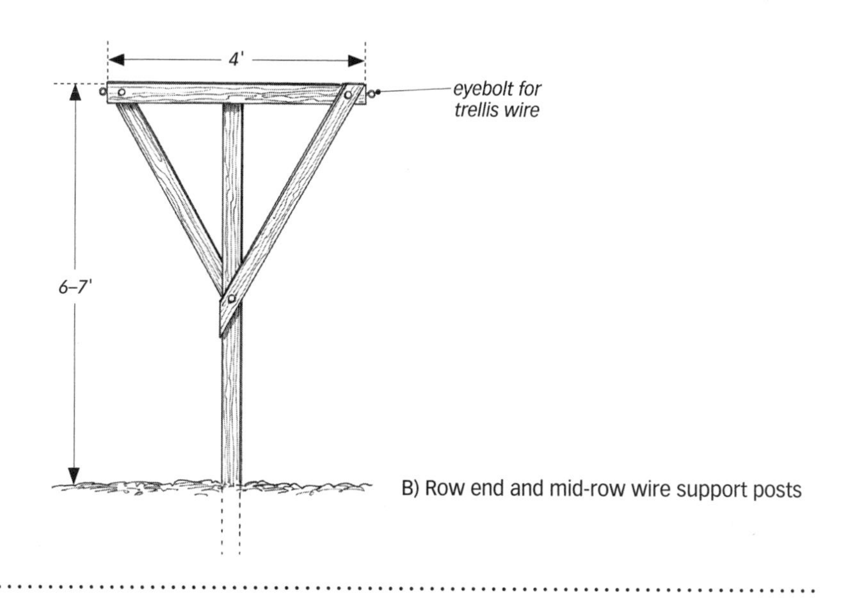

B) Row end and mid-row wire support posts

The open Tatura design (figure 12.15) improves on early Tatura systems by lowering the canopy and reducing the trees' intrusion into the alleys. Trees are maintained about 6 to 9 feet tall. The advantage over freestanding trees is that there is more fruiting surface per acre. The disadvantages are that the trees need trellis wires and poles, and there need to be many more trees per acre. From a pest and disease control

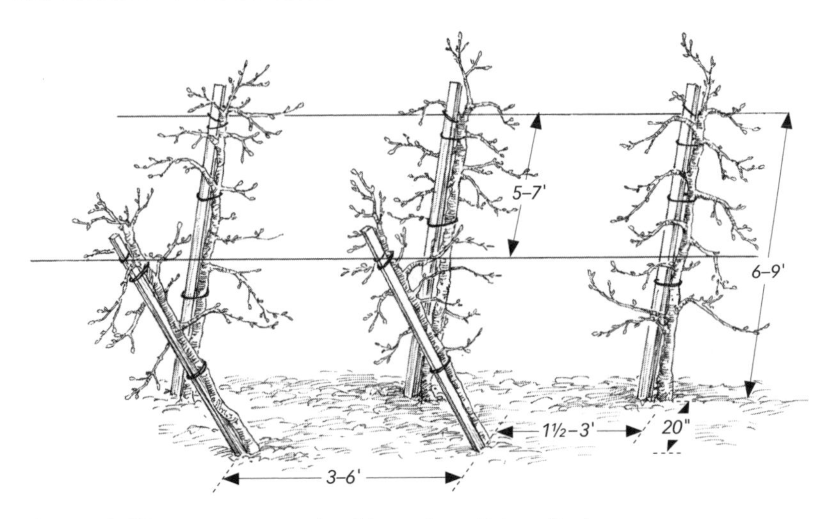

Figure 12.15 · **Pear Trees Trained to an Open Tatura Design**
The trees are planted in double rows spaced about 20 inches apart and are offset between rows to maximize the distance between trees. Poles and wires are used to train and support the trunks.

perspective, the V-shaped rows can make it difficult to scout and apply pesticides to the interiors of the crop rows, particularly as the orchard matures. It can also be challenging to manage weeds within the double rows of trees without using conventional herbicides.

Double Leader

This training system is especially good for commercial and home organic pear production (figure 12.16). While several variations exist, they all develop two leaders from each tree and train the leaders vertically, forming two trunks. Essentially, this is a high-density version of the old palmette training system, which uses greater numbers of smaller trees and spaces the leaders closer together. The two leaders are formed about 10 inches above the ground in a U shape and are trained vertically using horizontal wires at a density of about 2,400 leaders per acre. Unlike the V-hedge and Tatura designs, the U-shaped trees are oriented parallel to the crop rows in the double leader system and create narrow walls.

By doubling the leaders, you use half the number of trees per acre to produce the same number of leaders as with single-trunk axis or spindle systems. This saves you a huge amount of money in establishment costs.

Growers that double the number of leaders also have an advantage in areas where fire blight is a problem — you can lose one of the two leaders and still remain in production. Organic growers that use this system have the advantage of easier pest and disease management, due to the flat, narrow fruiting surface, than with freestanding or V-trained trees.

Harvesting the fruit is also very easy and is ideally suited to U-pick orchards, where the trees are maintained to a height of about 8 feet. Grower-picked trees can be trained to around 10 to 12 feet tall to increase yields. The taller trees lend themselves well to mechanization, where pruning the tops of the trees is done from tractor-drawn platforms, from which workers prune or harvest two rows simultaneously. This mechanization approach applies to all of the spindle and axis systems described in this chapter and can provide big savings in labor in commercial orchards.

Before planting, install your support wire posts the same way you would install a tall spindle system for apples (see page 432). There should be roughly five wires between 18 inches and 9 to 11 feet above the ground. You can install the wires after planting. Begin with feathered trees or whips. For yellow European varieties, use OHxF 40, 69, or 217 rootstocks or similarly vigorous pear rootstocks. Use OHxF clones

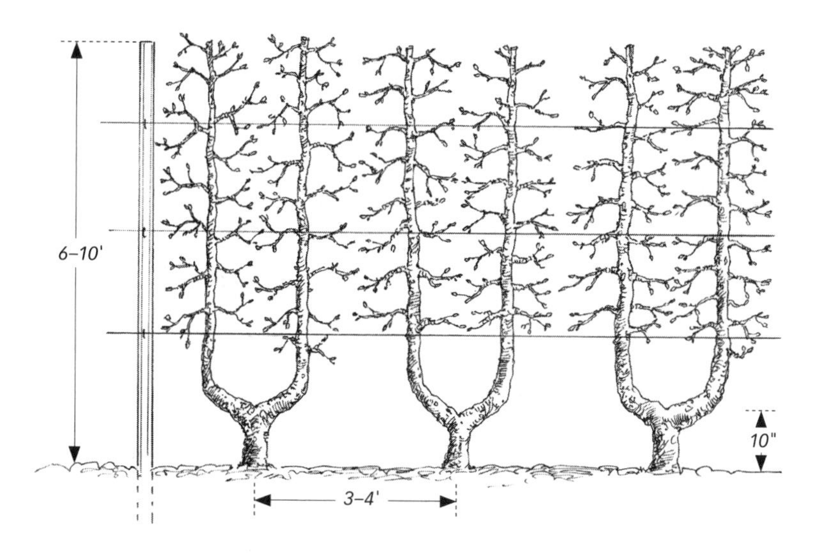

Figure 12.16 · **Pear Trees Trained to a Double Leader Design**

Developing two trunks for each tree reduces establishment costs. The flat, narrow canopy provides for excellent pest and disease control and easy harvest.

277 or 97 for Asian pears and for red European pear varieties. Plant the trees 3 to 4 feet apart in rows that are about 11 feet apart.

You will need to develop two leaders about 10 inches above the ground. If lateral branches are not available, head back the leader to a height of 12 to 14 inches above the ground in order to force new lateral shoots. When forming the double leaders, create a rather gentle U shape at the top of the single shared trunk. Take great care to avoid bark inclusions (see pages 414–415). For a spacing of 3 feet between trees, the paired leaders for each tree should be 18 inches apart. For a spacing of 4 feet between trees, the leaders should be 24 inches apart. Although not absolutely necessary, temporary posts are very helpful during training to ensure straight, evenly distributed leaders. Loosely tie the leaders to the horizontal wires or clip them into place using the U-shaped wire clips described for the tall spindle system for apples. When the leaders develop enough strength to support the crop load, you can remove the support wires.

Pruning established trees mostly involves removing crowding branches and heading back vigorous laterals to form fruiting spurs. Pear spurs live for 8 to 10 years, and it is a good practice to occasionally remove or head back older spurs to form new fruiting wood.

For high-density, organic pear production, the double leader design appears to offer more advantages and fewer disadvantages than the V-hedge or Tatura systems. The V-hedge does allow you to produce four leaders per trunk, reducing the numbers of trees needed per acre as compared to single and double leader designs. The spindle and axis systems described for apples are worth experimenting with for pears.

Training Loquat Trees

Loquats are tropical to subtropical evergreen trees or shrubs that can grow to be 30 feet tall, although in orchards we keep the trees about 10 feet tall. The trees naturally spread, and you can train them to central leaders, open centers, or multiple-stemmed shrubs. Loquat trees are typically spaced 20 to 25 feet apart in rows that are also spaced 20 to 25 feet apart.

Loquats bear their fruits in clusters at the ends of short side branches. After harvest, some growers head back the fruiting laterals by about 2 inches to stimulate new fruiting wood. The naturally dense evergreen canopy remains in place year-round and can create challenges with pest and disease control. Use the training and pruning practices we discussed for apples to shape and maintain the trees. All cuts will essentially be summer pruning. Use thinning cuts to keep the canopy open, and maintain the plants at a height of about 10 feet. Head back fruiting laterals several inches after harvest.

Training Mayhaw Trees

Mayhaw trees naturally develop open canopies and are among the easiest pome fruits to train. Once they are trained, pruning is generally limited to occasionally thinning out the top to maintain the desired tree height and an open canopy. You also need to prune suckers from the base of the tree and water sprouts in the canopy.

The trees can be trained to either modified central leaders or open centers. In either case, you use similar training and pruning methods as those used for apple and pear (see page 425). In commercial settings, the trees are typically trained to wide, flat-topped shapes (figure 12.17) and resemble flowering crab apples. The small fruits are harvested using

Figure 12.17 · **Mayhaw Trees in a Typical Commercial Orchard**

Trees are freestanding and can attain 30-foot height and spread. They can be trained to modified central leaders or open centers.

mechanical shakers and are collected on tarps under the trees. In training young trees, select your first branch at least 18 inches above the ground and 36 to 48 inches above the ground if you plan to harvest mechanically. In new mayhaw orchards, trees are usually planted about 25 feet apart in rows that are spaced 25 to 30 feet apart. Early recommendations were for closer spacing, but the higher density caused crowding.

Training Medlars

Medlars require very similar methods of pruning and training to apple and pear trees. They are naturally spreading and are traditionally grown as freestanding, single-trunk trees. They are planted 15 to 20 feet apart in rows that are spaced 15 to 20 feet apart. In landscapes, the trees typically grow to be 13 to 20 feet tall and broad. In an orchard setting, keep them no more than 12 to 14 feet high. Medlars can be trained to modified central leaders or open centers, and you can stake freestanding trees for 3 to 4 years after planting to produce straight trunks.

The low-density designs described above for apple and pear work well for medlar. Medlars can be trained to trellises, although they are most often grown freestanding. By using the training methods described for apples and pears, the trees might be planted at higher densities, perhaps as close as 7 feet apart in rows that are spaced 15 feet apart. Medlars' ability to adapt to high-density spindle and axis systems remains to be determined.

Early training is important because mature medlars do not tolerate heavy pruning well. For pruning, follow the recommendations for apple and pear.

Training Quince Trees

Quince trees generally resemble apple and pear trees and can be trained to either central leader or open center designs. Although quince trees have a strong apical dominance that makes them well suited to central leaders, their susceptibility to fire blight means that open center training is best in areas where fire blight is a problem. If one of three to five scaffold limbs in an open center tree is lost to blight, the tree can still produce fruit. Losing the main trunk of a central leader tree is more serious.

The biggest difference between quince trees and apple and pear trees is that apples and pears develop fruiting spurs that can remain productive for 8 to 10 years. Quinces bear their fruits at the tips of current-season shoots. The goal is to produce new fruiting shoots each year.

Quinces are usually grown as freestanding trees or tall bushes and are planted in low densities, about 15 feet apart in rows that are 20 to 25 feet apart. Trees are typically 8 to 15 feet tall. While there may be promise in growing quinces in high-density axis and spindle systems, similar to apples and pears, very little research has been published on the subject. For now, high-density quince plantings should be considered experimental.

Training Saskatoons

Saskatoons are grown as freestanding bushes, although some growers place two trellis wires, one on either side of the row and about 4 feet above the ground, to keep the bushes from leaning out into the alleys. Freestanding bushes that are spaced 20 feet or more apart make excellent specimens in an edible landscape. For fruit production, the bushes are usually grown in fairly dense hedgerows with the plants spaced 1 to 5 feet apart in rows that are 10 to 12 feet apart. Plants are spaced very closely together in commercial orchards to obtain early yields using mechanical harvesting. This close spacing can create crowding and may necessitate thinning out bushes several years after planting. An in-row spacing of 4 to 5 feet is more common and is recommended for home and market orchard production.

It's important to prune saskatoons because it rejuvenates the bushes and helps ensure high and consistent yields and healthy plants (see box on page 446). We usually prune saskatoons in early spring, before new growth starts. For the first 3 years after planting, prune only to remove dead, damaged, or diseased stems. At the same time, remove stems that droop close to the ground. Cut these off at ground level using sharp pruning shears or long-handled loppers. Saskatoon stems are normally not headed or thinned. Instead, remove the stems at ground level.

Depending on the plants' growth, regular pruning begins about 3 to 6 years after planting. Remove one-third to one-fourth of the stems, targeting those that are more than about 4 years old and those that show declining fruit production. Also, remove weak new stems and any stems

Goals of Saskatoon Pruning

- Replace all of the fruiting wood every 3 to 4 years.
- Keep the plants short enough to comfortably harvest, either by hand or with over-the-row harvesters. For fruit production, the bushes are kept 10 feet tall or shorter.
- Maintain open shapes and plantings that allow good light penetration and air movement.

that lie close to the ground. The next year, remove one-third to one-fourth of the oldest stems and repeat this practice yearly. Eventually, you develop a system where the oldest stems are 3 to 4 years old. For saskatoons, the best fruit production usually occurs on vigorous 2- to 4-year-old stems.

For ornamental landscape plantings, modify these steps to meet your needs. In general, you will prune to remove dead and damaged stems and to thin out the stems to create an attractive specimen shrub. The mature heights of selected cultivars are given in chapter 5, although in commercial cultivation the bushes are kept about 7 to 10 feet tall, depending on whether they will be hand or mechanically harvested. Strive for an open, spreading, vase-shaped shrub with a spread that is about equal to its height. Occasionally remove old stems to maintain a healthy and vigorous shrub.

Training Sweet Cherry Trees

Until recently, sweet cherries were most often trained to vase shapes and allowed to develop into tall trees, with most fruit produced far above the ground. Improved training systems and dwarfing rootstocks now allow us to grow smaller trees with 80 percent or more of the fruit harvested without ladders. The smaller trees and more open designs allow for improved pest and disease management.

Sweet cherries bear their fruits at the base of 1-year-old wood and on short spurs located on 2-year-old and older wood. Unlike the short-lived spurs on tart cherries, the fruiting spurs on sweet cherries live for 10 to 12 years, but the best-quality fruit comes from young to middle-aged spurs.

The type and amount of pruning needed for sweet cherry trees depend on the rootstock and the training system. With trees that grow on vigorous rootstocks, like Mazzard and Mahaleb, you are faced with the challenges of bringing the trees into production quickly, reducing growth, and maintaining heavy crop loads. With dwarfing and highly productive rootstocks, such as Gisela 5 or Gisela 6, the trees come into production quickly and tend to overbear, leading to loss of vigor and small fruits. With these highly productive rootstocks, the challenges are to increase vigor and reduce crop load.

Types of Pruning

The pruning practices we discussed at the beginning of this chapter work well for vigorous sweet cherry rootstocks. For trees on highly productive rootstocks, we need to prune differently. Oregon State University has developed an approach to pruning highly productive cherry rootstocks using thinning, stub, and heading cuts to maintain high vigor and reduce crop loads (figure 12.18).

Thinning Cuts

Use thinning cuts to remove weak and drooping wood. This practice is common to all of the pruning systems described in this book and is essential in maintaining open canopies that allow good air movement and light penetration. Maintaining high light penetration is important with the vigorous rootstocks to encourage new shoot development low in the tree as part of regular fruiting branch replacement. Also thin out leaders and lateral branches in the tops of the trees to allow light penetration. For vigorous cherry rootstocks, thinning also removes wood that is most likely to produce small fruits. Follow the practices described earlier in this chapter.

Stub Cuts

Stub cuts are a form of heading cut used on fruiting branches. According to the Oregon State University system, about 20 percent of all fruiting branches are cut back each year in order to renew the branches and spurs and increase tree vigor. Over a 5-year period, you will renew all of the fruiting branches and most of the spurs. The length of the renewal stub depends on the amount of light it receives. Well-exposed fruiting branches near the top of a tree can be stub-cut to be 3 to 5 inches long.

Stubs for less-exposed branches lower in the tree may be left as long as 2 feet if the canopy is dense.

Heading Cuts

Make heading cuts during the dormant season by removing the outer third to half of each fruiting shoot that formed during the previous growing season. This practice increases branching, causes the fruiting spurs to develop farther apart, reduces crop load, and increases leaf size and photosynthetic capacity. The number and size of leaves are important factors in the fruit's size and quality. To produce large, high-quality cherries, you need about five leaves for each fruit. Leaves that develop at the base of new shoots are generally larger and have greater photosynthetic capacity than those near the shoot tips.

For organic orchardists, a serious drawback with this system is that the heading cuts produce very dense canopies that can increase humidity and shading in the canopy, limit spray coverage on the inside of the canopy, and increase disease and pest problems. If you use this approach to prune highly productive sweet cherry trees, also increase thinning cuts to limit the number of branches in the canopy. You will have smaller yields but fewer pest and disease problems.

Training Systems

While sweet cherries can be trained to simple open centers or, with more difficulty, central leaders, the trees' vigorous growth complicates management. Newer designs help maintain shorter trees, greater yields, and improved pest and disease management.

Steep Ladder

In the steep ladder system (figure 12.19), a freestanding tree is grown on either standard or dwarfing rootstocks. This is the system of choice for many sweet cherry growers in the Pacific Northwest and is an adaptation of the open center or vase training system design that has long been popular. Trees are planted 16 to 20 feet apart in rows that are 18 to 24 feet apart, producing low- to moderate-density orchards on vigorous rootstocks, usually Mazzard. Production does not begin until the fifth or sixth year after planting.

By using dwarfing rootstocks such as Gisela 5 or Gisela 6, you can increase early yields while reducing tree size. These trees are spaced

Figure 12.18 **· Pruning Practices for Sweet Cherry Trees on Highly Productive Rootstocks**

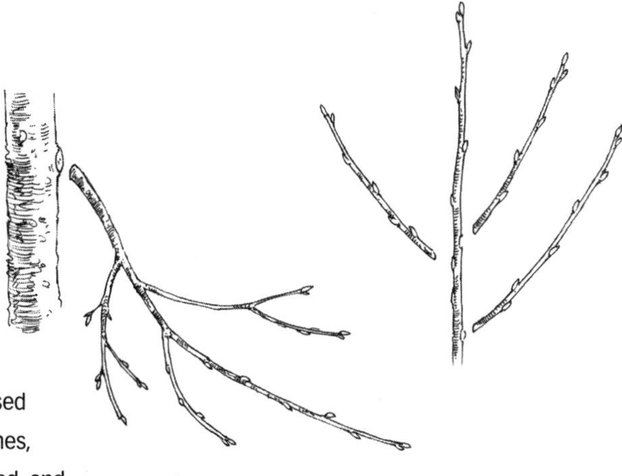

A) Thinning cuts are used to remove drooping branches, small diameter fruiting wood, and branches near the tops of leaders.

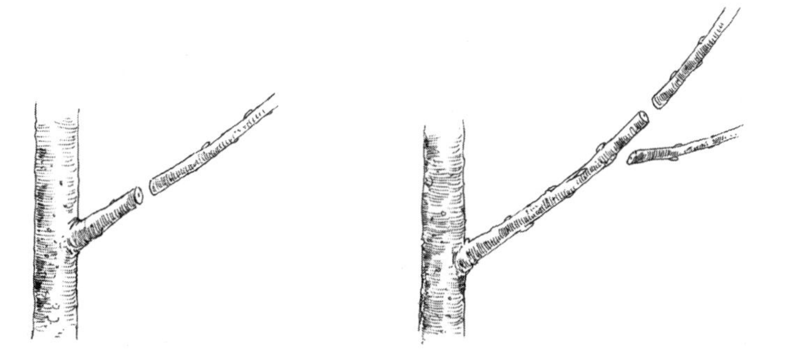

B) Stub cuts produce new fruiting wood. Renewal stubs near the tops of trees can be as short as 3 inches. For lower, more shaded branches, stubs can be up to 24 inches long.

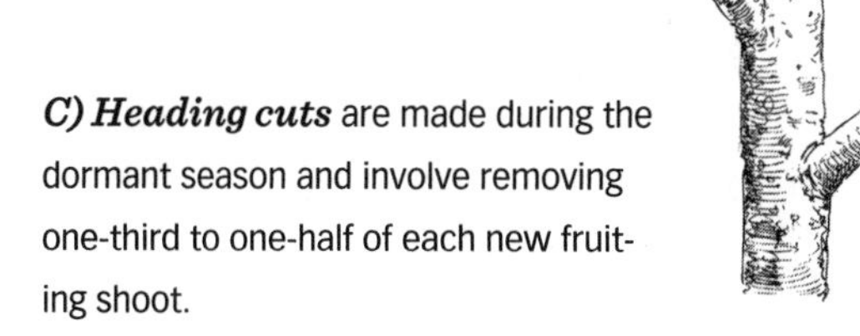

C) Heading cuts are made during the dormant season and involve removing one-third to one-half of each new fruiting shoot.

closer together than trees on vigorous rootstocks. California growers have found, however, that it can be difficult to maintain vigor on the dwarfing rootstocks, and small fruit size has been a problem as orchards mature. The pruning methods we discussed above should help maintain vigor and fruit size on productive rootstocks.

The steep ladder design has many advantages, particularly for larger-scale commercial growers. It is suited to good or poor soils and frosty sites, adapts to a range of vigorous and dwarfing rootstocks, and works with different cherry varieties. The system does require relatively high management skills and is slow to come into bearing (at least on vigorous rootstocks), and harvest costs are higher than with smaller trees.

Using the steep ladder system. When developing a steep ladder sweet cherry tree, you will create a temporary modified central leader that has four permanent scaffold limbs. You will treat each of these limbs as a separate spindle or trunk. Start with either feathered trees or whips and select four primary scaffold branches that are distributed evenly around the tree about 30 to 36 inches above the ground. Remember to keep the central leader for now. Head the scaffolds so that each develops two permanent secondary scaffold branches. After two to four growing seasons, remove the central leader to form an open center tree. Use heading and thinning cuts to maintain a conic shape. Figure 12.19 shows the steps in developing a steep ladder sweet cherry tree.

Large Sweet Cherry Trees

If you have a large, freestanding sweet cherry tree and want to keep it, pruning and training mostly involve thinning out excessive growth. Sweet cherries are notorious for producing multiple upright leaders and lateral shoots with narrow branch angles. Use thinning cuts to keep the canopy open. Spreaders and weights can be used to position branches and shoots to ensure branch angles wide enough to prevent bark inclusions and later limb breakage. Over several years, gradually thin out large branches. The goal is to lower the tree height and reduce its spread while developing new scaffold limbs that are more manageable and easier to reach. Train sweet cherry trees to an open center as in figure 12.10.

Figure 12.19 · **How to Train a Sweet Cherry Tree to a Steep Ladder Design**

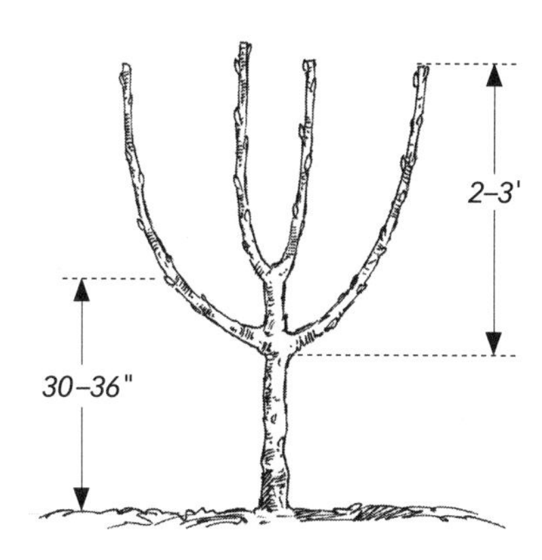

1) Develop three or four primary scaffold branches that are distributed evenly around the tree and 30 to 36 inches above the ground. Keep the central leader for now.

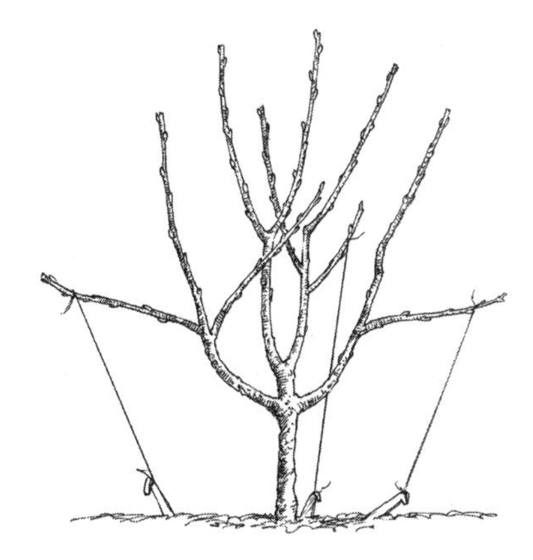

2) Head each primary scaffold branch to develop two secondary scaffold branches. For each primary scaffold, tie down the lowest of the two secondary scaffold branches, as shown.

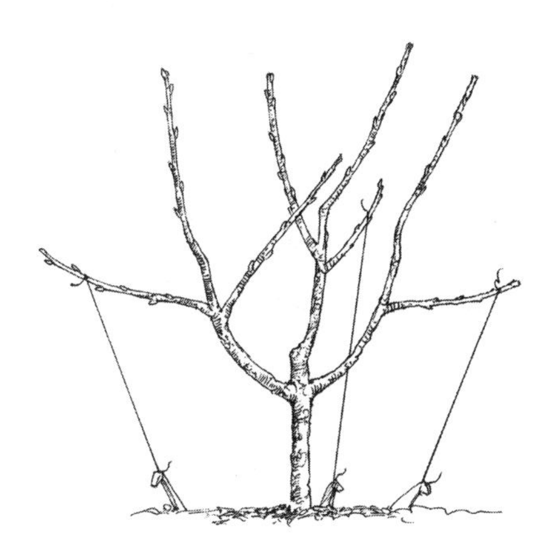

3) After two to four growing seasons, remove the central leader to form an open center tree.

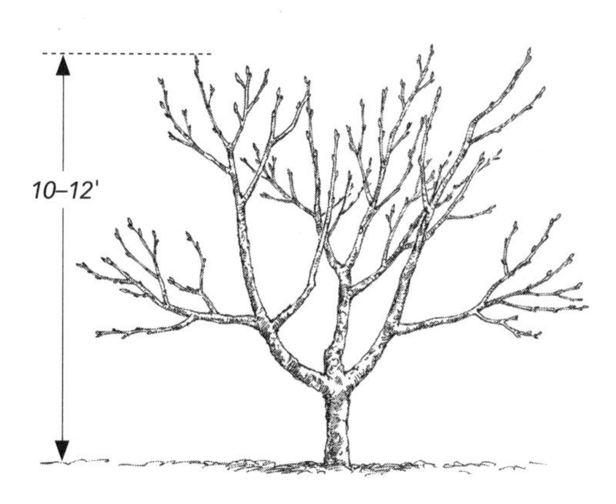

4) Use heading and thinning cuts to maintain a conic shape.

Vogel Central Leader

The Vogel central leader system uses dwarfing rootstocks, such as Gisela 5 or Gisela 6, to produce moderately high-density orchards that contain trees spaced 8 to 12 feet apart in rows that are 15 to 18 feet apart. Trees in this system are easy to develop and maintain, produce high early yields when used with highly productive varieties, and are easy to harvest. Because the trees are taller, the Vogel central leader design is better suited to frosty sites than the Spanish bush. This design tends to produce lower yields than do the other sweet cherry systems we are covering here. For home and market fruit growers, however, the Vogel central leader can be a good fit and is a replacement for the old open center training systems for cherries.

Because the trees grow 10 to 12 feet tall, the design is not ideal for U-pick orchards. By heading the leaders back to weak lateral branches, however, you can keep the trees shorter. Use the pruning practices described above for highly productive rootstocks in order to maintain commercially acceptable fruit size.

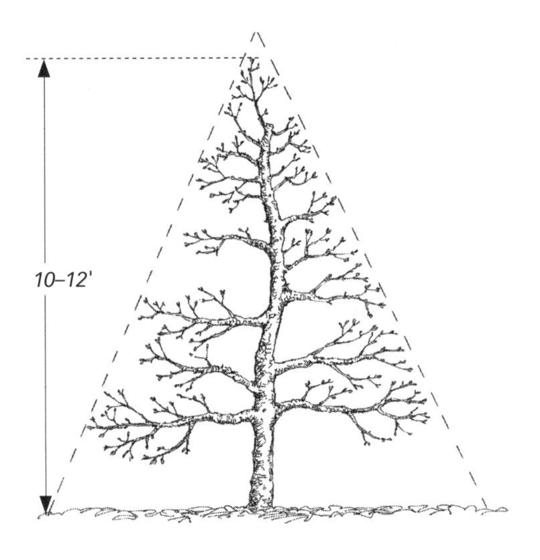

Figure 12.20 · **Sweet Cherry Tree Trained to a Vogel Central Leader**
Use thinning cuts to keep the canopy open and develop new fruiting wood. Head back branches that become too long or are half the trunk diameter or more at that height. Keep branches horizontal. Maintain tree height by thinning to a weak lateral.

Using the Vogel central leader system. Start with whips and head them back to 30 to 36 inches above the ground. If you want your primary fruiting branches to be higher or lower than this, adjust the height at which you head the leaders. Instead of heading the leader, you can leave the two uppermost buds on the leader and rub off the next five or six buds below them to reduce competition for the leader. Then leave every second or third bud. The goal is to develop about four fruiting branches 18 to 24 inches above the ground. Use clothespins and weights to train the branches horizontally, as we discussed earlier. If both top buds produce shoots, prune off the stronger shoot and keep the weaker shoot.

Repeat these procedures for the new leader during the first and subsequent dormant prunings. You will not need to head back the leader again unless it grows to be more than about 32 inches long. Your goal is to develop a series of fruiting branches that are distributed more-or-less evenly around and along the leader, rather than to form distinct whorls of branches every 18 to 24 inches.

As the trees mature, thin out drooping and crowding branches. Head back laterals that start to turn upwards, crowd neighboring trees, or intrude into the alleys. For each fruiting lateral, in summer pinch off the tips of new shoots when they are about 6 inches long. Thin out or head back branches that are more than one-half the diameter of the section of trunk where the branch is located. Based on Oregon State University recommendations, you may need to renew as many as 10 to 15 fruiting branches each year in mature trees. Figure 12.20 shows a Vogel central leader sweet cherry tree.

Spanish Bush

The Spanish bush system (see figure 12.21) uses dwarfing rootstocks to produce high-density orchards that provide high early yields with low labor costs. Most of the fruit can be harvested without ladders. In colder climates and on poorer soils, which limit cherry tree growth, use more vigorous rootstocks.

While this particular design was developed in Spain, other variations exist. Essentially, they are high-density variations of open center trees. You need more skill to manage the Spanish bush than you do to manage the Vogel central leader, and the Spanish bush is

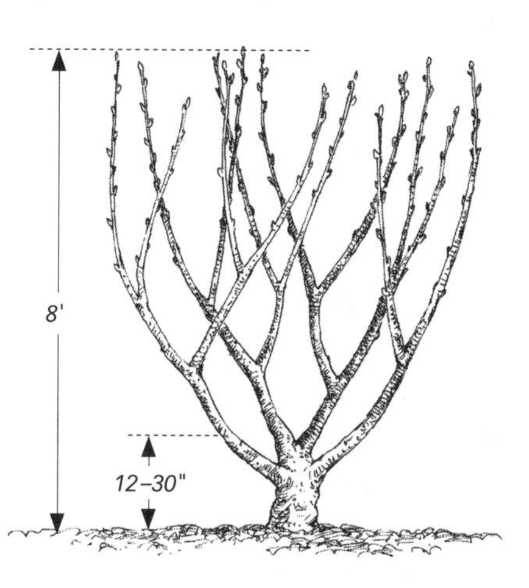

Figure 12.21 · **Sweet Cherry Tree Trained to a Spanish Bush**

Develop four primary scaffold branches. Head these to produce two secondary laterals each. Head each secondary lateral to produce two tertiary laterals each. Use weights, ties, or spreader bars to train the scaffolds to a wide, uniform bowl shape. Use thinning and stub cuts to keep the canopy open and develop new fruiting wood.

riskier on frosty sites because the trees are shorter. On sites where spring frosts are not serious problems and for growers who are willing to learn the system and put in the required management effort, the Spanish bush can be well suited to home, market, and U-pick orchards. The system is also worth testing in larger commercial orchards. Plant the trees 6 feet apart in rows that are 15 to 18 feet apart.

As with other systems that use highly productive rootstocks to produce early yields, small fruit size can be a problem as orchards mature. In trials in the northeastern United States, Spanish bush–trained trees produced relatively low yields, and the dense canopy created problems with shading and low-quality fruit. Use the pruning practices described above for trees on highly productive sweet cherry rootstocks. Also make extra thinning cuts to remove excessive numbers of fruiting branches and keep the canopies open.

In an Australian modification to the Spanish bush, called the Kym Green's bush (KGB), sweet cherry trees are grown on vigorous rootstocks and are spaced roughly 6.5 to 9 feet apart in rows that are 13 to 15 feet apart. The trees are roughly 11 to 12 feet tall. When newly planted, they are headed back to develop scaffold limbs. Over the next 2 years, each scaffold limb is headed back four times to produce 25 to 30 limbs. When the trees are mature, two to four large limbs are removed each year during dormant pruning, and renewal limbs are trained up. In both the Spanish bush and KGB systems, you need to frequently head back laterals during the first four years after planting.

While Spanish bush and KGB training systems produce much fruiting wood, they create dense canopies that make it very difficult to scout and treat pests and diseases under all but the poorest cherry-growing conditions.

Spindle and Axe

Many different spindle and axe training systems are being used and tested with sweet cherries around the world. These are generally similar to the Vogel central leader design, but the trees are spaced more closely — at densities of more than 1,000 trees per acre. In this book, we recommend wider spacings and a maximum of about 670 sweet cherry trees per acre.

A long-term orchard trial in New York produced the highest yields on a vertical axis system, with lower yields on slender spindle and

V-systems. Spanish bush-trained trees produced less than half the yields of those trained to the vertical axis, and central leader trees produced the lowest yields.

Using the spindle and axe systems. Precocious, dwarfing rootstocks are generally used for vertical axis and spindle cherry systems, although more vigorous rootstocks are used where soils and climate limit cherry tree vigor. As with other high-density systems, establishment costs and management input are high. The trees are well suited for producing hand-picked fresh fruit, and around 80 percent of the fruit is harvestable without ladders. The narrow rows also make it relatively easy to control pests and diseases.

For trees on Gisela 5, Michigan State University recommends planting the trees 8 feet apart in rows that are 15 feet apart. For trees on Gisela 6, plant the trees 9 to 10 feet apart in rows that are 16 to 17 feet apart. Where soils and climate produce vigorous sweet cherry trees, the wider spacings can provide less crowding over the long term.

While vertical axis and spindle training for sweet cherries show promise, adoption by commercial growers in North America has been slow. Widespread use of these techniques in North America may depend on rootstock breeders developing improved dwarfing rootstocks that maintain acceptable fruit size. Consider testing these designs in experimental plantings at this time. The following guidelines for slender spindle sweet cherry training are based, in part, on work done at Cornell University and recommendations by Michigan State University.

The vertical axis and slender spindle systems develop conic-shaped trees that are very narrow at the tops and provide good light penetration to the bottoms of the trees. Vertical axis trees are generally taller than those trained to slender spindles, which accounts for their greater yields. The increased height of the trees, however, complicates pest and disease management and increases harvest costs. Except for height, the two systems are very similar for cherries. For organic orchards, the lower tree height is better for most growers.

Managing growth. Cherry trees can be grown without support posts or wires, but because they need straight, uniform trunks and fruit crops develop early, it is advisable to use supports. Wooden stakes or metal conduit stand alongside each tree and support the young trunks, keeping them straight. The support posts can be driven into the ground or they may be supported by one overhead wire 10 to 14 feet above the ground.

Growers often divide the canopy of spindle- and axis-trained trees into two management zones: the lower fruiting zone and the upper fruiting zone. The lower fruiting zone includes seven or eight branches between about 2 feet and 4.5 to 6 feet above the ground. Begin with a well-feathered tree that has abundant lateral branches. Select the branches you intend to leave, distributing them evenly around the trunk and spacing them evenly along the trunk within the lower fruiting zone. Prune off the unwanted branches, and rub off unwanted shoots as they begin developing in the spring.

Horizontal branches are the key to the axis and spindle systems, but we do not use trellis wires to tie the branches. Train the lower zone branches horizontally or just above horizontal. You can tie larger branches down to stakes that are driven into the ground between the trees, or tie weights to the branches. For new shoots, use spring-type clothespins clamped around the leader with the tails of the clothespins pressing the lateral shoots down beginning when they are about 4 inches long. Leave the clothespins in place for about 2 weeks.

As the shoots grow longer, they naturally turn up at the tips. To prevent this, clip the clothespins to the ends of the developing branches and keep moving the clothespins further out about every 10 days as the branches grow. Add more weight, if necessary, to keep the branches flat in the lower fruiting zone. The upper fruiting zone is formed similarly to the lower, but the branches are trained to an angle below horizontal using weights and tie-downs.

Sweet cherries have strong apical dominance and often develop long sections of "blind wood" without fruiting laterals. To form the laterals in the lower and upper fruit zones, some growers head back the leaders

Divided Canopy Training Systems

Tatura and other divided canopy training systems are sometimes used for cherry production and can be very productive. The move today, however, is toward central leader trees and narrower rows. Divided canopies with cherry trees tend to become very dense and are likely to increase pest and disease problems. For that reason, I do not recommend them for organic growers.

to 50 percent of their length during the first three dormant prunings to force lateral buds to develop into fruiting branches. This practice, however, slows tree development and reduces yields during establishment. Rather than heading the leader to force laterals, leave the terminal bud but rub off about six buds below it with your thumb just as the buds are swelling in spring. Then leave about every third bud along the leader to develop lateral branches, as we discussed earlier. Use this spacing as a guide, but make sure you develop branches that are evenly distributed around and along the trunk. Mature trees appear similar to those shown in figure 12.13.

Managing mature trees. Once the cherry trees mature, pruning mostly involves thinning out excess wood to keep the canopy open and develop renewal fruiting wood. To accomplish this, thin out several of the larger branches each year in both the upper and lower zones. When cutting back large branches, leave branch stubs about 6 inches long. This practice helps limit the incidence of bacterial canker in the stubs and protects the main trunk from infection. It is especially important to use stub pruning in humid regions, where bacterial canker is a serious problem. Also use heading cuts to keep branches from intruding into the alleys or crowding adjacent trees. If you are growing the trees on Gisela 5, Gisela 6, or other highly productive rootstock, use the pruning practices described on pages 447–449.

Training Tart Cherry Trees

Tart cherry trees are naturally smaller than sweet cherry trees and are trained somewhat differently. Tart cherry trees bear their fruit mostly on 1-year-old wood and on spurs that are concentrated on 2- to 3-year-old wood. Due to these characteristics, the fruiting wood needs to be renewed relatively frequently compared with sweet cherries.

Because the fruit is generally used for processing, rather than sold fresh in markets, growers often use mechanical harvesting techniques to reduce labor costs. Many common training systems are designed around mechanical harvesting, in which machines grasp the trunks and shake the trees. Researchers are testing designs that use smaller trees trained to bushes and harvested with over-the-row blueberry harvesters. They generally do not use dwarfing rootstocks, and the trees are typically trained to modified central leaders for both hand and machine harvesting.

In Michigan, which is the leading producer of tart cherries in North America, fruit specialists recommend spacing trees 14 to 18 feet apart in rows that are 20 to 22 feet apart for standard-sized trees bearing amarelle-type (yellow flesh) fruits. Amarelle cherries make up nearly all commercial tart cherry production in North America. Morello cherries (red flesh) are sold in niche markets and produce somewhat smaller trees that can be spaced 15 feet apart within and between rows.

'Meteor' and 'North Star' are cold-hardy genetic dwarfs that can be kept 6 to 8 feet tall and 8 to 12 feet wide. Plant these varieties 10 to 12 feet apart in rows that are 16 to 18 feet apart. Dwarf tart cherries are ideal for home orchards and may be useful for market orchards and to diversify U-pick orchards. It is unlikely the trees will produce enough fruit to meet the needs of grower-picked commercial orchards.

For hand and mechanical harvesting, a modified central leader design works well (figure 12.9). Although you can use whips, if you start with well-feathered planting stock your trees will produce more quickly. If you use whips, rub off unwanted buds as described above for sweet cherry (see page 452), rather than heading back the leaders. Heading back the leaders of tart cherry trees tends to stunt the trees.

The mature trees will have four or five scaffold branches that attach to the trunks between 2.5 and 7 feet above the ground. Some growers prefer to have the lowermost branch face southwest. This practice helps shade the lower trunk and offers some protection against sunburn, to which tart cherry trees are susceptible. As with other central leader training systems, avoid developing two or more branches at the same height on the trunk. When the leader grows to the desired height, cut it back to a weak lateral.

The pruning techniques used for tart cherry trees are similar to those used for sweet cherry trees, but the buds of tart cherry trees are shorter-lived, and fruiting wood should be renewed about once every 5 years. Once trained, tart cherries require less pruning than most other stone fruits; you only need to renew fruiting branches and thin to prevent shading and crowding. If you harvest your fruit mechanically, head back willowy branches to stiffen them. Long, thin fruiting wood is too limber to release fruits easily when shaken.

Training Peach and Nectarine Trees

Peaches and nectarines can be grown both freestanding and supported on trellises in divided canopies. The Tatura trellis has proven especially popular with some peach growers and produces high yields. The dense canopies and difficult-to-reach interiors of the trees greatly complicate pest and disease management, especially for organic orchardists. For the purposes of this book, we will focus on using freestanding peach and nectarine trees.

Open Center Training

Open center training remains the standard for home orchard and commercial peaches and nectarines throughout North America. This design is standard in commercial California and Texas orchards and works well in warmer climates, where cold injury and stem cankers are not serious problems. In some cases, the steep ladder design we discussed for cherries (see page 448) is also used for peaches and nectarines to enhance color and fruit size.

For open center training, plant peach and nectarine trees 18 to 24 feet apart in rows that are 18 to 24 feet apart. In dry areas where trees are not irrigated, space the trees 24 feet apart within and between rows.

It is easy to adapt the training and pruning practices described for apple, pear, and sweet cherry trees to peaches and nectarines. The most popular open center design uses three or four main scaffold branches that are evenly distributed around the trunk to form a symmetrical bowl. Develop the lowest branch about 18 to 30 inches above the ground, and space other branches about 4 to 6 inches apart vertically along the trunk. Choose shoots with wide branch angles and/or use clothespins and weights to ensure wide branch angles. Leave the central trunk in place for now, and use it as an anchor point for spreader bars to train the scaffolds.

As the scaffold limbs grow, select two or three strong secondary branches on each of them, with the lowest at least 15 inches above the base of the scaffold. Try not to use heading cuts during this stage of training. Once you have established your secondary scaffolds, remove the central leader where it joins with the uppermost main scaffold branch. Trees that are trained to open centers usually mature in 5 to 6 years.

Use light to moderate annual pruning to maintain a strong, open framework. Because peaches and nectarines bear their fruit on 1-year-old wood, you need to renew the fruiting wood annually. Otherwise, fruit production moves upward and outward, producing an overly wide, top-heavy tree that has little fruit close to the ground. Thin out secondary scaffolds as needed to keep the trees within a desirable height range and to maintain fruit production throughout the trees. When pruning to develop outward-pointing branches, avoid making bench cuts (figure 12.1), if possible, because they tend to weaken the scaffold limbs. Use spreaders, weights, and ties, rather than pruning, to position branches. Figure 12.10 is similar to an open center peach tree.

Perpendicular-V Training System

The perpendicular-V training system is adapted from the open center, vase-shaped designs, but it uses only two scaffold limbs. The scaffold limbs are oriented perpendicular to the crop rows. The design creates an open canopy with good light penetration and air movement. It is relatively simple to control pests and diseases in this system. Although perpendicular-V trees yield somewhat fewer fruits than trees trained in the quad-V, the perpendicular-V system appears to be a better choice for

Figure 12.22 · **Peach Trees Trained to a Perpendicular-V System**

Each tree has two scaffold branches perpendicular to the row. Note how the scaffold branches are headed back occasionally during establishment to develop the shape. Use thinning cuts to develop renewal fruiting wood along each scaffold branch.

high-density organic peach and nectarine orchards because it maintains a more open canopy.

Perpendicular-V peach orchards typically have 375 to 580 trees per acre, and trees are spaced 5 to 7 feet apart in rows that are 16 to 20 feet apart. In areas with vigorous tree growth, unless space in your orchard is limited and you are experienced in growing peaches and nectarines, I recommend a spacing of 6 feet apart in rows 18 feet apart. In areas with cold winters and where tree vigor is limited, you can space peach and nectarine trees as closely as 5 feet apart in rows that are 16 feet apart. Whenever possible, orient the tree rows north–south to reduce sunburn on the scaffolds.

Develop two strong scaffold limbs, opposite to each other and that are perpendicular to the tree row. Together, these branches form a vase shape with a 50-degree angle. The limbs should arise from the trunk about 18 inches above the ground. Secondary scaffold branches are not

Figure 12.23 · **Peach Trees Trained to a Quad-V System**

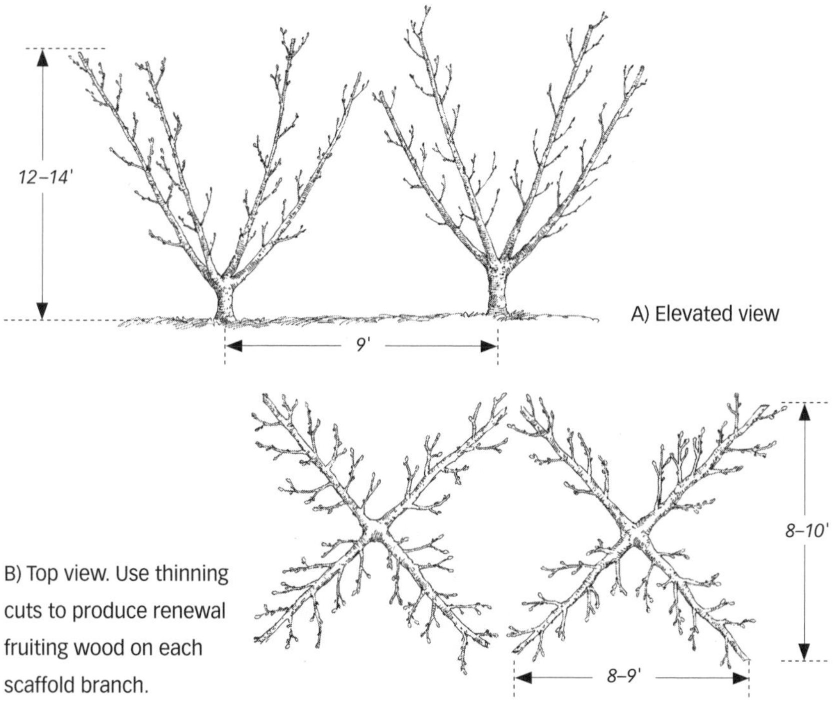

12–14'

9'

A) Elevated view

8–10'

8–9'

B) Top view. Use thinning cuts to produce renewal fruiting wood on each scaffold branch.

used in the perpendicular-V system; only fruiting wood is allowed to develop on the main scaffolds. Figure 12.22 shows peach trees trained to a perpendicular-V design.

Quad-V System

The quad-V system is another adaptation of open center training. It uses four permanent scaffold limbs that are spaced approximately 90 degrees apart, perpendicular to and parallel to the crop rows. The design is somewhat more difficult and labor intensive to establish than the system for perpendicular-V trees. Quad-V orchards use fewer trees per acre than perpendicular-V orchards, and they create greater early yields by doubling the number of fruit-bearing scaffolds per tree. Plant the trees 9 feet apart in rows that are 18 feet apart, for a density of 269 trees per acre. This design can be used for home and commercial high-density peach and nectarine production, but it requires greater management skill and labor than the open center or perpendicular-V systems. See figure 12.23 for how peach trees are trained to a quad-V design.

Training Apricot Trees

Apricots develop fruit both on 1-year-old limbs and on spurs found on older wood. In general, apricot trees require similar pruning and training practices as peach trees, and the amount of pruning they need lies somewhere between that needed for apples and that needed for peaches. Apricots are usually grown as freestanding trees. These trees tend to spread, so relatively wide spacing is the rule.

When you are training apricot trees to open centers, follow the recommendations for peaches. Space the trees about 18 to 24 feet apart in rows that are 18 to 24 feet apart. In the southeastern United States, fruit specialists recommend spacing apricots 20 to 24 feet apart within and between rows.

Apricots can also be trained to modified central leaders. Follow the recommendations for training apples, but use more thinning and heading cuts to ensure an open canopy. A tree spacing of 18 to 24 feet apart in rows that are 18 to 24 feet apart works well.

For higher-density plantings, you might consider experimenting with the perpendicular-V or quad-V training systems, as described above for peaches.

Training Plum and Prune Trees

Plums are among the easiest stone fruits to train, and they adapt well to both central leader and open center designs. For higher-density orchards, plums (especially European varieties) can be trained to palmettes parallel to the crop rows. Plums have a tendency to develop erect shoots with narrow branch angles and bark inclusions. Early pruning and branch spreading take care of this problem.

Like peaches, plums bear their fruit laterally on 1-year-old shoots. They also set fruit on relatively long-lived spurs, like apples and sweet cherries. Japanese plums tend to bear fruit more heavily along shoots than their European cousins and are more often trained like peaches.

As a general rule, encourage about 10 to 24 inches of new shoot growth for young trees and at least 10 inches of new shoot growth for mature trees. If shoots exceed those lengths, you may need to prune less, spread the branches more, and/or reduce the amount of nitrogen fertilizer you use. If shoot growth falls below these levels, prune more heavily and have your foliage tested to determine nitrogen status.

A modified central leader system and the pruning practices that work for apples also work well for upright European plum varieties such as 'Stanley' (figure 12.9). With upright varieties, pay careful attention when removing unwanted new shoots that form narrow branch angles. Select new shoots that point outward. Use branch spreaders, when necessary, to ensure suitable branch angles. European plums that have been trained early and well to a modified central leader design require relatively little pruning throughout their lives.

Training Bush Cherries and Plums

Bush cherries and plums are typically planted about 5 feet apart in rows that are 10 to 12 feet apart, depending on the mature size of the bushes. These crops are usually grown as freestanding bushes. The training and pruning techniques described for saskatoons generally produce satisfactory results (see page 445). The goal is to maintain productive, open canopies and to occasionally renew the canes.

European and Japanese plums adapt well to open centers (figure 12.10). For most European varieties, follow the recommendations for open center apples and pears described at the beginning of this chapter.

Japanese and European French plum varieties are trained more like peaches. When training new trees, select three to five laterals for your scaffold, with the branches starting about 18 to 36 inches above the ground and equally spaced around the trunk. The system is very similar to that described earlier for open center peaches (see page 459), but you may want to leave a few more secondary branches.

With Japanese plums, pay particular attention to thinning out water sprouts and unwanted lateral branches. Japanese varieties tend to produce too much fruit, leading to small size fruit and more disease problems. For these reasons, you must prune Japanese plums more heavily than you would European varieties. Use thinning cuts to open up the canopy. Use stub cuttings (described for sweet cherry, page 447) to renew fruiting branches that are more than about 5 years old. By renewing fruiting branches, you can help reduce shading and maintain large fruits without using excessive hand thinning. You can also renew branches of European plums by removing older branches heavily set with fruiting spurs.

FRUIT THINNING AND HARVESTING

- Types of Thinning | 468
- Thinning, Crop by Crop | 473
- Harvesting | 482

Many temperate fruit crops develop excessively heavy fruit loads, which creates several problems. Some fruits, particularly apples and apricots, can develop a heavy crop one year and a very light crop the next year or sometimes two. Our goal is to produce annual, sustainable yields of large, high-quality fruit.

In many crops, having too many fruits on the tree reduces the size of the fruit. Excessive numbers of fruit also interfere with light distribution, decreasing the number of flower buds formed in some crops for the following season and creating poor fruit color. Crowding of fruits interferes with air movement and spray applications and allows adjacent fruits to touch one another, increasing pest and disease problems. Trees require large amounts of energy to develop fruits, and excessively heavy crops weaken and stunt trees, making them more susceptible to pests and diseases and increasing the likelihood that limbs and trunks will break.

Fruit crops naturally thin themselves. For most crops, flowers that have not been pollinated fall off without developing fruits. In pome fruits, very young fruits that have only a few seeds drop off in what is called "June drop." June drop can be alarming for new fruit growers because it looks as if the entire crop will be lost or the tree has some terrible disease. In reality, you need only a small percentage of the total possible number of fruits that could develop on a tree to mature in order to obtain high yields and excellent-quality fruit. In the case of apples, up to 90 percent of young apples must be removed following a heavy fruit set in order to maintain good size and quality and consistent yields. Stone fruits also drop poorly developed fruits right after bloom and sometimes drop fruit later in the season, when the tree's reserves cannot support the crop load. For all crops, fruit that is damaged by pests and diseases may fall.

Berry crops, including saskatoons, are usually only thinned through regular pruning. For some varieties that tend to overbear, you can prune the bushes rather heavily to reduce fruiting wood and remove some flower clusters.

For apples, pears, loquats, quinces, apricots, nectarines, peaches, and plums, fruit thinning in addition to natural fruit fall is required to maintain good quality, high yields, and healthy trees. Cherries, especially sweet cherries grown on highly productive rootstocks, also benefit from thinning. When to thin depends on the crop and the thinning method used.

Proper pruning techniques remove excess fruiting wood and are an important step in managing the crop loads of many tree crops. As we discussed in chapter 12, for example, special pruning techniques help reduce the crop load in sweet cherries that are grown on highly productive rootstocks like Gisela 5. Another example is the special pruning technique used for apples grown in the tall spindle training system. In both cases, careful pruning removes older wood that produces large numbers of small fruit.

Thinning Recommendations for Pome and Stone Fruits

Table 13.1

Crop	Thinning Required	Fruit Spacing
POME FRUITS		
Apple	yes	1 or 2 fruits per cluster or 4–8 inches apart
Loquat	yes	3–5 fruits per cluster
Mayhaw	no	n/a
Medlar	no	n/a
Pear (Asian)	yes	1–3 fruits per cluster or 4–8 inches apart
Pear (European)	yes	1–3 fruits per cluster or 4–8 inches apart
Quince	No	Usually managed by pruning
Saskatoon	no	n/a
STONE FRUITS		
Apricot	yes	3–5 inches apart
Bush cherries and plums	no	n/a
Cherry (sweet)	sometimes	Usually managed by pruning
Cherry (tart)	no	n/a
Nectarine	yes	3–6 inches apart
Peach	yes	3–6 inches apart
Plum (European)	yes	2–4 inches apart
Plum (Japanese and hybrid)	yes	3–4 inches apart

Excessive pruning, however, delays the onset of fruiting and reduces yields. Pruning too heavily also stimulates excessive, upright, and unproductive wood that complicates training procedures, increases pest and disease problems, and requires more pruning and branch bending to correct. While pruning is an important step in managing crop load, it is not the only step. For some crops, additional thinning to remove some of the young fruits is needed.

Types of Thinning

CROP LOADS CAN BE REDUCED BY REMOVING FLOWERS; very young, green fruit; or older fruit. Thinning can be done by hand, with handheld devices, or by large, tractor-mounted machines.

Hand Thinning

By picking off some of the fruits, you leave more room and resources for the remaining fruits. Start by removing double fruits (fruits that are fused together), fruits that are smaller than average for the tree, and damaged or deformed fruits. Then remove additional fruits to achieve the densities shown in table 13.1 and described below.

While you can follow the guidelines for spacings in table 13.1 exactly — for example, spacing apples 4 to 8 inches apart — a better approach is to thin according to size and use the spacings as averages. Fruits that are small at the time of thinning will likely remain small at maturity. Leave larger fruits when possible, even if it means leaving a smaller spacing between some fruits and a larger spacing between others. Use the spacings given in the table as a guide for the average number of fruits that should be along a given branch. When thinning apples, for example, for every 48 inches of fruiting wood, you should leave 6 to 12 of the largest fruits. Space the individual fruits far enough apart that they do not push each other off the branches or rub against each other excessively as they enlarge. For apples, leave no more than two fruits per cluster and leave some fruiting spurs without fruit to ensure adequate return bloom the following season.

Pole Thinning

Pole thinning speeds up the thinning process and is especially useful where hand thinning requires a ladder. You can attach about 6 to 10 inches of rubber hose to the end of a broomstick or similar pole, or tape a piece of cloth around the end of a pole to serve as padding. The rubber or cloth softens the blows and helps prevent damaging the branches. Plastic or rubber-covered baseball bats also work well for thinning.

Pole thinning is less accurate than hand thinning but much faster, and with practice, you can obtain good results. Use the padded end of the pole to knock off individual fruits, but avoid breaking off the fruiting spurs. Also use the poles to thin out fruit clusters. Strike a cluster once or twice to break it up and remove some of the fruit.

Mechanical Thinning

Mechanical thinning is used in pome and stone fruit orchards. Early thinning methods included ropes mounted to rapidly spinning cylinders or disks, or high-pressure water jets. In these methods, the spinning ropes or water jets knocked some of the fruits off the trees. They did not produce uniformly effective results in orchards, however, and have given way to shakers and string thinners.

Regardless of the mechanical thinning method you use, you will need to follow up with hand and pole thinning to achieve optimum fruit sets. The advantage is that mechanical thinning reduces the amount and cost of hand thinning required.

Shakers. In fruit thinning trials, researchers found that shaker devices can be quite effective, and the method is now being adopted by orchardists, particularly for peaches. These devices also hold promise for apple and plum orchards. Several versions have been developed by the U.S. Department of Agriculture and private sources and are being tested across the United States. Essentially, they are modified citrus and raspberry harvesters with long rods mounted to oscillating spindles or drums. As the tractor travels along the alley, the rods reach into the canopy and shake the branches. The devices are probably best used 40 to 55 days after full bloom to remove green fruits, although they have been tested during bloom and can effectively remove flowers. The type of crop,

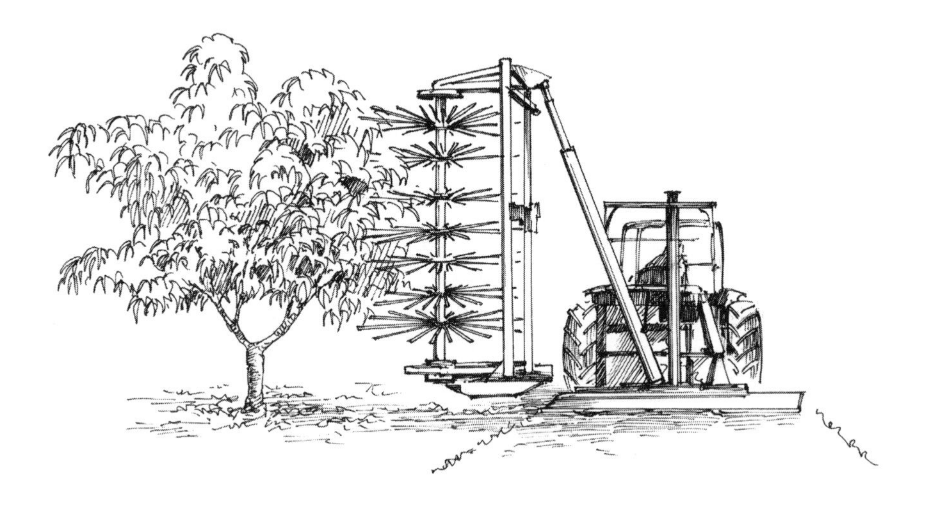

Figure 13.1 · A shaker device used to thin green fruits and sometimes to thin flowers

variety, and canopy design, as well as other factors, affect the thinning process. You will need to run some tests in your orchard to determine the best time to thin using a shaker. The shakers have primarily been developed and tested for use in narrow canopy designs, particularly perpendicular-V and quad-V training systems for peaches. Figure 13.1 shows a tractor-mounted shaker used for thinning fruit trees.

String thinners. Trials in Ontario, Pennsylvania, and Washington State have shown quite good results using a specially designed, tractor-mounted string thinner. The Darwin string thinner consists of long, spinning, vertical cylinders equipped with hundreds of strings. The tractor is driven about 2.5 to 3 miles per hour through the alleys, with the strings striking the flowers. The best time to thin is when about 20 percent of the flowers have opened. Typically, about 30 to 50 percent of the flowers and flower buds are removed. You only need about 10 to 15 percent of the flowers that set to produce sustainable crops of apples, pears, peaches, and nectarines. Various designs are emerging with long single-spindle thinners, shorter and independent upper and lower spindles, and horizontal over-the-tree thinners. Figure 13.2 shows several tractor-mounted string thinner designs for orchard thinning.

The vertically mounted tractor-mounted string thinner works best with narrow-row, flat canopies. For other training systems, researchers

Figure 13.2 · **Thinning Flowers Using Tractor-Mounted String Thinners**

Some models allow the thinner to be operated vertically along the rows or horizontally over the tops of the trees.

A) Thinner positioned vertically or at an angle.

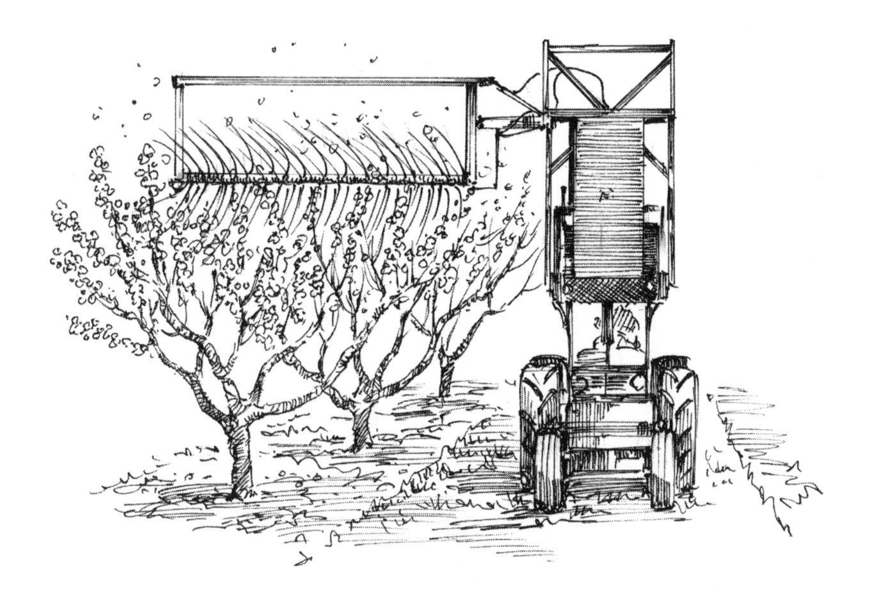

B) Thinner positioned horizontally above the tree canopy.

are testing handheld string trimmers that have been modified for use in fruit tree thinning, and one version (Bonner) is marketed commercially by a German firm.

Advantages and disadvantages. Mechanical thinning shows promise and is being adopted commercially, but we are in the early part of the learning curve using these methods. While mechanical thinners effectively remove flowers and green fruits, they can damage the trees, increasing their susceptibility to diseases. We do not yet know how much damage the thinners will cause and how they will affect the long-term health of the tree. The design of the canopy also influences the effectiveness of thinning techniques and their damage to the trees. Pennsylvania peach growers have found that branches that project into the alleys suffer greater damage from string thinning than those within the crop rows. Trials in West Virginia have shown that large, rigid limbs are more susceptible to damage than thinner, more flexible branches.

Growers with high labor costs for thinning should consider testing mechanical thinning. Do your homework and evaluate the devices that are available. At present, the string thinner approach seems to be somewhat more popular with commercial orchardists than the shakers, but both designs continue to evolve. The most effective approach will probably be to use a combination of the two methods: string thinning to remove flowers, followed by shaker thinning to remove green fruits. This technique reduces hand thinning more than either thinning method alone, and it allows you to judge how many fruits to remove in midseason after the fruits have set.

Test the devices on small blocks within your orchard. Compare fruit size, yields, quality, and tree health over several years before adopting the approach throughout your entire orchard. You may find that you need to adjust your training methods and canopy designs to optimize thinning and reduce tree damage.

Chemical Thinning

Chemical thinning has a greater impact on pome fruit size and fruit bud formation than does hand or pole thinning and is a standard practice in conventional apple orchards. Unfortunately, despite decades of research, no effective and reliable chemical thinners have been found for organic stone fruits. Chemical thinning is discussed under specific fruits, below.

Thinning, Crop by Crop

While the general principles of thinning apply to most stone fruit and pome fruit crops, the specific methods vary slightly from crop to crop. We cover these specifics in the following sections.

Apples

In pome crops, if a flower receives abundant, compatible pollen and weather conditions are favorable, many seeds form. The presence of these seeds stimulates growth, causing a large and uniformly shaped fruit. If a flower receives only a fraction of the pollen that it requires, only a few seeds form in the fruit. Fruits that have a small number of seeds are typically small and often misshapen. Some of these fruits fall naturally during June drop. Those that remain continue to compete for nutrients, reducing the size of well-formed fruits on the same tree.

Hand thinning. In apples, the centermost blossom is called the "king" blossom. King blossoms usually open first in a cluster and produce the largest fruits. When hand or pole thinning, try to target the smaller surrounding fruits; ideally, you want to leave one apple on every other spur. Spurs that do not have apples are more likely to develop fruit buds for next year's crop. As mentioned above, however, keep the larger fruits when possible and leave 6 to 12 fruits for every 48 inches of fruiting wood.

Chemical thinning. Growers in conventional apple orchards have applied chemical sprays during and shortly after bloom for many years to cause some of the flowers and fruits to fall. For a long while, the materials of choice were carbaryl insecticide, usually combined with synthetic plant growth regulators like naphthalene acetic acid (NAA). None of these materials can be used in certified organic orchards.

Using synthetic chemicals is an extremely effective and relatively inexpensive way to thin fruits. Not only are the chemicals effective, but they can be applied over an extended time during bloom and early fruit development. This long period allows you to see how many fruits have been formed and to accurately remove just enough fruit to ensure annual crops of large fruits.

Organic apple orchardists have more limited chemical thinning options. Researchers have tried salt (sodium chloride) and vinegar, but

Choosing the Right Thinner

As of this writing, fish oil was not registered as a thinning agent in California, and researchers there found that soybean and cottonseed oil were more effective than fish oil. Note that not all liquid lime sulfur products are approved for thinning by certified organic fruit growers. Check all of your oil and sulfur products to ensure that they are approved for thinning before applying them.

they have not produced consistently effective or reliable results. The most popular thinning agents at this time are a combination of oil (usually fish oil) and liquid lime sulfur (LLS). Fish oil and LLS sprays proved so successful in Washington State trials that some conventional growers have adopted the practice.

A typical thinning spray consists of 2.5 percent fish oil and 2 to 4 percent liquid lime sulfur. Use the lower rates when you expect fruit set to be lower. Use the higher rates if you expect a particularly heavy fruit set. According to Cornell University, Crocker's Fish Oil (Quincy, Washington), was used in their thinning trials, although other fish oil products may be equally as effective. In thinning trials, fish oil was somewhat more effective than mineral crop oils. California researchers reported that soybean and cottonseed oil were more effective thinning agents than was fish oil.

When to spray. Washington State fruit specialists recommend applying a fish oil/LLS spray during bloom. Unfortunately, at that time you do not yet know what the fruit set will be and you run the risk of thinning too heavily. For example, cold, wet weather can interfere with pollination and fertilization. A late frost can damage flowers and developing fruits, as can hail and heavy winds. In the generally warm, arid or semiarid fruit growing regions of western North America, applications during bloom appear suitable. The objectives in this region are to thin out some of the flowers and to reduce pollination and the percentage of fruit set. For eastern fruit growers, Cornell University suggests waiting to make the first application of fish oil/LLS until petal fall, when danger of frost has passed and you can better determine how much fruit has set. Apply a second spray 4 to 7 days after the first, and a third spray may be needed.

The downside of the Cornell approach is that it appears to be less effective than making applications during full bloom, and there is a greater risk that the lime sulfur will damage the leaves and fruit. Lime sulfur is also toxic to beneficial organisms and repeated applications of LLS can increase pest problems. The present theory for why the oil/LLS sprays work is that:

- The liquid lime sulfur inhibits photosynthesis and puts a stress on trees when their reserves are low and demands for those reserves from flowers and fruit are high. This stress causes some of the developing fruits to drop.
- The liquid lime sulfur directly damages the flowers and reduces the percentage of flowers that are successfully pollinated and develop into fruits.
- The oil increases the plant uptake and efficiency of the LLS.

Cornell fruit specialists note that dark, cloudy weather for about 2 days or more before or after applying chemical thinners increases the thinning response. If you expect cloudy weather just before or after applying oil/LLS, consider reducing your rate of application.

Temperatures also influence the effectiveness of chemical thinners. In general, higher temperatures increase the rates at which chemical reactions take place. Night temperatures of 65°F (18°C) or greater and daytime temperatures of 85°F (29°C) or greater during the 3 to 5 days after applying sprays can increase the thinning response. If you observe or expect temperatures in these ranges, consider using the lower application rates.

Loquats

For commercial production, you must thin loquat fruits to produce fruits of acceptable size. Loquats also tend to develop an alternate year bearing pattern like apples. You can thin clusters to help reduce alternate bearing and remove flowers and fruits to increase loquat fruit size.

Loquats set about nine or ten fruits per cluster. By leaving three to five fruits per cluster, you will increase fruit size and the percentage of marketable fruits. In Chinese trials, enclosing the fruits in bags (bagging) after thinning reduced the amount of sunburned and rusty fruit from about 48 percent to less than 1 percent and greatly increased the percentage of premium-quality fruits. Bagging tree fruits in fabric or weather-resistant paper to protect the fruits from pest, disease, and environmental damage and increase the percentage of premium-grade fruit is a common practice in parts of the Orient.

There is a trade-off to thinning loquats, however. Worldwide, purple spot is the most serious physiological disorder affecting loquat, causing the rind to discolor and dry out. In Spanish trials, thinning to leave five fruits per cluster when the fruits were ⅜ inch in diameter increased the incidence of purple spot from about 2 percent to about 6 percent. Leaving three fruits per cluster increased purple spot to 12 percent of the fruits, and leaving one fruit per cluster increased purple spot to 36 percent of the fruits. The cause of purple spot is not yet known, but it correlates with the amount of mineral nutrients in the fruits. Thinning alters the concentrations of mineral nutrients in the fruit. In short, be careful not to overthin.

You can use several approaches to thin loquats. One approach is to thin out the flower clusters, usually at or shortly after full bloom. Clip off about 50 percent of the flower clusters, leaving the remaining clusters evenly distributed throughout the tree. Alternatively, you can clip off about half of the flowers from each of the clusters.

You can also thin the fruits shortly after petal fall but before the normal drop, when the fruits are about ⅜ inch in diameter. By removing fruit clusters that set late during the bloom, you harvest larger fruits in a shorter period. Larger commercial growers desire a concentrated harvest. For home and market production, you might want to keep some of the later fruits in order to extend the harvest. Using this approach, remove some clusters entirely or remove some of the fruits from each cluster. When removing some of the fruits from the clusters, leave roughly three to five fruits per cluster. In Japan, the 'Tanaka' variety is thinned to one fruit per cluster and 'Mogi' to two fruits per cluster.

Some growers use chemical thinning methods for loquats, particularly those in Taiwan, but the results have been variable and the materials used are not allowed in organic production.

Pears

Pears generally require less thinning than apples, although some thinning is generally required with European varieties to produce large fruits. Asian varieties require heavier thinning than European varieties, and the following recommendations work for both crops. Although some growers thin by clipping off all but two or three flowers per cluster during bloom, thinning normally starts immediately after June drop, when you can estimate the crop load.

Some fruit specialists recommend removing all pear fruits from trees for the first few years after planting in order to allow more energy for tree development. This practice is most likely to be useful on trees that are grafted to dwarfing rootstocks.

For average or light crops, you can usually leave one to three fruits per spur. 'Bartlett' and 'Bosc' normally set three to five fruits per spur and can often mature them if the crop load is not too heavy. 'Bartlett' also usually self-thins enough by itself to produce marketable crops.

When pears set an unusually heavy crop, you may want to remove all but the single largest fruit per spur. Another approach is to space the pears an average of 4 to 8 inches apart along the branches. As with apples, this works out to be 6 to 12 fruits for every 48 inches of fruiting wood.

When crop loads are light to moderate, you can wait until harvest to pick the largest fruits from each cluster first and leave the remaining fruits on the trees for another 7 to 10 days to develop more size. If you choose to harvest in two stages, be careful not to leave 'Bartlett' fruit on the tree too long, as this variety is prone to a physiological disorder called water core. Water core results in soft, watery, discolored flesh near the core, decreasing the market value of the fruit.

The organic chemical thinning practices we discussed using on apples may also apply to pears, but more research is needed before making recommendations for how to chemically thin organic pears.

Quinces

Quinces generally do not need to be thinned unless the trees set an unusually heavy crop. Some precocious varieties of quince, such as 'Aromatnaya', begin bearing fruit very young and can benefit from thinning during the first few years to develop a vigorous tree with good structure.

In mature trees, crops are usually managed through pruning to remove long, weeping branches. Thinning, if needed, is done by hand.

Mayhaws, Medlars, and Saskatoons

Mayhaws produce small, crab apple–sized fruits and are not thinned. Medlars are normally not thinned. In the case of an unusually heavy crop, you might reduce the number of fruits somewhat by hand thinning shortly after June drop. Saskatoons are not thinned.

Apricots

Apricots tend to overbear and produce small fruits. Just after June drop, when the developing fruits are about 1 inch in diameter, thin them by hand or with a pole. Over the past 30 years, recommendations for the spacing between thinned apricots have ranged from 1 to 2 inches to 6 to 8 inches. A good compromise is to leave an average of one fruit every 3 to 5 inches (6 to 10 fruits per 30 inches of fruiting wood); this will produce large fruits and economical yields. If possible, leave only one fruit per cluster.

As yet, no effective and reliable chemical thinners are available for organic apricot production. Remember that apricots are easily damaged by sulfur, and liquid lime sulfur (LLS) should not be applied to apricot trees.

Cherries

Although trials are still underway around the world, thinning sweet cherries remains controversial. For most varieties on vigorous rootstocks, flower or fruit thinning is not required. For self-fruitful varieties on highly productive rootstocks, small fruit size has been a problem. While chemical thinners have been effective in reducing fruit loads, the reduction in crop load has generally not increased fruit size.

Washington State University has been conducting experiments using the tractor-mounted string thinners discussed earlier to thin sweet cherries. The best results generally come when the trees are thinned when the buds have swollen to about ¼ inch in early spring. The results show some promise for thinning, but they are too preliminary to make recommendations.

For self-fruitful varieties grown on Gisela 5, Gisela 6, and other highly productive rootstocks, follow the pruning recommendations in chapter 12 to help manage crop loads.

Tart cherries do not require thinning.

Peaches and Nectarines

Peaches and nectarines require thinning to develop large fruits. Thinning is also required in commercial orchards for fruit destined for processing in order to achieve uniform sizes. Hand thinning has been the standard and is still required in commercial orchards. Researchers are testing both mechanical and chemical thinners, and mechanical thinners have proven very effective in reducing the amount of hand thinning needed. For large commercial orchards, reducing the amount of labor required to thin fruit often makes a large economic difference.

Shakers and string thinners. Commercial growers are adopting shakers and string thinners for peach production. In research trials, these methods produced more consistent and effective results than did the chemical thinners tested, and they increased fruit size and greatly reduced the amount of follow-up hand thinning that was needed. In Pennsylvania trials, the drum shaker reduced crop load more than all of the other treatments, produced fruit of a similar size to the fruit that was string thinned, and required about 15 worker hours per acre for follow-up hand thinning. The string thinner also increased average fruit size and reduced the variability in fruit size. About 22 worker hours per acre were required for hand thinning after using the string thinner. Blocks that were thinned only by hand required about 40 worker hours per acre of thinning.

So far, researchers have noted little damage to the trees when using mechanical thinners, except for damage done by string thinners to large limbs sticking out into the alleys. Most of the trials have involved relatively flat canopies, primarily quad-V and perpendicular-V training systems. The overhead string thinner can be used with any training system, but it will not be effective in thinning fruit in the lower canopy. For large peach and nectarine orchards, the most effective approach will probably involve using a combination of a vertically mounted string thinning at 20 percent bloom followed by shaker thinning 40 to 55 days after full bloom.

Chemical sprays. Chemical thinning in research and commercial orchards has proven less reliable with stone fruits than with apples and pears, although recent research shows some promise for developing effective chemical thinners that can be used in organic orchards. Vegetable oil emulsions applied to the dormant buds have produced some thinning and increased fruit size. In Pennsylvania research trials, a caustic fertilizer (ammonium thiosulfate) and various surfactants normally used in pesticide applications were tested and found to produce less consistent results than mechanical thinners. Organic oils applied with synthetic surfactants have shown promising results, but these materials cannot be used in organic orchards. While ongoing research shows some promise, no effective chemical thinners are yet available for organic peach and nectarine production.

Hand thinning. Hand and pole thinning, whether used alone or as a follow-up to mechanical thinning, remains the standard for peach and nectarine production. Begin hand and pole thinning after June drop, which is usually about 5 to 8 weeks after full bloom. Start with the earliest-maturing varieties.

Spacing. Table 13.1 suggests spacing the fruits an average of 3 to 6 inches apart (6 to 12 fruits per 36 inches of fruiting wood), although you can find other recommendations for wider spacings up to 8 inches apart. Much depends on the intended use of the fruits and the variety. Although counterintuitive, small-fruiting varieties require wider spacing than varieties that naturally set larger fruit. You might want to space fruit intended for fresh use somewhat farther apart than fruit intended for processing. Vigorous, well-exposed 1-year-old wood produces the largest and best fruits. Remove most or all of the fruits from thin and shaded wood.

In general, try to leave 30 to 40 full-sized leaves for each peach or nectarine on the tree. With a little counting and practice, you can visually estimate the numbers of fruit to leave. If the load is unbalanced and only part of the branches bears heavy crops, you might need to do little or no thinning. The goal is to adjust the crop load for the entire tree.

In commercial orchards, hand thinning is complex and based on total crop load, leaf surface, tree spacing, and tree health. The science behind the process is that it takes a certain amount of leaf surface to set and fully mature a high-quality peach of a given size. The process is very sophisticated in the clingstone peach industry, where the fruit is destined for canning.

Experienced growers of commercial processing peaches, for example, can accurately develop uniform peaches of a particular size by leaving a certain number of fruits per tree. Growers determine how many fruits to leave based on the size of the fruit at a specific reference date, typically 10 days after pit hardening. Pit hardening is determined by slicing through the fruits with a sharp knife. With a little practice, you can detect the hardening pit.

If the peaches are smaller than a certain diameter at pit hardening, you will need to thin heavily. Slightly larger fruits require less thinning and fruits that are above a given size require no thinning. According to *Modern Fruit Science*, diameters of 33, 36, and 38 mm, respectively, at the time of pit hardening serve as starting points. With experience, growers may adopt different threshold diameters based on climate, cultural practices, and varieties.

Plums and Prunes

Plums need to be heavily thinned in order to develop large, uniform-sized fruits with good color. Thinning also helps ensure that the trees are vigorous, and it reduces limb breakage due to heavy crops. Hand and pole thinning work well for plums. Leave European varieties an average of 2 to 4 inches apart (9 to 18 fruits per 36 inches of fruiting wood). Japanese plums set more heavily, and an average of 3 to 4 inches between fruits (9 to 12 fruits per 36 inches of fruiting wood) is better.

Some varieties, including 'Burbank' and 'Friar', often set heavily when cross-pollination and weather conditions are good. These varieties may need somewhat heavier thinning. Varieties that tend to self-thin throughout the growing season include 'Climax' and 'Santa Rosa'. Wait until just after June drop to thin by hand.

Commercial growers have long used mechanical trunk shakers designed for harvesting cherries to thin plums. Thin about 7 to 10 days after pit hardening to remove green fruits. Fruits will continue to drop naturally after this point, so be careful not to thin too heavily.

The mechanical shaker described for apples and peaches shows promise for plums as well, although little research has been published on the practice. Trials, however, show that large, rigid branches suffer more injury from the shakers than do smaller, more limber branches. For commercial orchards, you may need to adjust your pruning and training practices to facilitate mechanical thinning. Because plum trees are generally not trained to flat walls, thinning with a tractor-mounted string thinner is not a particularly effective method.

Researchers have tested chemical thinners for plums, but the practice has not been widely adopted by commercial growers. The available materials cannot be used in organic orchards.

Bush Cherries and Bush Plums

Bush cherry and bush plum varieties do not require thinning.

Harvesting

For home orchardists, few things beat tree-ripened fruit. For market and larger commercial orchards, however, determining when to harvest in order to get the best-quality fruit to your customers becomes more complicated. Some fruits also do not lend themselves to ripening on the tree.

There are a few things to consider when harvesting fruit. The first is not to mix fruit that has been damaged by pests, diseases, or weather with healthy fruit. There is great truth to the adage that one rotten apple spoils the barrel. Rot organisms can spread from infected to healthy fruits during picking, transport, sorting, packing, and storage.

Some fruits, like apples, also naturally produce ethylene gas as part of the ripening process. Ethylene causes fruits to ripen more quickly. Damaged apples and other ethylene-producing fruits produce more ethylene than normal and can cause the stored fruits to overripen.

Similarly, be cautious about salvaging dropped fruits from the orchard floor. For market and commercial orchards, it is often best to cull the fruit, which may contain insect pests or diseases. Dropped fruit is also typically bruised from the fall. In some cases, you can carefully examine dropped apples and cull infested or diseased fruit, then press the good ones for cider.

Cleanliness is critical. Before harvest, ensure that your bins and anything else that the fruit will touch are clean and disinfected.

Be gentle! Bruised fruit loses much of its appeal and marketability. Damaged fruits are also more susceptible to storage rots, and some generate excessive amounts of ethylene during storage. Even with mechanically harvested fruits, the goal is to reduce the damage as much as possible. For hand harvesting, pick the fruit into small containers or a picking bag worn about your waist. If you are transferring the fruit to bins, gently pour the fruit onto the bin floor or onto the fruit already in the bin. Do not drop the fruit into the bins.

For most fruits, cooling the fruit as quickly as possible after harvest slows the ripening process and helps reduce storage diseases. In commercial packing houses, fruits that benefit from refrigeration are placed into large walk-in coolers where tarps, tunnels, and fans are used to force cold air around the fruit to cool it quickly. We will discuss some exceptions below.

If you wash your fruit or cool it with cold water during packing, make sure that the fruit is thoroughly dry before placing it into storage. High humidity during storage is desirable for most refrigerated fruits. Droplets and films of water on the surface of the fruit are not.

Harvesting Pome Fruits

Determining when to harvest pome fruits is usually more difficult than with stone fruits. 'Red Delicious' apples, for example, develop water core if allowed to ripen on the tree. European pears are usually harvested before they are fully ripe and allowed to ripen off the tree. The situation is especially complicated if the fruit will be kept for long periods in refrigerated or controlled atmosphere storage.

Apples

The developmental stage at which apples should be harvested depends greatly on the variety, its intended use, and how it will be stored. Fruit that is picked too early is typically undersized and has poor taste and texture. Overripe apples become soft and mealy, and red varieties can develop dark, greasy skins. Fruit that is overripe also does not store or ship well. 'Red Delicious' develops scald disorder during storage if picked too early but develops water core if left on the tree too long.

If immediately consuming or processing, you want the fruit to be fully ripe; firmness, texture, and taste are good indicators of ripeness. If the fruit is to be stored inside a refrigerator for an extended period or in controlled atmosphere storage, you will want to pick the fruit after it is mature but not fully ripe.

Gauging ripeness. Many methods have been used to determine the ripeness stage of apples, including the number of days from full bloom, the air temperature, the sugar concentration, and the skin and ground color. Results from these methods vary too much due to weather and cultural practices to be reliable. Measuring ethylene production is more accurate, but it requires advanced equipment and training. Two methods have emerged that are reliable and easy.

As fruits ripen, they soften. Handheld firmness meters quickly and accurately determine fruit firmness. Testers cost anywhere from one hundred to several hundred dollars and are available from orchard supply companies. To use the meter, remove just the peel of the apple over a small area. Special peelers make this process quick, easy, and reproducible. Some meters come with interchangeable tips to allow them to be used on different fruits. Simply press the meter against the flesh of the apple and take the reading. Take readings from several representative samples to determine an average firmness. For apples, a firmness of 16 pounds at harvest for storage apples and 10 to 14 pounds for fresh sales are about average. You will probably use different values, depending on the variety and whether the fruit will go to controlled atmosphere storage, refrigeration, or fresh markets. By correlating readings with the starch test described below, you can develop a chart of suitable firmness values for your varieties and marketing strategy.

When using a firmness meter, use uniform-sized fruits and avoid those that are larger than average. Testing only large apples will give you misleading readings. Likewise, do not test apples that have developed water core.

A simple and reliable method that has long been used by commercial growers is the iodine test. As apples form and grow, they are first composed mostly of starch. As they mature, the starch is converted to sugar. Iodine reacts with starch to form a blue color, but it does not react with sugar. To use this method, collect several apples of each variety. Cut the apples in half to obtain a cross section perpendicular to the stem. Spray a standard iodine solution, available from orchard supply companies, on the cut surface. Starchy areas will appear blue. Areas that have converted to sugar remain yellowish-white. Rate your apples by comparing the staining pattern with pictures on standardized charts that are available for popular varieties.

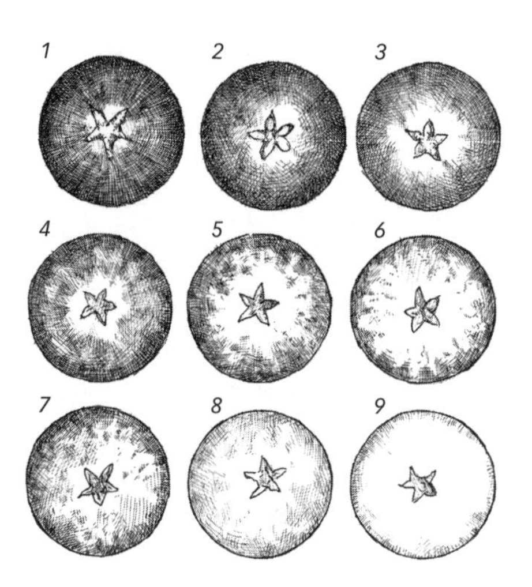

Figure 13.3 · This sample starch-iodine index chart for 'Liberty' apple is based on a chart published by the University of Massachusetts. Charts for different apple varieties are available from several university and commercial sources.

Typical ratings range from 1.0 (immature) to 6.0 (fully ripe), although different systems have ranges of 1 to 9. Although the scale differs somewhat between apple varieties, ratings of about 2.0 to 2.5 indicate the time to harvest for long-term controlled atmosphere storage. Ratings of about 3.0 to 4.0 indicate the time for long-term refrigerated storage. Ratings of 4.5 to 6.0 indicate the fruit is ready to eat. Start making your iodine tests twice weekly beginning 2 weeks before your expected harvest date, based on days from full bloom. Figure 13.3 shows iodine test patterns and harvest maturity for 'Liberty' apples.

Picking and storing. When picking apples, lift the fruit upward and twist. You should break the fruit stem off at the spur. If the stems pull out of the fruit, rot organisms can enter. Broken spurs attached to the apples are the result of improper harvest techniques, and they reduce the following year's crop.

Once harvested, cool the fruit as quickly as possible to between 32 and 34°F (0 and 1°C). Increasing the temperature from 32 to 40°F (0 to 4°C) doubles the rate of ripening. Going from 40 to 50°F (4 to 10°C) doubles the rate again, as does going from 50 to 70°F (10 to 21°C). In other words, fruit can be held at 32°F eight times longer than it can be held at 70°F.

Details on commercial apple packing and controlled atmosphere storage are beyond the scope of this book. If you use an off-farm packing house or CA storage, be very careful to avoid mingling organic fruit with fruit coming from uncertified orchards. If your organic fruit is mingled or stored with fruit from uncertified orchards, you will no longer be able to market it as organic. Packing houses that deal with organic fruit typically maintain separate rooms for organic and conventionally grown fruits.

Loquats

Loquat fruits typically reach maturity for harvest about 90 days after full bloom. The best indicator of ripeness appears to be a fully developed skin color typical for that variety, although the fruits become softer as they ripen. It is critical that you time the harvest properly because fruits that are not ripe when picked do not ripen properly after harvest and can be unpleasantly acidic. If using the fruits in baking, harvest them slightly before full ripeness.

Remember that loquats are very tender and are easily damaged during harvest. Rather than picking the fruits by hand, use a sharp knife or pair of clippers to cut them off individually from the clusters. After harvesting, place in shallow containers to prevent bruising and crushing.

Loquats remain usable for up to 10 days at room temperature. Cooling extends shelf life, but be aware that these fruits are susceptible to injury during chilling if kept at too low a temperature. Properly stored, the fruits can be kept for up to 2 months. Work in Lebanon showed that a storage temperature of 41°F (5°C) preserved the fruits much better than a temperature of 54°F (12°C).

Loquats are generally packed into small containers around 2.2 pounds or less. If picked and placed in wooden or paper containers, wrap the containers with cellophane film to preserve freshness and reduce shriveling. Research has shown that using wooden boxes increases fruit bruising. Clear plastic clamshell packaging is a good alternative.

Mayhaws

Mayhaw fruits are small, and commercial growers use trunk shakers and tarps on the ground to collect the falling fruits. If harvesting by hand, pick the fruits as you would apples.

Most mayhaws are destined for processing, in which case the fruit should be at or near full ripeness for harvest. If the fruit will be refrigerated for an extended period before processing, pick them after they are mature but before they are fully ripe.

Medlars

Medlar fruits are hard, acidic, and not considered edible until they have been softened on the tree by frost or allowed to partially decompose during storage in a process called "bletting." As the fruit ripens, it softens and the skin wrinkles and darkens. Once the fruit is suitable for eating, the flesh takes on a brown color and develops a texture that has been likened to applesauce, baked apples, and baked sweet potatoes. The aroma is rather earthy or musky, and depending on whom you ask, the flavor is described as being similar to applesauce, figs, kumquats, loquats, and Persian dates. At that point, the fruits can be eaten fresh, like applesauce, or processed into jelly or custards.

In regions where medlar is traditionally grown, the fruits are harvested in November, often after having been exposed to several frosts. Once harvested, the fruits are placed in a cool, dry location to continue ripening, usually for at least 2 weeks.

Pears

Determining the point at which pears should be harvested, particularly for commercial harvests, is somewhat more difficult than for apples. As the fruit ripens, the underlying "ground color" of the skin gradually changes from green to yellow, the aroma and sugar concentrations increase, and the fruit softens. Although charts are available to help judge the ground color, this evaluation remains subjective and is less useful on red varieties. Likewise, basing harvest maturity on aroma is highly subjective.

Firmness meters described for apples work well for pears. If the fruit will be stored in refrigerators, the firmness ranges from 11 to 23 pounds, depending on variety. Harvest Asian pears at about 13 pounds firmness, 'Anjou' at 15 pounds, 'Bosc' at 16 pounds, and 'Bartlett' at 19 to 22 pounds. As with apples, the firmness of fruit ready for sale will be lower, ranging from 7 to 10 pounds for Asian varieties, to 11 to 13 for 'Bosc', 10 to 14 for 'Anjou', and 15 to 18 pounds for 'Bartlett'.

For other varieties, keep a record of firmness measurements and correlate them with storage and quality evaluations. You will soon be able to develop a chart of firmness standards for your varieties and markets.

Firmness readings are sometimes combined with measurements of the sugar concentrations in the flesh (called total soluble solids or TSS) as measured with refractometers. Good-quality optical and digital refractometers are available from orchard supply companies at a relatively low cost.

European Pears

European pears are usually picked before they are fully ripe in order to develop maximum quality and storability. This is particularly true of 'Bartlett'. 'Anjou' and 'Bosc' are allowed to ripen a bit more before harvest but are still not fully ripe. Pears that are destined for canning are allowed to nearly ripen on the tree before harvest.

Like quinces, firm pears are still susceptible to bruising and must be handled gently. Once harvested, pears are stored at 30 to 31°F (−1 to −0.5°C). Sugar in the fruit prevents the flesh from freezing at temperatures above about 28°F (−2°C). Varieties differ greatly in their storability. "Fall pears," such as 'Clapp's Favorite', 'Red Clapp', and 'Starkrimson', can be kept in storage at 30 to 31°F for only about 2 to 6 weeks. "Winter pears" can be stored much longer. 'Red Angelo', for example, can be held at 30 to 31°F for up to 10 months. The most popular pear variety, 'Bartlett', can be stored for about 3½ months.

For commercial orchards, temperature control is critical. If the storage temperature rises to between 34 and 35°F (1 to 2°C), some varieties, such as 'Bosc', will fail to ripen properly, when removed from cold storage.

European pears are best ripened off the tree at temperatures of 60 to 70°F (16 to 21°C) and 80 to 86 percent relative humidity. At higher temperatures, the fruits may not ripen properly, and when ripened at low humidity, the fruits can become shriveled.

Pear fruits can be kept in controlled atmosphere storage, similar to apples, where temperatures are kept near freezing, ethylene is removed from the air, and the oxygen and carbon dioxide concentrations are manipulated. High concentrations of CO_2, however, cause pears to develop postharvest breakdown.

Asian Pears

Unlike their European cousins, Asian pears can be tree ripened and eaten right from the trees.

Home and market orchard growers do not really need firmness meters and refractometers. Given some experience growing pears at your location, you can estimate fairly closely by the calendar date when to harvest. Use changes in ground color and the ease with which the pears detach from the spurs to fine-tune your timing. As pears reach harvest maturity, they detach easily from the spur. Immature fruits will cling to the spurs and should be left where they are to ripen a bit more. As with European pears, flavor and softening are good indicators of maturity.

Quinces

In warm, long-season climates, quince fruits can remain on the trees long enough to soften and become sweet enough to eat out of hand. In most of North America, the practice is to leave the fruits on the trees as long as possible, but not to the point that they are exposed to frost. At this stage, the fruits are very firm, acidic, and astringent. They are, and will remain, quite gritty. Although they are firm, they still bruise easily and must be handled with the same care given to apples.

The fruits usually need to be stored for a few weeks in order to mature them to the point where they can be eaten. One practice is to store the fruits in single layers in a cool, dry location to continue ripening. Once ripe, the fruits can be kept refrigerated for a few weeks.

Quinces are sometimes eaten fresh, but they are more commonly made into jams and jellies or baked. Their flavor and aroma are very strong, and a small amount of quince can be added to processed apples and pears to provide flavor and aroma.

Saskatoons

Saskatoon fruits are allowed to fully ripen before harvest. The fruits will become an even blue color, soften, and be sweet and flavorful. At this stage, the fruits are rather delicate, and you must be gentle. If harvesting by hand, pluck, strip, or roll the ripe berries off the clusters, leaving unripe fruits on the cluster to ripen for a later harvest.

Commercially, saskatoons are hand harvested or mechanically harvested using over-the-row harvesters similar to those used for raspberries and blueberries. After harvest, store the berries at 32 to 34°F (0 to 1°C) and high humidity.

Harvesting Stone Fruits

Unlike apples, pears, and medlars that can be stored for months, stone fruits generally have short storage and shelf lives — typically three weeks or less. For long-term storage, freeze, can, or dry the fruits.

Apricots

Harvest apricots when they begin to soften and develop their characteristic aromas. If shipping, pick the fruits when they have fully colored but are still firm. Store the fruits at 35 to 40°F (2 to 4°C) for up to 3 weeks.

Cherries

Unlike for apples and pears, there are no objective measurements for determining the harvest maturity for either sweet or tart cherries. Also unlike apples and pears, cherries do not continue to ripen after they are harvested. Combined with the fact that cherry fruit size and yields continue to increase in the latter stages of ripening, most growers prefer to delay harvest until the fruits are fully ripe. At this point, the fruits should be quite soft and flavorful.

Varieties that develop very firm fruits provide a longer harvest window than soft-fruited varieties. Regardless, allowing for no more than a 10- to 14-day harvest is a good practice.

Ripe cherries are easily bruised and must be handled gently and cooled quickly. Fresh-market fruits are harvested by hand. By using the training methods and rootstocks we discussed earlier, pickers should be able to harvest about 80 percent or more of the fruits without using ladders. By keeping your pickers' feet on the ground, you can greatly speed up harvest; this is especially desirable for U-pick orchards, where customers and ladders are a poor combination from a liability standpoint.

For commercial orchards, most tart cherries and sweet cherries destined for processing are mechanically harvested using trunk shakers equipped with padded grips to reduce damage to the trees.

Cherries are highly perishable but can be held in cold storage for up to 14 days at 32 to 35°F (0 to 2°C). By maintaining high humidity but preventing free moisture from accumulating on the fruit, you can help extend shelf life.

Peaches and Nectarines

For home use, allow peaches and nectarines to fully ripen on the trees in order to obtain maximum flavor and the desirable texture and firmness. You can use several indicators to judge harvest maturity. As with other fruits, the underlying ground color changes from green to either yellow or white, depending on the flesh color of the variety. The fruits will round out, losing the flat side typical of peaches and nectarines. The flesh softens, and the fruits become aromatic, sweet, and juicy.

Fully ripe fruits are too soft to allow much handling or shipping and typically arrive at the market bruised, overly soft, and with unattractive skins. For that reason and the fact that the first fruits at the market can bring premium prices, some growers tend to pick the fruits too early, when their size, texture, and flavor resemble those of a baseball.

For distant markets, pick fruits about 3 to 6 days before they are fully ripe. You will be able to judge this with a little practice. For markets close to home, pick the fruits when they are still firm and just before they are fully ripe. Peaches and nectarines bruise very easily and must be handled with extreme care throughout the harvest, packing, and shipping steps.

Unlike cherries, where an entire tree is harvested at one time, you only pick the peaches and nectarines at the optimum harvest maturity. Small fruits remain on the trees to continue ripening. This means each tree will be picked at least several times, and generally more.

If the fruits will be processed, commercial growers often use mechanical harvesters to reduce labor costs. Mechanically harvested trees must be trained to specific canopy structures to allow the fruits to fall into catch devices and to avoid damage to the tree limbs caused by the mechanical harvester. Mechanically harvested fruit is transported very quickly to the processing facility to prevent bruises from developing.

It is very important that peaches and nectarines are rapidly and continually cooled. To ensure freshness, keep the temperature at about 32°F (0°C) from packing to the consumers' tables.

Plums and Prunes

The harvest maturity of plums is largely determined by changes in color at the tips of the fruits and by the firmness of the flesh.

The level of firmness appropriate for harvest depends on the variety and the amount of time before the fruits reach consumers. An average firmness of 8 to 10 pounds works for most plum varieties that will be cooled quickly after picking and before shipping. If you cannot precool the plums before shipping, pick them earlier, when firmness is in the 13- to 15-pound range. Fully ripe plums are very soft, measuring around 4 pounds of firmness. For home fresh use and local fresh markets, harvest plums when they are fully colored and still firm. Plums destined for processing can be allowed to fully ripen on the trees.

Like peaches and nectarines, plums continue to ripen after harvest. Being even softer than peaches and nectarines, plums intended for shipping are picked mature but still firm and before they achieve their full color.

With the exception of loquats, plums and prunes are the most fragile fruits covered in this book and must be handled with great care. When picking, use a lifting and twisting motion to detach the fruits from the tree. For fresh market and shipping plums, the stems should remain attached to the fruit. Ripe plums in good condition can be stored for several weeks at 30 to 32°F (−1 to 0°C).

Prunes are allowed to fully ripen on the trees before harvest, usually when they measure at about 4 pounds of firmness. The skins should be smooth and fully colored. On a commercial scale, prunes are often machine-harvested. They are dried inside ovens, or the prunes can be sun-dried.

Bush Cherries and Plums

Bush cherries and plums are harvested fully ripe, when they have developed full color and flavor and have softened. They can be stored for 1 to 2 weeks at 30 to 32°F (−1 to 0°C).

CHAPTER 14

MARKETING AN ORCHARD BUSINESS

- Selecting Products | 496
- Selecting a Farm Design and Selling Venue | 498
- Setting Prices and Terms of Sales | 510
- Creating an Enterprise Budget | 512
- Publicizing Your Enterprise | 515
- Packaging and Selling | 517
- Developing a Business Plan | 519

Some years ago, I was part of a joint university-industry project designed to identify challenges to establishing and operating specialty crop businesses, such as orchards. We found the main obstacle was not getting a farm established or producing a crop; it was marketing the product. It does not matter how well or how cheaply you produce something if there is no market for it. If you cannot sell it profitably, the business will fail. Trailing far behind and roughly tied for second in terms of challenges were labor and capital to start and maintain the enterprise until it was profitable. The easiest part of the process was growing a crop.

While poor marketing can quickly kill a business, good marketing can make it a great success. The nice thing is that marketing is not new or a mysterious unknown. People have been marketing farm products successfully since ancient times. Industry groups, government agencies, and universities often conduct workshops and seminars on the subject. You must, however, be ruthlessly objective about your own situation. You cannot make marketing decisions based on how much you like living in a particular area or how much you like growing orchard crops. To be successful, you must create high-quality products that people want, let the people who want to buy those products know that you have them, and sell the products at competitive prices.

One of the first problems that new business owners get into — and a commercial orchard is simply a business — is that they think marketing is just another word for selling. Selling is only a small part of the marketing process! It's important to always remember that you are not selling apples or peaches. You are selling customer satisfaction. To be successful you must sell that satisfaction at a profit.

Successful marketing involves many things, some of the most important being:

- Selecting products that are, or can be made to be, in demand
- Identifying (often called targeting) a market population
- Selecting an orchard site and farm design that will allow you to create your products and deliver them to customers profitably
- Developing a business plan
- Setting prices and terms of sales
- Developing good public relations for your enterprise
- Advertising and promoting your products
- Appropriately packaging the products for the intended market
- Selling the products
- Delivering the products to the customer in good shape and on time
- Practicing good customer service before and after the sale

The second big mistake that new marketers often run into is that they wait until their products are ready to ship before they worry about marketing. For a new orchard, marketing should start before you even decide what fruits you are going to grow. According to Dr. Levitt of Harvard University, "A firm should make what the customer wants, not what the company's machinery is set up to make."

Evaluate your own capabilities — and here, again, you must be ruthlessly objective. Can you produce fruit or value-added products that are in demand, in marketable quality and quantities, and do it profitably? How can you market your goods in a way that will give you an edge over competitors? In other words, how can you provide more customer satisfaction than other orchards that are selling the same varieties and products? It might involve selling your product in more convenient sizes, with better packaging, personal service, or faster or less expensive delivery. Be very cautious about undercutting competitors' prices. Most orchards already have a slim profit margin. Price wars usually hurt everyone in the business and help no one.

Selecting Products

MARKETING SPECIALISTS OFTEN TALK ABOUT the four "Ps" of marketing: Product, Place, Price, and Promotion. The first thing you should do in planning your orchard is to gather information on markets (where, how, and to whom you sell your products) and competitors. Identify what orchard products are available, where the markets are for those products, and who your competitors would be. This task might sound overwhelming at first, but do not worry — it is not as hard as it sounds, and help is available.

Start by making a trip to the produce section of several local grocery stores. Visit a large chain store; a small, locally owned supermarket; and a specialty health food store and talk with the produce managers. Also visit farmers' markets and roadside fruit stands in your area. Talk to buyers at fruit cooperatives and packing houses in your area. Make a list of the fresh produce that is available. Ask the following questions.

- What fresh produce and value-added orchard products are most in demand? Are they seasonal or in demand year-round? (Value-added products are orchard fruits that have been packed or processed in some way to increase their value, compared with raw fruit, such as jams, ciders, or fruit baskets.)
- Where do the various products come from?
- Are purchases made through a central company buyer, produce brokers, or local farmers?
- What size and kind of packaging do the outlets prefer?
- What quality and packaging standards must be met for the produce and value-added products?
- What quantities do the outlets buy?
- What products do they have trouble getting enough of, or would love to carry if they could find a supplier?

State and provincial departments of agriculture, local chambers of commerce, and agricultural economics departments at universities often have marketing specialists that can provide information on the kinds and sizes of successful orchard operations in your area. Cooperative Extension agents and specialists can provide information about successful and unsuccessful farming efforts in the area. What is especially nice is that much of this consultation will be free of charge.

The best sources of information on marketing orchard products are successful fruit growers who are established in that business. Select a time of year when the farmers are not overwhelmed with work and make appointments to talk with them. I have generally found that farmers are willing to talk to people who are considering setting up their own farms. Established farmers work hard for years to establish reputations for providing high-quality products, and they want to be sure their good reputations are maintained. One farmer putting out lousy produce in an area hurts the reputation of everyone else who sells the same thing. A word of caution, however: Do not try to take advantage of prospective competitors — doing so can blight your reputation from the start. If you are planning to move out of the area and are just getting background information on prospective enterprises, be sure to let the growers you talk to know that. If you are planning to market in the same area, you should tell

Identifying a Market Population

In marketing your goods, you cannot be all things to all people. North America is filled with diverse cultures and interests. Not everyone will be interested in what you produce. Instead of using a shotgun approach to try to sell to everyone, select certain groups that offer you the greatest potential return for the least effort. This process is called "targeting a market population." You must answer the following questions:

- Is there a market for my product?
- Who makes up the market?
- What part of that market, if any, can I successfully target?

Also evaluate your own skills and personality. If you like growing fruit but do not particularly enjoy interacting with customers, selling through a packing house or cooperative will probably be more enjoyable for you than selling the products yourself through a U-pick orchard or on-farm fruit stand.

them that as well. Packing house and food cooperative representatives are also excellent sources of information, as are representatives of farmers' markets.

Consumers' preferences vary by market population, and these preferences will influence the fruit crops and varieties that you choose. If you plan to sell directly to a packing house or supermarket, you will probably need to focus on crops and varieties that have high consumer recognition and demand. For example, popular apple varieties at the moment include 'Fuji', 'Gala', 'Honeycrisp', and a handful of others. One problem is that varieties in greatest demand are often not the best varieties to grow organically because they have less resistance to disease than other varieties. If you plan to sell through local food cooperatives, farmers' markets, and roadside stands, you will often have greater flexibility in choosing the crops and varieties that you market.

Selecting a Farm Design and Selling Venue

WE WENT INTO DETAIL ON SELECTING AN ORCHARD SITE and designing an orchard in chapters 2 and 3. The emphasis in those chapters was on fruit growing, but you also must consider marketing as well.

Beautiful views, lake access, or seclusion are not good reasons to buy a commercial orchard site. Before you design your orchard, know what you are going to grow, what value-added products you plan to make, and how you are going to market them.

Say, for example, you want to grow apples, pears, and peaches and sell them through a U-pick operation. You may design the orchard differently than if you planned to harvest the fruit yourself. For a U-pick orchard, you want small trees that customers can harvest without stepping on a ladder. "What's a ladder got to do with it?" you ask. Just imagine if a customer falls off a ladder and is seriously injured or killed. Are you willing to risk the potential lawsuit? The same reasoning applies if you have barns, sheds, or equipment that will attract children — you want to make sure they won't get hurt. If you are planning a U-pick farm, you should eliminate every possible hazard that you can. You will also need to provide roadside advertising, customer parking, a checkout stand, and possibly transportation between the parking area and fields. Keep it safe. Keep it convenient. Keep it pleasant.

If you plan to harvest the fruit yourself and sell it at a roadside stand on your property, you can design the farm for maximum production efficiency. You will have to provide cold storage, parking, the stand, and fencing to keep children and pets away from dangerous areas. For U-pick and roadside stand operations, you might want to leave room to put in a small, fenced-in playground where children can safely play while their parents pick or buy produce. One successful organic orchard in Michigan has a petting zoo and provides tours for local schoolchildren. A cool, attractive picnic area might help boost sales, especially impulse sales to tourists. You might want to modify the barn and farmyard to make customer visits recreational experiences, such as with a cider stand, hay wagon rides, or decorated pumpkin displays in the fall. According to one study, a satisfied customer will tell from one to three people about the buying experience. A dissatisfied customer will tell seven people. Even if a customer does not buy anything, make sure the experience is pleasant.

If you plan to sell your products off-farm and customers will seldom or never see your farm, then design your orchard for maximum production efficiency and convenience to yourself. Make sure you have appropriate storage, processing, and packaging facilities. Provide plenty of convenient parking and turnaround space for delivery trucks and employees. In short, tailor the farm design for your products and market.

Despite the many variations available for marketing orchard products, there are only two basic systems. You can sell your product directly to the final consumer (retail) or you can sell to a middleman who will then sell to the final consumer (wholesale). When dealing with brokers and large retail outlets, your goods can actually change hands several times before they get to consumers.

On-Farm Sales

One of the simplest marketing strategies is to sell your goods on the farm itself. The distinct advantages are that you control everything involved with the sale and are close to your work. You are normally paid immediately, and since there's no middleman, whatever money you take in is yours — before paying suppliers and taxes, of course. On the other hand, your farm is not likely to be on Main Street, and you must ensure that you bring in enough customer traffic to sustain the enterprise. The possibility of personal injuries and the issue of liability are more of a concern when you bring customers onto your farm, and curious customers and their children can disrupt farming operations by wandering around and getting into things they shouldn't. Do not let these considerations scare you away from on-farm marketing, however, because good planning will reduce their impact to a minimum.

If you do plan to market directly, make sure your farm is designed to accommodate customers and that the design is suitable for your products and the clientele you are marketing to. For example, if you are selling a few cartons of plums or apricots to neighbors, your entire sales facility might be the back porch or kitchen table. For a higher volume of customers or a more diverse range of products, a roadside stand or salesroom might be more appropriate. If your produce is sold U-pick, you will need a stand or room where the produce can be weighed, packaged, and paid for.

U-Pick

Also called Pick Your Own or PYO, U-pick is well suited to fruits and vegetables, particularly crops like peaches that have some obvious characteristic indicating when they are ripe. With this approach, the customer has the responsibility of harvesting and transporting all of the goods sold. For that reason, your expenses for harvesting, packaging, storing, and transporting will be lower than if you do those tasks yourself. You also get paid immediately and do not have to worry about shipping and postharvest spoilage. One disadvantage is that you need a sufficiently large local population (whether permanent residents or visitors) made up of people who are willing to pick their own produce.

Other drawbacks include the need for easy vehicular access, a large parking area, and increased risk of liability should a customer be injured. You must also direct and supervise pickers who may not be particularly adept at harvesting your crop, or who want to go through your fields and "high grade," picking only perfect fruits. You may suffer some crop damage from customers, and not all of the crop may be harvested. Bad weather during the harvest can drive away customers and leave your crop rotting in the field. A client of mine once lost an entire crop of U-pick cherries because of heavy, prolonged rains. If it had been a farmer-harvested operation, at least part of the fruit could have been salvaged. I must hand it to him, though — even when we were standing in the middle of his orchard, up to our knees in mud and in a drenching rain, he never stopped smiling. He just loved farming and knew that the sun would shine again.

A U-pick operation can be profitable, but it is not for everyone. If you are not comfortable with direct retail sales and do not work well or enjoy working closely with people, then you may be better off selecting another marketing strategy or have someone else in your operation interact with customers.

Consumers are typically drawn to U-pick farms looking for fresh, high-quality produce at bargain prices. In some cases, customers want enough produce to freeze, can, or otherwise preserve, but they cannot afford supermarket prices. For many customers, however, harvesting their own food is also a form of recreation, where they can get out into nature with friends or family or enjoy some quiet time alone in the great outdoors. Others appreciate seeing where and how their food is produced.

Successful U-pick orchards are seldom found in isolated, hard-to-reach locations. You must be where customers can and will be willing to visit your farm. Locations within about 20 miles of a city or large town are best. Your chances for success also increase if you are in an area that has other established U-pick operations, preferably selling a variety of crops that pull people into the area throughout the summer and fall. The diversity of products available is a powerful magnet that draws potential customers into the area.

When designing a U-pick orchard, provide easy access from a main road and make sure you have a large, well-marked parking area. It is a good idea to fence the parking area or otherwise isolate it so that people cannot drive into your farmyard or fields. Locate the parking area close to the fields, if possible. Roads and paths should be wide, clearly marked, and laid out for maximum customer convenience. Use easy-to-read, clearly worded signs to direct customers to the portions of the orchard that you want harvested. Rope off other areas, if necessary, to keep people out of them.

U-pick customers often like to wander over a large area, selecting only perfect produce. You, on the other hand, want all of your crop harvested and will need to direct and closely supervise the customers to make sure they are doing a thorough job of harvesting. If it is difficult to know when to harvest a crop, such as when to harvest apples for storing or processing, plan on educating your customers. You cannot, of course, insist that they take rotten or otherwise unmarketable produce. Open only one part of the orchard to harvest at a time or assign customers to pick specific areas. When those areas are thoroughly harvested, open up new areas. Just remember that people have gone out of their way to visit your farm. Keep directions and supervision upbeat and pleasant. Word-of-mouth advertising is the most important promotional tool for U-pick operations. Do everything that you reasonably can to make the visit pleasant.

For parents who are not able or willing to closely supervise their children when picking, you might offer a free babysitting service or provide a supervised playground that will keep the children occupied and prevent them from wandering into potentially dangerous areas. Even with the cost of a playground and hiring a babysitter, the reduced risk of personal injury and increased customer appreciation should more than pay you back. Along the same lines, you can make harvesting food a recreational experience for customers by adding a picnic area or petting zoo.

As I mentioned earlier, keep safety foremost in your mind when designing your U-pick farm. Large livestock, dangerous equipment, and farm ponds should be out of sight or otherwise inaccessible. Bees are important in orchards but do not mix well with customers. Remove hives from the fields before harvest and place them at least a quarter mile from customer areas. Strictly follow pesticide regulations to avoid exposing workers and customers to danger. Keep pesticides and sprayers out of sight and reach of customers. If you are asked if you use pesticides, be totally honest. Keep detailed records of your pesticide use, whether organic or nonorganic.

In terms of orchard design, it's a good idea to use small trees and bushes that allow most or all of the fruit to be harvested without ladders. Trellises and other narrow training systems work very well. Keep planting blocks relatively small with short rows to facilitate customer picking. Few of us would feel comfortable with customers driving around our orchards. For customers picking a carton or more of fruit, allow them to leave the cartons at the ends of the tree rows, and use your own farm vehicle to transport the fruit to the sales stand.

Although alfalfa is an excellent alley crop for orchards, walking on alfalfa stubble is less than pleasant. For a U-pick orchard, consider maintaining the alleys in a relatively fine-textured grass sod. Mow the alleys and under the trees, ensure the orchard is attractive, and do some thorough weeding before you open up the orchard to pickers. Avoid doing any cultivation that leaves large clods or powdery dust for customers to walk through. Remember to keep the entire orchard experience pleasant for customers.

Roadside Stands

Marketing produce and value-added products from roadside stands is a popular option for small and medium-sized operations. In some cases, roadside stands are combined with U-pick and farmers' market operations to provide customers with alternatives and to increase sales. With a roadside stand, you can either market only your own produce or broker produce from other growers to increase the diversity of your goods. Two or more growers might also operate a single stand in a cooperative venture. Besides raw produce, you might offer preserves, honey, dried herbs, cookbooks, cider, or other value-added products. Make sure that any consumable, value-added products meet food preparation and safety regulations for your area.

As you can see, roadside marketing offers a lot of opportunities. Be cautious, however. Local ordinances vary tremendously, so be sure that you will be able to sell what you want where you want. In one county where I farmed, for example, unless a roadside stand was located on a commercially zoned site, it had to be on the farmer's property and carry only produce grown on that farm.

As with U-pick operations, there are several things that you must get right from the start for a roadside stand: location, location, and location. The best stand in the world will not last long if it is isolated and there is little customer traffic. Main, heavily traveled highways are best.

Roadside stands provide excellent opportunities for creative marketing. When I lived in upstate New York, my family loved going to a roadside stand that put on a large Halloween display every fall. The owners would create the agricultural version of a wax museum with hundreds of figures made from pumpkins, squash, apples, cornstalks, and other produce. There were pumpkin marshals and pumpkin outlaws gunning it out on Main Street. Pumpkin and cornstalk witches flew through the

The Value of Diversity

If you are selling directly to consumers through U-pick or a roadside stand, a diversity of products is usually better than having a single crop and variety. Providing a diversity of products brings customers to you throughout the summer and fall. The same applies if you are selling fruit to a local packing house. Quite often even large, well-established growers prefer to raise at least several different varieties, and often two or more different crops. As with buying stocks and bonds, diversification reduces your risks and helps even out your cash flow.

The downside to diversification is that you must become an expert at growing several crops and varieties. If you are new to the orchard business, remember to start small and focused. As your skills develop and you gain experience, you can safely expand your offerings.

treetops. Squash children played on the lawn. I know the family that owned the stand had a lot of fun putting on the show, and customers came back year after year to enjoy it, putting late-season dollars into the farmer's pocket. The owners realized they weren't selling pumpkins and squash — those products were available from other sources closer to town. These entrepreneurs were selling fun and memories.

Off-Farm Sales

Selling produce off the farm provides markets for orchards that are not well located for U-pick or roadside stand sales, and it can increase marketing options for growers with U-pick and roadside stand outlets. Although your labor inputs are higher than for U-pick, harvesting the fruit yourself generally results in increased yields of marketable fruit over U-pick operations, and you can ensure the fruit is harvested at the proper ripeness for your particular markets. Farmers' markets are among the most common outlets for retail off-farm sales. You can also market directly to consumers through subscriptions in a program called community supported agriculture (CSA).

Farmers' Markets

Farmers' markets provide opportunities to sell produce, get to know other food producers in the area, and develop a reputation with local customers. This is especially useful for small orchards and those new to farming. Market orchardists often find this is the only marketing outlet they ever need. You will find organized farmers' markets in practically every city and in many small towns. The markets range from a few people selling tomato seedlings off card tables to huge operations covering many acres.

Several things make selling goods through a farmers' market attractive. First, there is virtually no overhead or capital involved in marketing. You sell everything from the back of a pickup or from a booth made with a folding table and canopy. Signs and promotional aids usually consist of your offerings and prices displayed on hand-lettered poster board. Second, customers are coming to see you with the idea of shopping in mind. By keeping track of what you sell and what produce is popular at other nearby stands, you quickly learn what goods are in demand, and when they are in demand. Many farmers' market organizations do, however, charge participants a flat fee to operate, in addition to a percentage of the profits.

There are three basic types of customers at farmers' markets. The first come with a shopping list, just as if they were at a supermarket. They have specific purchases in mind and little interest in anything else. As you might gather, novelty and impulse-type goods are of little interest to these shoppers; their primary concerns are freshness, overall quality, and price.

The second type of customer comes to the market out of curiosity, seldom with any specific purchases in mind. These customers browse for a while, look skeptical, occasionally buy a few small impulse items or some fruit to snack on, and leave without putting much money into anyone's pocket.

The third type of customer is looking for a recreational experience, just as many people visit shopping malls to window shop and spend time with friends or family. Particularly in tourist areas, this group of shoppers can be very important to sellers at a farmers' market. Impulse items are the big sellers here, such as fresh fruit for eating out of hand or value-added snack, delicacy, and gift items. For these spur-of-the moment shoppers, the experience is everything, while prices are seldom important.

To learn about farmers' markets in your area, contact chambers of commerce and Cooperative Extension and Ministry of Agriculture offices, or search online. Be sure to ask about the markets' rules. Some markets limit sales only to fresh produce while others allow you to sell virtually anything.

CSAs

You can use a contract strategy for selling retail directly to consumers. This approach is called community supported agriculture (CSA). Customers agree to purchase certain amounts of produce at specified prices and at certain times. The practice is sometimes called "subscription agriculture."

The CSA model can be used anywhere you have neighbors who want to purchase your produce, and it can be especially profitable when marketing to an urban population. The consumers want fresh, high-quality produce in certain amounts at certain times. Delivery to their homes may be included in the contract or subscription. They benefit by obtaining fresh produce with certain quality standards, sometimes with the convenience of home delivery. The grower benefits by having much of the market uncertainty eliminated and knowing that she or he can expect a given amount of cash flow at some particular time.

The keys to a successful CSA marketing strategy are reliability and impeccable quality. Word-of-mouth is your greatest asset in this approach, and you must ensure your customers are satisfied. Keep them updated on when they can expect deliveries and if there are any problems. Make deliveries on time, and ensure the produce meets (or exceeds!) the customers' expectations. Get to know your customers, and maintain records of what they want in terms of produce, quality, and delivery. Establish and maintain good relations with each of them. Not only are they the foundation for your present sales, they are the best source of advertising for future customers. In addition to established customers providing referrals, lists are published on the Internet and in print of CSA farms for different regions. Consumers looking for a CSA supplier may use such a list to find you.

The details of setting up a CSA marketing strategy are beyond the scope of this book. Many government, nonprofit, and commercial sources of information on CSAs can be found online.

Wholesaling

A big problem with U-pick, roadside stand, and farmers' market sales is that you depend upon the customers coming to you. Bad weather can keep them away in droves. If your goods are perishable, the losses can be substantial. By selling wholesale directly to local purchasers, you take marketing one step further by going to the customers yourself. Restaurants, grocery stores, specialty food stores, and food processors are common outlets for direct local sales. In most cases, you will be selling at wholesale prices, which is less than you would normally sell directly to the final consumer at a roadside stand or farmers' market. What you lose in price, however, you hope to gain in volume and steady sales.

In many cases, you will find it to your advantage, as well as to the customers', to contract grow. In this type of arrangement, you agree with the customers in advance as to what you will supply, what quantities they will purchase, what the purchase prices will be, what quality standards you need to meet, and how and when the produce will be delivered. Your customers are not going to want to make any spontaneous or impulse purchases. The last thing that restaurateurs and store owners like is to be surprised by their suppliers.

When you sell directly to businesses, you are dealing with hard-nosed professionals who insist on reliability and quality. It can be a difficult market to break into, and you may face stiff competition from other commercial fruit growers, and not only those from your immediate area. It is no accident that states like Florida and California dominate the produce industry. The growers in these areas have established reputations for quality, reliability, and professionalism. With today's rapid transportation, peaches picked in California on Monday morning can easily be on a Boston supermarket shelf Tuesday afternoon.

To beat out the competition, you must offer something that other growers cannot. Tree- or vine-ripened freshness, lower transportation costs, flexible delivery schedules, the lack of a middleman, and the appeal of buying locally are some of your strongest tools in the local market. Do not try to compete by undercutting the big suppliers' prices. Competition in the produce industry is fierce, and prices are about as low as they can be and still keep growers and brokers in business. Underpricing only hurts you in the long run. Set reasonable prices, and concentrate on quality and reliability.

To get into direct local sales, start by visiting with the buyers personally. Be sure to set up an appointment well ahead of time to ensure they will be able to fit you into their busy schedules. Be honest and forthright with your prospective customers. Explain that you produce orchard crops and would like to supply their business. Clearly explain the advantages they will enjoy by doing business with you, but do not make promises that you cannot or will not keep. If you make a sale, great! If not, keep everything on a pleasant, upbeat basis. Circumstances change and that "no" may eventually become a big "yes." You will probably find that large supermarket chains are not particularly good prospects for locally grown produce. These firms usually purchase huge volumes of produce through reliable, well-established brokers that represent hundreds of growers and grower cooperatives. With literally tens of thousands of items on their shelves every day, local store managers usually cannot take the time to deal with producers individually. Smaller and locally owned grocery stores, locally owned specialty food and health stores, restaurants that are not part of national or regional chains, and local food processors are probably your best opportunities.

If you are in a centralized fruit-growing region, such as California, Washington, and upstate New York, you may have access to fruit packing houses, fruit brokers, and fruit-grower or food cooperatives. The packing houses may be privately owned or owned by growers' cooperatives. They take in fruit from growers and clean, sort, package, and store it. The fruit may be sold immediately, placed into refrigerated rooms for sales within a few weeks, or placed into controlled-atmosphere rooms for sales months after the harvest. In some cases, the packing house buys the fruit. In other cases, the grower can rent storage space and take the fruit out later for marketing himself or herself. Before choosing a packing house, ensure that it is set up to handle certified organic fruit. Organic fruit must generally be processed, packed, and stored separately from noncertified fruit.

Fruit brokers buy from packing houses and individual growers and cooperatives. They serve as intermediaries between the growers and consumers. Because brokers usually represent many packing houses and growers, they provide the volume sales needed to meet the needs of large supermarket and restaurant chains. Cooperatives are local or regional in nature and are made up of growers who pool their fruit and other produce. In a sense, they act much like brokers, but the grower retains more control over prices and how the fruit is sold. An individual orchardist may use only one or all of these outlets.

These outlets provide excellent and relatively stable markets for your fruit, you will not have promotional expenses, and marketing is very easy. On the downside, you are now selling your produce as a commodity and the prices you receive for your fruit will be quite inflexible. Wholesale marketing can be very demanding. Quality standards are high and unforgiving, production and shipping schedules are often tight, and you may have to modify your production practices to meet buyers' demands.

Wholesale markets generally mean big business. It is unlikely that these will be your primary outlets when you first get into sales, unless you start on a fairly large scale. An exception might be if you become part of a cooperative marketing group in which anywhere from a few to hundreds of farmers pool their produce in order to meet wholesale quantity requirements.

Blending Your Marketing Approach

Your marketing plan doesn't need to be all or nothing; you can combine different marketing strategies to meet your needs. For example, you might choose to sell some of your grower-picked fruit directly from your farm, perhaps at a farm stand, and the remainder to a local food cooperative. This blended approach provides you with increased flexibility and stability. Instead of a food cooperative, you might sell fruit that is sound, but not cosmetically perfect, to a local cider press operator or other processor. By doing so, you can sell more of your fruit than you would if you were only selling it for fresh consumption.

Regardless of the primary marketing strategy that you employ, be sure to have a backup market. It often makes the difference between a successful orchard operation and one that is not successful.

Setting Prices and Terms of Sales

BUSINESS OPERATORS THAT SIMPLY GUESS AT PRICES generally do not last long. If you guess too low, you will cut into your profits and may even sell at a loss. If you guess too high, you lose sales. It's better to take a systematic approach.

Follow the competition. As a consumer, particularly in these economically difficult times, prices are important to you. You do not want to pay more than you have to for goods and services, and you probably compare prices at different stores and shop for bargains. Your customers will do the same thing. This is the basis for using your competitors' prices to set your own. You normally do not want to be too far out of line with everyone else's prices unless you provide some special service or other benefit that makes customers consider your price a bargain. There are several variations on the practice of pricing your goods according to competitors' prices.

You can set your prices the same as the competition. This is a simple, straightforward way of setting prices. You simply produce your fruit or value-added products, take a look at what those same things sell for in the market you are targeting, and charge the same prices. Simplicity is about the only thing going for this approach. Most importantly, it does not take into account your costs of production. It does not make any sense to work hard, produce a quality product, and sell it for $0.50 per pound when you have to spend $0.75 per pound to create it. The same problem applies to all pricing strategies where you base your prices on what other people charge.

You also have to look at who you are comparing prices with. For instance, you visit the local supermarket and find that conventionally grown 'Fuji' apples are selling for $1.89 per pound and organic 'Fuji's are selling for $2.29 per pound. You immediately go back to your roadside stand or farmers' market and price your organic 'Fuji's at $2.29 per pound, which is the full retail price for organic. There is a problem with this strategy. One of the reasons people shop at roadside stands, farmers' markets, and other direct outlets is because they are looking for bargains. Customers may need to travel farther to visit your stand than they would to shop at the neighborhood store, and they are certainly missing the convenience of one-stop shopping that they would get in a supermarket.

Of course you can advertise your produce as being fresher than that in the supermarket (not necessarily true, by the way), locally grown, or somehow better than supermarket produce, but the bottom line is that customers often expect to pay less than full retail price for direct market goods. The obvious solution is to set your price somewhere between what growers who sell to the supermarket receive (wholesale price) and what the supermarket that sells the product receives (retail price). You make more than your competitors and your customers find a bargain.

You may choose to set your prices lower than the competition. This strategy gets many new enterprises into trouble. Customers, of course, love price wars, and you can gain a definite marketing advantage if you consistently price lower than the competition and still make a profit. Discount department stores have done this for years. Making a profit is the hard part, and price gougers soon go out of business. The established competition may have to lower their prices for a while, but once you are gone the prices will come back to sustainable levels.

You might choose to set your prices higher than the competition. It may be that your product has some special advantage that you believe makes it more valuable than the competition's. Organically grown produce sometimes does bring premium prices. Goods sold through specialty shops, direct mail, or other niche markets often bring premium prices. This strategy can be successful, but be sure that what you sell is worth the extra cost to your customers. In some way, they must perceive that they are gaining some special advantage by buying from you. Consistent quality is an absolute must with this pricing strategy. Before you undertake this approach, get to know your customers and make an objective appraisal of how much they are willing to pay for what you sell.

Assess your costs. A more logical approach to pricing than following the competition is to set your prices according to what it costs you to produce your goods. Occasionally you might sell some item at below cost to bring in customers, hoping they will buy other goods at profitable prices. Below-cost goods sold this way are referred to as "loss leaders" and are common in retail stores and nurseries. They normally are not used in wholesale marketing.

Loss leaders aside, however, you must charge what it costs you to produce your goods, plus whatever profit you want to make. This approach is called "cost-plus pricing." Say, for example, that you have to spend $1.00

per pound to grow and market organic cherries, and you want your costs to average 50 percent of the sales price.

$$\text{Cost} = 0.50 \times \text{selling price}$$

$$\text{Cost} \div 0.50 = \text{selling price} = 1.00/0.50 = \$2.00 \text{ per pound}$$

Another variation on pricing according to your costs is to first estimate how much of a particular item you can sell, then figure out what it will cost you to produce the item and how much total profit you want to make. This is called the set profit approach. For example, you are growing nectarines and you figure you can sell 10,000 pounds per year, based on your previous sales or a request from a local fruit broker. Your enterprise budget shows that it will cost you $1.00 per pound to produce the nectarines, and you want to make $5,000 from that particular operation. It will cost you $10,000 to produce the nectarines. To make the $5,000, you add the cost of production to the desired profit, then you divide the required sales by the amount of product you estimate you can sell. In this example:

$$\text{Desired Profit} + \text{Cost of Production} = \text{Required Sales}$$

$$\$5{,}000 + \$10{,}000 = \$15{,}000 \text{ required sales}$$

$$\$15{,}000 \div 10{,}000 \text{ pounds} = \$1.50 \text{ per pound selling price}$$

Creating an Enterprise Budget

WHETHER YOU USE THE COST-PLUS OR THE SET PROFIT APPROACH, you must know how much your goods cost you to produce. Enterprise budgets help you identify your costs and prospective returns and design and refine your production and marketing programs accordingly.

An enterprise budget is simply an estimate of all of the income and expenses associated with a business. By calculating how much it will cost you to produce a ton of plums, for example, and estimating how much money you will receive for the plums when you sell them, you can estimate how profitable your orchard business will be.

If the enterprise budget shows your profit will be lower than you want or need, you can start playing the "what if" game with production and marketing practices. "What if I change to a higher-density planting? What will that do to my profits?" you might also ask, or "What if I sell some fresh plums at a roadside stand, instead of selling all of my fruit to the packing house?" These changes can be done in minutes with no cost but a little time.

Many computerized budgets are available online or in spreadsheet form that allow you to change a single figure and look at the effect on profit. It is much easier and less costly to change your operation during the planning stage than it is to change your actual operation after you've established an orchard.

There are two general types of cost that you need to consider when making an enterprise budget: "fixed costs" and "variable costs." These concepts may sound difficult but are actually very simple. To help you understand them, let's look at how they apply to owning a car. Say, for example, that you purchased a car last year and agreed to pay for it over a period of 2 years. That means that every month you are going to have to make a payment to the bank. The law requires that you also have to keep the vehicle licensed and insured, so you have to pay registration fees and insurance premiums. These costs are "fixed." You pay them whether you use the car or not, and you legally do not have the right not to pay these costs and still keep the car available for your use. On the other hand, you do have control over some costs. You decide whether or not to purchase gas and oil, have a faulty brake repaired, or have the car washed. Because these costs vary, according to your use of the car, they are called "variable."

The same terms and concepts apply to farming. Fixed costs include things like mortgage and other loan payments to the bank, insurance, property taxes, and depreciation of equipment and buildings. Whether you actually farm or not, these costs are not going to change. Variable costs include such things as fertilizer or seed, labor to plant or harvest a crop, and equipment repair. These expenses only occur when you farm, and you control how much they are going to be.

One kind of fixed cost that is a little more difficult to understand is "depreciation." This term simply refers to the way in which things lose value over time. Say, for example, that you purchase a tractor for $20,000 and you expect it to last for 20 years. Because it will be used in your orchard for many years, it is not reasonable to count the entire $20,000 against your business the year that you buy the tractor. Instead, you might elect to include a fixed cost of $1,000 in your enterprise budget for each of the next 20 years, thereby accounting for the $20,000. There are different ways of calculating depreciation. In general, the goal is to have an estimate at any given time of the worth or value of an item. In the tractor example, you might expect a rather large decrease in value (for example, what you could expect to sell it for) during the first year in service, then a gradual decline for the remainder of its expected life. Using this strategy, you might take $5,000 in depreciation the first year, then $789.47 during each of the following 19 years. At age 20, you estimate that the tractor will have no value and will need to be replaced.

By accounting for the depreciating value of equipment, buildings, irrigation systems, and other similar items over their expected working lives, you gain a clearer picture of the profitability and net worth of your business at any one time. Suppose you were to count the entire $20,000 as a single business cost the year that you make the purchase. During that year, your profitability would appear disastrous, even though you've made a reasonable investment that will pay returns in the form of the tractor's usefulness on the farm for many years. During the second year, your profitability might look better but your net value will appear lower than it really is because the $15,000 value still left in the tractor does not appear on your books.

The process of amortizing establishment costs is really very similar. Do not let the terms frighten you. "Amortization" means spreading a cost out over a period of time. Depreciation is simply a form of amortizing where you spread out the cost of an item over its expected working life. The term "establishment costs" refers to how much money it will take you to get the orchard going. Say that your new orchard will cost about $10,000 per acre to establish. If you establish a 10-acre apricot and peach orchard, your establishment cost is going to be $100,000. If you decide that you want to amortize the cost of establishing the orchard over 10 years, you would include a fixed cost of $10,000 in your enterprise budget each year for the next 10 years.

Many tree fruit enterprise budgets are available online and from universities, Cooperative Extension offices, and ministries of agriculture. A simple Internet search should supply you with all of the help you need. Two particularly valuable resources are the Ontario Ministry of Agriculture, Food, and Rural Affairs Budgeting Tools website and the British Columbia Ministry of Agriculture, Food and Fisheries Planning for Profit website, which provide simple enterprise budgets for apples, apricots, nectarines, peaches, pears, plums, and tart and sweet cherries.

Publicizing Your Enterprise

IT STANDS TO REASON that if people do not know about your business, they are not likely to become your customers. For most business owners, the obvious answer is to advertise. While advertising may become an important part of your marketing effort, it can be expensive. Fortunately, there are ways to become known that will not cost you anything. When I started a bonsai nursery many years ago, it was a novel enterprise in the community. My local newspaper, which reached about 25,000 people, was so interested that they sent a reporter to interview me. The result was a large front-page photo of me training a bonsai tree, along with a quarter-page article. I could never have afforded an advertisement like that, yet the tremendous publicity I got out of the article cost me nothing but an hour of pleasant conversation. Particularly in smaller communities, local newspapers are hungry for unusual stories. If your enterprise is novel in some way, you could easily end up with an article like the one I enjoyed.

In addition to traditional newspapers, chambers of commerce frequently put out newsletters that feature stories about new members and enterprises. Some chambers and other university and community groups also offer agricultural tours for the public. One chamber of commerce that I belonged to put on a county-wide farm tour each year. If you were promoting a new orchard, how would you like to see two or three busloads of people, anxious to see and hear about what you do, pull into your farm yard? It is hard to buy that kind of publicity, yet you can often get it for free.

Ten years ago, retail businesses could operate well without an Internet website. That is no longer the case. If you plan to market retail from a farm stand or U-pick orchard, you should have a simple website that tells who you are, where you are located, what you have to sell, and when you are open. You can easily find many examples, both good and poor, of websites for retail orchards and farms. Model your website after the best that you can find. Although you may need a commercial website designer's assistance, creating and maintaining a simple and effective farm sales website is fairly inexpensive. Shop around, and obtain referrals from other retailers that have effective websites. If you sell only wholesale to local packing houses and cooperatives, you do not need to have a website.

Regardless of your marketing strategy, you will need to use some form of promotion to make your enterprise successful. This portion of your marketing program may be as simple as paying a personal visit to a packing house buyer or placing a hand-lettered sign by your driveway or roadside stand. Or it might involve formal advertisements in newspapers or on the radio. You will probably find that word-of-mouth advertising by satisfied customers is the most effective promotion. That is why it is important that every customer feels good about having worked with you, whether he or she buys anything or not.

While most enterprises depend on advertising for success, it is also an area where you can waste money. Before you start advertising, specifically identify your target population — the people who are most likely to buy your products. Then examine the various ways that you can reach that specific population with your message.

Where and how you sell your goods will greatly influence how you advertise. For example, if you plan to sell directly to local packing houses, food cooperatives, restaurants, and stores, it would be pointless for you to invest in radio or newspaper ads. Your best advertising strategy would be to make direct, personal contact, possibly preceded by a letter of introduction and a request for a meeting. The same strategy applies if you plan to sell your goods through a broker on the wholesale market. With this type of marketing, you target specific buyers. If you are new to an area and want to wholesale-market locally, you can identify potential clients by using an online or printed telephone directory or by visiting the chamber of commerce. Other established fruit growers are an excellent source of information.

If your goal is to market directly to local consumers, you will probably find that satisfied customers are your best promoters. To get started, roadside signs (where local and county ordinances allow) are one way to stimulate interest. For a roadside stand or U-pick operation alongside a major road, a series of signs can be effective in promoting sales, especially for out-of-town visitors. Only people who drive by your signs are going to be exposed to the message, however. To develop local interest, newspaper and radio advertisements may be the most cost-effective advertising strategy. Remember to target your population. If you want to sell U-pick cherries to homemakers, your best success may come by advertising during a local radio program listing items for sale and trade. Similarly, rather than advertising in a traditional newspaper, you might find it more effective to place your ad in a "Nicklesworth" type of paper that consists almost exclusively of classified ads for buying and selling. In both cases, you are promoting your enterprise at a time and in a place where potential customers are looking for things to buy. Not only do you reach the target population, but you do so when those people are in buying moods.

When you sell at a farmers' market or growers' cooperative, you usually pay a fee to them. This fee helps cover the cost of advertising the market, which advertising directly benefits you. The positive aspects are that you can often get more bang for your advertising buck and leave the details of producing the ads to someone else. On the negative side, your success is largely tied to the success of the particular market or cooperative.

Packaging and Selling

Because of the large array of orchard fruits and value-added farm products and marketing strategies, it is impossible to go into detail on packaging in this book. You should be aware, however, that there are industry packaging standards for most of the goods that you will be producing. In some cases, the standards are formalized and may even have the force of law. In other cases, packaging standards may be informally and voluntarily accepted within an industry. The bottom line is that industry members have adopted standardized packaging methods because they are the most cost-effective way to get products to market in good condition.

To get a quick overview of the packaging demands you will face, talk to established farmers who produce the crops you are interested in growing. For U-pick farms, roadside stands, and farmers' markets, know whether you will be selling your produce by the piece, by weight, or by volume. If you are selling by weight or volume, find out if your state or province requires that your scales or other measuring devices be certified. Be aware that some common terms have specific meaning when selling produce. The terms box, lug, flat, and crate all mean specific amounts when applied to fruit sales. If you will be selling on the wholesale market to brokers or directly to stores or restaurants, you should also talk to the people that will be buying your goods. Find out exactly what they expect in terms of packaging and labeling.

Whatever packaging and delivery system you use, packages should be clean and strong and present a professional appearance. Recycling old cardboard boxes and paper sacks may save you a few pennies, but it certainly will not do much to build customer confidence in your professionalism. When designing a system appropriate for your operation, take advantage of the expertise that commercial packaging suppliers have to offer and clearly identify your product and the name of your business. Competition is fierce, so be sure to give yourself every possible marketing advantage. After spending time, money, and effort to create a fine product, do not present it to your customers in shoddy paper bags or used cardboard boxes bearing the names of your competitors.

In many cases, you will find that clearly marking containers with your farm name and logo is an inexpensive and effective means of advertising. I had occasion to visit a large fresh produce brokerage firm in Oregon a few years ago. As I walked through the facility looking at huge stacks of boxed and bagged produce stored in the cold rooms, the manager, knowing that I was from Idaho at the time, pointed out several pallet loads of potatoes in cardboard boxes. The boxes were all clearly marked with the name and logo of the Idaho grower who had supplied the potatoes. My host explained to me that several of his largest customers insisted on receiving only potatoes from that particular grower, citing consistently high quality. The customers were confident that as long as the potatoes arrived in a box with that name and logo, they did not have to worry about quality. That grower had it made. By maintaining a high level of quality and good customer relations, as well as by clearly identifying his product and setting it apart from his competitors, he had customers coming to him.

Selling

If you have taken care of the steps that we have already discussed, selling will be easy. You will have produced something for which there is a viable market, your product will meet quality and delivery standards, you will have established contact with your prospective customers, and you will have clearly set forth sales terms. There will not be any surprises for you or your customers. You have something they want, and you are willing to provide it for a fair price. The only thing left is the exchange. There are literally hundreds of books and training programs available on how to sell, and you may benefit from reading some of them. You are, however, already an expert on the subject by having made thousands of purchases of your own.

Make the exchange quick and professional. Always provide receipts. For wholesale purchases, work out the details, such as delivery and payment schedules, in advance. Meet your end of the bargain and provide clear, concise invoices. Keep your records and accounting meticulously accurate. If you need to charge sales tax, then do it. Never bend the law for anyone. Doing so can land you in deep trouble with the authorities and will only damage your reputation with customers. They know that if you are willing to cheat the government, you are probably also willing to cheat them.

Developing a Business Plan

According to an old adage, "If you do not know where you are going, then you will never get there." This is especially true of creating and running an orchard business. A business plan is simply your road map. It gets you started in the right direction and helps you make midcourse corrections when you run into detours and dead ends.

New small businesses, including farms, traditionally have high failure rates. One commonly quoted estimate is that 80 percent of small businesses fail within the first 5 years. I believe two of the main reasons they go under is that the owners fail to adequately address marketing planning and business planning. All too often a new commercial farmer will say that her or his operation is too small to worry about doing market research or prepare a business plan. The farmer drifts along for a while, hoping that something good will happen. "Once I get well established," he or she thinks, "then I'll put together a professional business plan and really market my product." The problem is that the farmer usually continues to drift along, putting in much hard work for small or no returns.

If at all possible, the best time to prepare a business plan is before you buy your farm site. For those of you who have already established your orchard, take time to develop a business plan before expanding your orchard or continuing sales.

After deciding on a product and market, put together a business plan that will direct your efforts and resources so that you earn the best possible return for your investments of time and money. A solid, well-designed business plan written for your specific products and market will give you the direction you need for success. You will invest — not spend — your time and resources by putting them into profitable activities that will help you attain your goals.

A full treatment of orchard business planning is beyond the scope of this book. Many books, commercial software, and online resources are available, however, that provide detailed guidance on developing business plans for commercial orchards.

The goal of this book has been to help you grow organic orchard fruits effectively and in ways that are environmentally sound and sustainable. If you are a commercial fruit grower, hopefully you will have gained knowledge and ideas that will prove profitable. For all readers, whether you have a tree or two in your yard or thousands of trees in a commercial orchard, I hope that you will find great satisfaction and enjoyment in growing fruit trees, as I have.

APPENDIX A

Sample Planning, Preparation, and Planting Year Calendar

The following calendars show typical activities for organic fruit growers. Depending on your location, orchard size, management strategies, and crops, adjust the calendars to meet your needs.

PLANNING YEAR

Midsummer to Fall

- ☐ Test soil for nutrients, pH, exchangeable acidity, organic matter, texture, and nematodes.
- ☐ Test water supply for quality, flow rate, and capacity.
- ☐ Conduct water percolation tests in fields.

PREPLANTING YEAR

Late Winter to Early Spring

- ☐ Apply lime and sulfur to adjust soil pH, if needed. Base application rates on soil tests.
- ☐ Apply fertilizers and soil amendments, if needed. Base application rates on soil tests.
- ☐ Choose fruit varieties and training systems.
- ☐ Design field layout and irrigation system.
- ☐ Determine the number of plants needed.

Mid-spring

- ☐ Order plants for next spring.
- ☐ Correct soil drainage problems.
- ☐ Mark out planting blocks, roads, and buffer strips.
- ☐ Plant a rotation, biofumigant, or green manure crop in planting blocks, maintaining good weed control.
- ☐ Establish roads and buffer strips.

Midsummer

- ☐ Till under and replant green manure or biofumigant crop.
- ☐ Continue weed control.

Fall and Early Winter

- ☐ Till under green manure crop if it cannot be left until next spring.
- ☐ Plant a winter green manure crop, if needed.
- ☐ Order irrigation and fencing materials.

PLANTING YEAR

Early to Late Spring

- ☐ Apply preplant fertilizers and soil amendments, if needed. Base rates on soil tests.
- ☐ Deep-rip entire planting blocks with a chisel plow, first at right angles to the crop row direction, then parallel to the crop row direction.
- ☐ Level the planting blocks with cultivation disks or a harrow.
- ☐ Lay out and mark planting rows.
- ☐ Install irrigation system, fences, and trellis posts.
- ☐ Test irrigation system.
- ☐ Plant fruit crops.
- ☐ Install trellis wires and tree support posts, if needed.
- ☐ Install rodent guards to trees.
- ☐ Apply mulches.
- ☐ Irrigate to keep soil moist but not saturated.
- ☐ Begin tree training.
- ☐ Begin pest, disease, and orchard floor management programs.

Summer

- ☐ Continue tree training.
- ☐ Continue irrigation.
- ☐ Continue pest, disease, and orchard floor management programs.

Late Fall to Winter

- ☐ Examine and adjust trellises, tree supports, and tree ties.
- ☐ Winterize irrigation system.
- ☐ Apply Bordeaux spray after leaves fall (do not apply to apricots).
- ☐ Continue rodent management program.

APPENDIX B

Sample Establishment and Production Years Calendar

The following calendars show typical activities for organic fruit growers. Depending on your location, orchard size, management strategies, and crops, adjust the calendars to meet your needs.

ESTABLISHMENT YEARS

Winter to Early Spring

- ☐ Arrange for contract labor, if required.
- ☐ Order pesticides, fertilizers, and other supplies.
- ☐ Inspect and maintain fences, trellises, and irrigation systems.
- ☐ Prune fruit crops.
- ☐ Apply compost and slow-release fertilizers.
- ☐ Apply dormant and delayed dormant fixed copper, lime sulfur, and horticultural oil sprays (don't use sulfur on apricots and allow 30 days between applying sulfur and oil products).

Mid-spring

- ☐ Continue tree training.
- ☐ Continue irrigating.
- ☐ Continue pest, disease, and orchard floor management programs.
- ☐ Apply quick-release fertilizers, if needed.

Late Spring to Early Summer

- ☐ Continue tree training.
- ☐ Continue irrigation.
- ☐ Continue pest, disease, and orchard floor management programs.

Mid- to Late Summer

- ☐ Continue tree training.
- ☐ Continue irrigation.
- ☐ Continue pest, disease, and orchard floor management programs.

Fall to Winter

- ☐ Examine and adjust trellises, tree supports, and tree ties.
- ☐ Winterize irrigation system.
- ☐ Apply Bordeaux spray after leaves fall (do not apply to apricots).
- ☐ Continue rodent management program.
- ☐ Arrange to rent beehives for first production year.
- ☐ Order baskets and other supplies for harvesting for first production year.

PRODUCTION YEARS

Winter to Early Spring

- ☐ Arrange for contract labor, if required.
- ☐ Order pesticides, fertilizers, and other supplies.
- ☐ Inspect and maintain fences, trellises, irrigation systems, and refrigeration equipment.
- ☐ Prune fruit crops.
- ☐ Apply compost and slow-release fertilizers.
- ☐ Apply dormant and delayed-dormant fixed copper, lime sulfur, and horticultural oil sprays (don't use sulfur on apricots and allow 30 days between applying sulfur and oil products).
- ☐ Apply prebloom kaolin and other prebloom to early-bloom pesticides.

Mid-spring

- ☐ Place hives in orchard.
- ☐ Thin flowers, if needed.
- ☐ Apply quick-release fertilizers, if needed.
- ☐ Continue irrigation.
- ☐ Continue pest, disease, and orchard floor management programs.

Late Spring to Early Summer

- ☐ Thin green fruits, if needed.
- ☐ Continue irrigation.
- ☐ Continue pest, disease, and orchard floor management programs.

Midsummer to Mid-fall

- ☐ Harvest fruit.
- ☐ Every 2 years, test soil samples from each planting block for pH.
- ☐ Conduct foliar nutrient tests.
- ☐ Continue irrigating.
- ☐ Continue pest, disease, and orchard floor management programs.

Late Fall to Winter

- ☐ Examine and adjust trellises, tree supports, and tree ties.
- ☐ Winterize irrigation system.
- ☐ Apply Bordeaux spray after leaves fall (do not apply to apricots).
- ☐ Continue rodent management program.
- ☐ Arrange to rent beehives for next year.
- ☐ Order baskets and other supplies for harvesting for next year.

RESOURCES

Useful Websites

Agricultural Marketing Service
United States Department of Agriculture
www.ams.usda.gov
Provides complete information on the National Organic Program, including a list of certifiers and the list of allowed and prohibited synthetic materials

Harris Manufacturing
www.wonderweeder.com
Maker of the Wonder Weeder

Mayhaw Recipes
Louisiana Mayhaw Association
http://mayhaw.org/original/recipes.html

National Climate Data and Information Archive
Environment Canada
http://climate.weatheroffice.gc.ca

Natural Resource, Agriculture, and Engineering Service
www.nraes.org
Publishes practical books related to agriculture, including the *On-Farm Composting Handbook* (NRAES-54. Edited by Robert Rynk, 1992)

NOAA Regional Climate Centers
National Climatic Data Center
www.ncdc.noaa.gov/oa/climate/regionalclimatecenters.html

Organic Materials Review Institute (OMRI)
www.omri.org

USDA Plant Hardiness Zone Map, 2012 update
United States National Arboretum
www.usna.usda.gov/Hardzone

USDA Plant Hardiness Zone Map, 1990
United States National Arboretum
http://planthardiness.ars.usda.gov/PHZMWeb/Downloads.aspx

BUDGETING TOOLS

Ontario Ministry of Agriculture, Food, and Rural Affairs
www.omafra.gov.on.ca/english/busdev/bear2000/Budgets/budgettools.htm#tree
Budgets for various commodities

Enterprise Budgets: Planning for Profit
British Columbia Ministry of Agriculture
www.agf.gov.bc.ca/busmgmt/budgets/tree_fruits.htm
Budgets for various commodities

Publications Cited in Text

Canadian General Standards Board. "Organic Production Systems General Principles and Management Standards." CAN/CGSB-32.310-2006.
Available on the Canadian General Standards Board website, www.tpsgc-pwgsc.gc.ca/ongc-cgsb

Canadian General Standards Board. "Organic Production Systems: Permitted Substances List." CAN/CGSB-32.311-2006.
Available on the Canadian General Standards Board website, www.tpsgc-pwgsc.gc.ca/ongc-cgsb

Childers, Norman F., Justin R. Morris, and G. Steven Sibbett. *Modern Fruit Science,* 10th ed. Horticultural Publications, 1995.

Westwood, Melvin Neil. *Temperate-Zone Pomology,* W. H. Freeman and Company, 1978.

INDEX

Page numbers in *italic* indicate illustrations; those in **bold** indicate charts.

A

access roads
- marking, 86
- orchard design and, 74
- planning blocks and, 64
- sod-forming grasses for, 307
- staking the orchard, *87*

access to orchard, utilities and, 40–41
acetic acid, 295–96
advertising, 515–17
An Agricultural Testament (Howard), 7
alfalfa, 305–6
alfalfa meal, 271
algae, 267
alley(s), 64
alley crops, 300–306. *See also* perennial alley crops
- annual, 248, 301–3
- benefits of, 289, 299, 300–301
- drawbacks of, 301
- green manure crops as, 276
- management of, 306
- nutrient management, 309
- permanent, **308**
- research and, 288
- soil amendments and, 243

alternate hosts, removal of, 331–32
aluminum sulfate, 94–95
ambrosia beetle, 399–400
amendments, 108–11. *See also* compost/composting; manure
- off-farm, 109
- other soil-building, 112–13
- for planting holes, 243
- sand and/or organic matter, 79
- soil standards and, 89–90

American hawthorn rust, 347
ammonium nonanoate, 297–98
ammonium sulfate, 281
amortization, 514
aphids, 14, 120, 381, 383, 384, 386, 387–88
apple, 122–31. *See also* apple varieties; supported apple trees; training apple trees
- average yields, new orchard, 61, *61*
- basics, 125
- cross-pollination, 126–27
- disease-resistant varieties, 128, 152–57, 343–44, 345
- harvesting, 483–86
- most common, 120
- picking, spurs and stems, 485
- proliferation, 364
- ripeness, gauging, 484–85
- rootstocks, 50, *50*, 128–31, 429, 434
  - large trees, 130–31
  - medium-sized trees, 130
  - small trees, 129–30
  - very small trees, 129
- soil conditions, 124
- temperature concerns, 123–24
- thinning, 473–75
- varieties, 124–28
  - choice of, 127–28
  - cold-hardy, 158–60
  - low-chilling, 163–64
  - medium-chilling, 161–63
  - pollination, 126–27
  - popular, 498
- yields, average, 61, **61**

apple and thorn skeletonizer, 395
apple maggot, 120, 121–22, 299, 374, 376, 396
apple mosaic, 364
"apple pears." *See* Asian pears
apple scab, 39, 119, 341–44
- causes and timing, 342
- hours of wetting for, **343**
- management, 342–44
- symptoms, 341–42

approved materials, 265–66, 267–80

apricot, 178–81, *179*
basics, 179
climate concerns, 180
fungicides and, 333
harvesting, 490
planting/cultivating/harvesting, 180–81
pollination, 181
rootstocks, 181
thinning, 478
training systems, 59–61, **60**, 462
varieties, suggested, 200–203
armillaria fungus, 366–67
Asian pears, 132–34
basics, 133
pollination, 134
temperature concerns, 133–34
varieties, suggested, 164–65
"available potash," 105
axis systems, 245, 424–25
azadirachtin products, 390

B

Bacillus thuringiensis (Bt), 12, 382–83
bacterial canker, 357–58
bacterial pathogens, 39
bacterial spot, 358–59
bactericides. *See* fungicides/bactericides
Balfour, Lady Eve, 7–8
bare orchard ground
maintaining, 323–24
small mammals and, 372–73
bare-root trees, *231*, 231–32
bark inclusion, *414*, 414–15
bats, 379–80
bees. *See also* honey production
fire blight and, 345
insecticidal soaps and, 381
pesticides and, 398, 403
pollination and, 192, 198, 378
safety and, 502
Spinosad and, 384
beetles and plant bugs, 385, 402–6
campyloma bug, 405–6
Japanese beetle, 404
plant bugs, 404–5
plum curculio, 402–3
bench cuts, 412, *412*
beneficial microorganisms, 240–42
product selection, 241–42
treatment of plants, 242
types of, 240–41
beneficial organisms, 377–78
insectary crops and, 306, 378
natives, management of, 377
releasing, 377
between-row spaces. *See* alleys
bicarbonates, 339
biodynamic preparations, 270–71
biofumigant crop, 115
bird netting, 374
birds
insect eating, 299, 374, 379
small mammals and, 373
wild, 379
bitter pit, 353–55
bitter rot, 352–53
black knot, 359–60
black mica, 104
black rot, 351–52
bletting, 138, 142
"blind wood," 456
blood meal, 271
blossoms
crop loads and, 468, 471, *471*
"king" blossom, 473
petal fall, 262
temperatures and, 26
bonemeal, 102, 243, 271–72
Bordeaux mix, 334–37, **335**
boring insects, 399–402
ambrosia beetle, 399–400
pacific flatheaded borer, 400
peachtree borer/lesser peachtree borer, 400–401
peachtwig borer, 401–2
shothole borer, 399–400
boron (B), **99**, 107–8, **257**, 272, 285
branches. *See* scaffold branches; whorls of branches
branches, positioning, 416, *416*
branching angles, 415
Brassica plants, 110–11
brown blight, 39, 355–57
brown rot, 340
budding onto a rootstock, 47, 48, *48*, 121
budget, 512–15

buds
- formation of, 333
- terminal, *427*

bud sport mutations, 120–21
bud swell, 390
buffer strips
- marking, 86
- non-crop areas, 63
- in sample plan, 74
- staking the orchard, *87*

bush cherry
- cherry-plum hybrids, 187, 209–10
- dwarf sour cherry hybrids, 188, 210
- harvesting, 492
- Nanking cherry (*Prunus tomentosa*), 187, 210
- thinning, 482
- training systems, 463
- varieties, suggested, 209–11
- Western sandcherry (*Prunus besseyi*), 187, 211

bush plums, 482, 492
business plan, 519–20
"butterflying," 239–40

C

calcium (Ca), 105–6, 107, **107**, **257**
calcium carbonate equivalent (CCE), 92, 93
calcium polysulfide, 337
calcium sulfate, 276–77
campyloma bug, 405–6
Canada
- approved materials, 265
- climate data for, 19, 22, 25
- metric areas, fertilizer for, 283
- organic certification program, 13

Carson, Rachel, 9
case study, orchard preparation, 84
caterpillars, 381, 383, 384
cation exchange capacity (CEC)
- calcium and, 105–6, 107
- soil organic matter (SOM) and, 289–90
- soil test and, 28, 88

cedar-apple rust, 347
central leader trees, 425–26, 427, *427*
certification, organic, 13
- 3-year recertification, 38

chelates, 272
chemical thinning, 472, 473–74, 480
cherry, 181–88. *See also* training sweet cherry trees; training tart cherry trees
- basics, 183
- bush cherry, 187–88, 209–11
- climate concerns, 182–83
- harvesting, 490
- pests and diseases, 183
- pollination, 185
- rootstocks, 50, *50*, 183–85, 447
- sweet cherry, 185–86, 204–7
- tart cherry, 186–87, 208–9
- thinning, 478–79
- training systems, 57–58, *58*, **58**

cherry mottle leaf, 364
cherry-plum hybrids, 187, 209–10
cherry slug, 398–99
cherry twisted leaf, 364
chickens, 299, 379
chilling requirements, 20
chilling units, 20, 26
cinnamaldehyde, 339
citrus products, 296
clay. *See* hard pans/clay layers; kaolin clay
climate considerations, 18–27. *See also* frost; temperature concerns
- apricot, 180
- frost, 37, *37*, 67, 367
- loquat, 148–49
- mayhaw, 143–44
- peach, 188
- pome fruits and, 119
- saskatoon, 150
- spring and summer temperatures, 21
- winds, 26–27
- winter temperatures, 19–20

clones/cloning, 121
clopyralid, 273
clothespins, spring-type, 416, *416*
clove oil, 297
cocoa bean hulls, 272
codling moth, 120, 122, 374, 376, 391–92
colloidal phosphate, 102, 279

community supported agriculture (CSA), 504, 506
companion crops
  green manure crops as, 276
  intercropping, 323
  nitrogen availability and, 309
  soil amendments and, 243
compost/composting, 270, 272–73, 278, 309
containerized trees, 232–33
controlled atmosphere (CA) storage, 15
Cooperative Extension, 40, 260, 386, 497, 506, 514
cooperatives, 40, 508–9, 517
copper (Cu), **257**, 334, 367
corn gluten meal, 271, 297
cost-plus pricing, 511–12
costs. *See also* financial considerations
  assessing, 511–12
  types of, 513–14
cottonseed meal, 273
cottonseed oil, 474
cover crops, 300–310. *See also* alley crops; perennial alley crops
  benefits of, 299
  insectary crops, 306–9
  living mulch, **320–21**
  nutrient management, 309–10
  suggested, **308**
  understory crops, 300
crab apples, 127, 332, 345, 347, 396, 429
Crataegus aestivalis and *Crataegus opaca*. *See* mayhaw (*Crataegus aestivalis* and *Crataegus opaca*)
crop load
  "bench" cuts and, 412, *412*
  nutrition and, 326
  pruning and, 467, 468
  thinning and, 467, 468
crop oils, 337
crop row orientation, 72–73
crop spacing, 73–74
cross-pollination. *See* pollination
crowding, 62, 64, 65, 466
crown gall, 242
CSAs, 504, 506
cultivar vs. variety, 122
cultivation, mechanical, 114–15, 322
cultural practices, 330–33
  alternate hosts, removal of, 331–32
  fruit crop nutrition, 332
  orchard floor, 330–31
customer demand
  for apple varieties, 498
  for Asian pears, 133
  keeping track of, 505
  marketing and, 495
  for organic farm products, 9, 496
  for quince, 140
cutworms, 120, 393
Cydonia oblonga. *See* quince (*Cydonia oblonga*)

D

Damson plum (*Prunus insititia*), 195
"deep ripping," *80*, 80–81
deer, 69, 71, 74, 146, 235, 369–72
Demeter-International, 270
depreciation, 513–14
diseased wood, 363, 413
disease(s), 39–40. *See also* pome fruit diseases; stone fruit diseases
  of both pome/stone fruits, 363–64
  cherry, 183
  loquat, 149
  management
    in dormant season, 338
    scouting, 327–30
    site selection and, 326
  mayhaw, 146
  peach, 189
  pear, 131–32
  pome fruits and, 119–22
  quince, 139–40
  root, 364–67
  saskatoon, 151
  tree prepping and, 240
disinfectant solution, 339, 346, 413
diversity of products, 504
divided canopy training systems, 423, 456
dolomite, 273–74
dormancy
  disease/pest control, 338
  temperatures, 26
dormant pruning, 409, 417, *417*
double leader pear system, 440–42, *441*

drainage, 78–85. *See also* soil types and drainage
adding sand/organic materials, 79
ditches/drain tiles/grading, 82
excessive, 84–85
raised beds, 82–84, *83*
waterlogged soil and, 31
draught-tolerant grasses, 305
drip irrigation, 68–69
dropped fruit, 482
dwarf fruit trees
dwarf sour cherry hybrids, 188, 210
rootstock and, 45, 46, 47, 51, 319, 446

E

earwigs, 390, 399, 407
ectomycorrhizal fungi, 241, 242
elemental sulfur, 94–96, **95**
elements, essential, 252–53
endomycorrhizae, 241, 242
enterprise budget, 512–15
entomosporium leaf spot, 353
enzymes, 274
Epsom salts, 274
ericoid mycorrhizae, 241
Eriobotrya japonica. See loquat (*Eriobotrya japonica*)
eriophyid mites, 390
erosion control, 310
espalier trellis design, *430*
essential elements, 252–53
establishment years, seasonal tasks, 524
ethylene gas, 482, 483
European pears, 134–36, *135*
basics, 135
low- and medium-chilling, 168–69
medium- and high-chilling varieties, 166–68
pollination, 136
temperature concerns, 135–36
varieties, 136
European plums (*Prunus domestica*), 193–95
greengage plums, 194
Imperatrice-type plums, 194
Lombard-type plums, 194
prunes, 194
varieties, suggested, 218–21
yellow egg plums, 194
evaporation pans, 35–36
eyespotted bud moth, 393

F

fabraea leaf spot, 349–50
farm design/selling venue, 498–509
on-farm sales, 499–504
U-pick farm, 498–99
farm ecosystem, 14–15
farmers' markets
customer types and, 505
orchard design and, 69, 73, *73*
packaging and, 517
product selection and, 496, 498
roadside stands and, 503
rules regarding, 506
feathered planting stock, *231*, 231–32, 425, 427, *427*
feather meal, 274–75
fencing
design for, 69
large mammals and, *370*, 370–72, *371*
in place before planting, 235
for sample orchard plan, 71–72, 74
fertilizer, 280–85
application rates, 281–84
less than an acre, 282
in mature orchards, 283–84
for metric areas, 283
per acre, 282
per-tree rate, 282–83
for small orchards, 283–84
excessive amounts of, 251
how/where to apply, 284–85
newly planted trees, 248
nutrients in, percentages, **268–69**
financial considerations, 122. *See also* costs
fire blight, 119, 340, 345–47
chemical and biological control, 346–47
management, 346
symptoms, 346
firmness meters, 484, 487
first year care/management, 245–48
alley crops, perennial, 248
irrigation, 246
weed management, 246–48, *247*

fish emulsions, 275
fish meal, 275
fish oil, 474–75
fixed costs, 513–14
flame weeders, 294
flat apple disease, 364
flat limb, 364
flies, sawflies, and midges, 396–99
  apple maggots, 121–22, 299, 374, 376, 396
  fruit flies, 397
  pear leafcurling midge, 399
flooding orchards, 66
flowers. *See* blossoms
flyspeck, 350–51
foliage, nutrient status and, 254–55, 260–61
foliar analysis. *See* tissue testing
"French axe" training system, 424
frogeye leaf spot, 351–52
frost
  damage from, 367
  orchard topography and, 37, *37*
  protection against, 67
frost heaving, 304
fruit. *See also* crop load; *specific crop*
  nutrient status and, 254–55
fruit brokers, 508
fruit flies, 397
fruit-grower cooperatives, 40, 508–9, 517
fruiting wood
  pruning and, 433, 467
  rootstocks and, 47
fruit pomaces, 275
fruit variety. *See* scion
fruitworms, 120, 393–94
Fukuoka, Masanobu, 8
fungal pathogens, 39
fungicides/bactericides, 333–40
  Bordeaux mix, 334–37, **335**
  cinnamaldehyde, 339
  copper (Cu), 334
  hydrogen peroxide, 339
  lime sulfur, 337
  microbial fungicides, 340
  neem oil, 338
  plant-derived oils, 339
  streptomycin, 340
  sulfur, 333
  Surround, 263, 340, 385–86
  tetracycline, 340

G

Genetic Resources of Temperate Fruit and Nut Crops, 125
glazing, 238
glyphosate, 116
grading fields, 82
grafting, 45, 47, *48*, *49*, 121, 238–39
granite dust, 103, 276
grass alley crops. *See* alley crops
grasses
  draught-tolerant, 305
  low-growing, bunch-forming, 303–4
  orchard grass, 305
grazing. *See* livestock in orchards
green lacewings, 390
green manure crops
  as approved material, 276
  biofumigant quality, 110–11, 117
  carbon-nitrogen balance, 108
  for nitrogen fixing, 100
  nutrient management and, 250
"green revolution," 7
greensand, 104, 276
green tip, 390
growers' cooperatives, 40, 508–9, 517
guanos, 270
gypsum, 276–77
gypsy moths, 394

H

habitat
  loquat, 148–49
  mayhaw, 143–44
  saskatoon, 150
hand harvesting, 483
hand thinning fruits, 468, 473
hardiness zone, 25. *See also* Plant Hardiness Zone Map
hard pans/clay layers
  identifying/breaking up, *80*, 80–81
  looking for, 78
  rototillers creating, 81, *81*
harvesting, 482–83
  pome fruits, 483–89

apples, 483-86
loquats, 486
mayhaws, 486
medlars, 487
pears, 487-89
quince, 489
saskatoon, 489
stone fruits, 490-92
Haughley Experiment, 7-8
heading cuts, 411-12, *412*, 449, *449*
hedging, 411-12
herbicides
advent of, 287-88
clopyralid, 273
nonorganic, 116
organic, 295-99
acetic acid, 295-96
citrus products, 296
clove oil, 297
corn gluten meal, 297
herbicidal soaps, 297-99
herbivore fencing, 235
high-density pear systems, 437-42, *440*
double leader design, 440-42, *441*
open Tatura design, 438, 439-40
V-hedge design, 438, 439, *439*
history, orchard management, 287-89
honey production, 323. *See also* bees
hoof and horn meal, 277
horizontal trellis design, *430*
horticultural oils, 380-81, 390
Howard, Sir Albert, 7, 8
humates and humic acids, 277
humidity and precipitation, 25, 26
humus, 289
hydrated lime, 93
hydrogen peroxide, 339

I

inclusion, *414*, 414-15
indole-butyric acid (IBA), 144-45
infrared weeders, 294
inorganic mulches, 311-13
landscape fabrics, 312-13
plastic films, 311-12
in-row crops, 299. *See also* companion crops
insect and mite pests, 386-407. *See also* pests
beetles and plant bugs, 402-6
biological management, 299
boring insects, 399-402
cultural practices and, 374-75
flies, sawflies, and midges, 396-99
moth pests, 391-96
scales, 406
sucking pests, 386-90
thrips, 407
insectary crops, 306-9, 378
insecticides, 380-86
horticultural oils, 380-81
insecticidal soaps, 381-82
microbial, 382-86
insect monitoring devices, 376, *376*
intercropping food/nonfood crops, 322-23
internodes, 59
interstem
planting and training, 238-39
tree size and, 47, 49, *49*
iodine test, 484-85, *485*
Iron (Fe), **258**
irrigation
first year care/management, 246
orchard design and, 74-75
site selection and, 34-36
systems, 65-69, 235

J

Japanese beetle, 404
Japanese medlar. *See* loquat (*Eriobotrya japonica*)
Japanese plums (*Prunus salicina*), 195-96
June drop, 331, 405, 466

K

kaolin clay, 263, 340, 385-86
kelp meal and extracts, 277
kelps (brown algae), 267
King, Franklin H., 7
"king" blossom, 473
K-mag, 277
Kym Green's bush (KGB) system, 454

L

laborers, 41
ladybird beetles (ladybugs), 390

landscape fabrics, 312–13
laterals, 254, 435, 442
laws
    environmental, 36, 78
    lawsuits, avoidance of, 498
    sales tax and, 519
    state and local, 41
layout. *See* orchard layout design
leader growth, 254–55
leafhoppers, 120, 381, 384, 386, 388–99
leaf miners, 120, 381, 384
leafroll, 364
leafrollers, 120, 393–94
leather meal, 278
leaves. *See also* foliage, nutrient status and; tissue testing
    moth pests on, 391–96
'Liberty' apple
    cross-pollination, 127
    disease resistance, 128, **154**, 344, 345
    starch-iodine test, 485, *485*
    as trap crop, 403
limestone, 92, 277–78
lime sulfur, 337
liming materials, 92–93
limonene (citrus oil), 296
liquid lime sulfur (LLS), 474–75, 478
little cherry, 364
livestock in orchards, 299, 403
living mulches, 316–19, **320–21**
    how to use, 318–19
    Swiss sandwich system, **317**, 317–18
    what to use, 319
The Living Soil (Balfour), 8
LLS sprays, 474–75, 478
location for orchard. *See* topography
loopers, 393–94
loquat (*Eriobotrya japonica*), 141–42, 147–49, *148*
    basics, 149
    climate and habitat concerns, 148–49
    harvesting, 148, 486
    pests and diseases, 149
    pollination and varieties, 149
    thinning, 475–76
    training systems, 56, **56**, 442–43
    varieties, 173–74
low-input sustainable agriculture (LISA), 9
lygus bugs, 120

M

macronutrients, 252, 253
magnesium (Mg), 106–7, **257**
magnesium sulfate, 274
Malus domestica, 120
mammals
    large, pests, 369–72, *370*, *371*
    small, pests, 371, 372–73
manganese (Mn), **258**
manure
    animal, 269–70
    carbon-nitrogen balance, 108
    phosphates and, 102
mapping orchard design, 62–65
    non-crop areas, 63
    planning blocks, 63–65
marketability of harvest, 483
marketing
    budget, creation of, 510–12
    components of, 495
    farm design/selling venue, 498–509
    prices/terms of sales, 510–12
    product selection, 496–98
    selling and, 494
    target population, 497, 498, 516
mass trapping, 403
mayhaw (*Crataegus aestivalis* and *Crataegus opaca*), 142–47, *143*
    basics, 145
    habitat and climate concerns, 143–44
    harvesting, 486
    pests and diseases, 146
    planting/training/harvesting, 145–46
    propagation, 144–45
    thinning, 478
    training systems, 56, **56**, *443*, 443–44
    varieties, 146–47, 170–72
mealybugs, 120, 381, 383, 388
mechanical cultivation, 114–15, 322
mechanical thinning, 469–72, *470*, *471*

medlar, 140–42, *141*
basics, 141
common (*Mespilus germanica*), 140
harvesting, 487
Stern's (*Mespilus canescens*), 140–41
thinning, 478
training systems, 56, **56**, 444
metric areas, fertilizer for, 283
mice guards, 373, *373*
microbial fungicides, 340
microbial insecticides
Bacillus thuringiensis (Bt), 382–83
Beauveria bassiana, 383
neem products, 384–85
pyrethrum products, 385
Spinosad, 383–84
Surround, 385–86
micronutrients, 252, 278, 281
microorganisms, 111–12
microsprinklers, 36, 68–69, 72
midges. *See* flies, sawflies, and midges
mineral nutrients, 97–108
boron (B), **99**, 107–8
calcium (Ca), 105–6
magnesium (Mg), 106, 107
nitrogen (N), 99–101
phosphorus, **98**, 101–3
potassium, **98**, 103–5
sulfur, **98**, 105
Ministry of Agriculture, 13, 260, 386, 506, 514–15
minor pome fruits
plant spacing for, **56**
training systems, 56–57
minute pirate bugs, 390, 399
mites, 120, 374–75, 381, 384, 386, 389–90
Modern Fruit Science, 190, 481
modified central leader, 427, *427*
moose, 69, 235, 369–70
moth pests, 391–96
apple and thorn skeletonizer, 395
codling moth, 391–92
cutworms, 393
eyespotted bud moth, 393
fruitworms/leafrollers/loopers, 393–94
gypsy moths, 394
leafminers, 395–96
Oriental fruit moth, 392–93
tent caterpillars, 394–95
mulches, 311–21
inorganic, 311–13
living, 316–19, **320–21**
organic, 313–15
weed fabrics and, 246–47, *247*
mushroom compost, 278
mutations, 120–21
mycorrhizal fungi (mycorrhizae), 240–42

N

Nanking cherry (*Prunus tomentosa*), 187, 210
naphthalene acetic acid (NAA), 473
National List of Allowed and Prohibited Substances, 12
National Organic Program (NOP), 10–11, 12, 403
National Sustainable Agricultural Information Service, 401
The Natural Way of Farming (Fukuoka), 8
nectarine, 191–92. *See also* training peach and nectarine trees
harvesting, 491
thinning, 479–81
varieties, suggested, 217–18
nectary crops. *See* insectary crops
neem products, 338, 384–85
nematodes, 374–75
nitrogen (N), 99–101
application of, 100–101, 251
basics, 252–53
nutrient status, **256**
soil organic matter (SOM) and, 290
nitrogen fixing. *See* alley crops; companion crops; green manure crops
non-crop areas, 75
non-fruit crops, 299
nonsynthetic materials
NOP approved, 265–66
NOP prohibited, 266
North American plums, 196–97
American plum (*Prunus americana*), 196

North American plums *(continued)*
Beach plum (*Prunus maritima*), 196–97
Canada or black plum (*Prunus nigra*), 196
Hortulan or wild plum (*Prunus hortulana*), 196
Munson's or wild goose plum (*Prunus munsoniana*), 196
Oregon, Sierra, or Klamath plum (*Prunus subcordata*), 197
nutrients. *See also* mineral nutrients; *specific nutrient*
fertilizer materials and, **268–69**
foliage, fruit and, 254–55
soil organic matter (SOM) and, 289–90
symptoms/treatments of deficiencies/excesses, **256–58**

O

off-farm amendments, 109
off-farm sales, 504–9
CSAs, 504, 506
farmers' markets, 505–6
wholesaling, 507–9
offset planting, 64, **64**, 65
oil products, 474–75
old orchard sites, 38
One Straw Revolution (Fukuoka), 8
On-Farm Composting Handbook, 273
on-farm sales, 499–504
roadside stands, 503–4
U-pick, 498–99, 500–503
open center tree design
apple, 426, *428*, 428–29
peach and nectarine, 459–60
open Tatura pear system, 438, 439–40, *440*
orchard design, sample, 69–75, **73**
analysis of, 72–75
planning worksheet, 70–72, *71*
orchard floor management
cultural practices, 330–31
fertilization practices and, 285
orchard heaters, 67
orchard layout design, 62–69
mapping out, 62–65
Organic Farming and Gardening, 8
Organic Food Production Act, 10, 12
Organic Materials Research Institute, 108
Organic Materials Review Institute (OMRI), 12, 297
organic mulches, 313–15
advantages of, 314
disadvantages of, 314–15
types of, 313–14
Oriental fruit moth, 376, 392–93

P

Pacific Coast pear rust, 347
pacific flatheaded borer, 400
packaging, 517–18
packing houses, 508, 516
palmette trellis design, *430*
parasitic wasps, 390, 399
particle film technology. *See* Surround
particle size distribution (PSD) analysis, 28
peach (*Prunus persica*), 188–91, *189*. *See also* training peach and nectarine trees
basics, 189
climate concerns, 188
harvesting, 491
"meadow orchard" system, 46
pests and diseases, 189
rootstocks, 191
thinning, 479–81
varieties, 190–91, 211–16
peach leaf curl, 360
peachtree borer/lesser peachtree borer, 400–401
peachtwig borer, 401–2
pear, 131–37. *See also* high-density pear systems; training pear trees
Asian, *132*, 132–34, 164–65, 488–89
disease-resistant varieties, 345
diseases and, 131–32
European, 134–36, 166–68, 488
harvesting, 487–89
pests and, 131–32
rootstocks, 137
thinning, 477
"pear decline" 132, 137, 348–49
pear leafcurling midge, 399
pear leaf spot, 349–50

pear psylla, 389–90
pear scab, 119, 344–45
pear slug/pear sawfly, 398–99
peat moss, 278
perennial alley crops, 248, 303–6. *See also* alley crops
  alfalfa, 305–6
  draught-tolerant grasses, 305
  low-growing, bunch-forming grasses, 303–4
  orchard grass, 305
  tall fescues, 305
  white clover, 304–5
perpendicular-V training system, *460*, 460–61, 462
pest control, 116–17, 375–86
  biological management, 299
  in dormant season, 338
  insect monitoring devices, 375–76, *376*
  scouting, 375–76
pesticides
  pesticide drift, 63
  U-pick farms and, 502
  using less, 10
pests, 39–40. *See also* insect and mite pests
  of cherry, 183
  of loquat, 149
  mammals, large, 369–72
  mammals, small, 372–73, *373*
  of mayhaw, 146
  of peach, 189
  of pear, 131–32
  pome fruits and, 119–22
  of quince, 139–40
  of saskatoon, 151
  tree prepping and, 240
petal fall, 262
petioles (leaf stems), 262
pheromone traps, 376, *376*
phosphorus (P), **98**, 101–3
  application of, 102–3
  nutrient status, **256**
  for planting holes, 243
  ways to add, 102
phytophthora, 365
Phytophthora root rot, 315
phytoplasmas, 364
phytotoxicity, 381
picking fruit, 485
pinching, 246, 415
pirate bugs, 390, 399
planning year, seasonal tasks, 521
plant bugs, 390
Plant Hardiness Zone Map, 20, 22–24, *23, 24*
planting blocks
  drainage and, 84
  fencing and, 69
  orchard layout design, 63–65, *64*
  right angles for, 86–88, *87*
  safety and, 66
  soil amendments and, 89, 96, 106
  staking out, 86, *86*
planting holes, digging, 237–39, *238*
planting rows, laying out
  with marked rope, 235, 236, *236*
  with tractor, 235–36, 237, *237*
planting stock. *See* feathered planting stock; whips
planting trees, 243–45. *See also* preparing for planting
planting year, seasonal tasks, 522
plowing, 114–15
plum, 192–99. *See also* North American plums
  basics, 192
  bush, 482, 492
  Damson (*Prunus insititia*), 195
  European (*Prunus domestica*), 193–95, 218–21
  harvesting, 491–92
  hybrid varieties, suggested, 227–28
  Japanese (*Prunus salicina*), 195–96
  Japanese and Japanese-hybrid varieties, 221–26
  North American, 196–97
  pluots, 228
  pollination, 197–98
  rootstocks, 198–99
  thinning, 481–82
  training systems, 59–61, **60**, 62, 463–64
plum curculio, 39, 120, 121, 374, 402–3
plum leaf scald, 361–62
plum pocket, 360
pluots, 228

pole thinning fruits, 468
pollination
apples, 126–27
apricot, 181
Asian pears, 134
cherry, 185
European pears, 136
loquat, 149
plum, 197–98
quince, 139
saskatoon, 151
pome fruit(s). *See also* harvesting pome fruits; minor pome fruits; *specific pome fruit*
climate considerations, 119
defined, 15
diseases
bitter pit, 353–55
bitter rot, 352–53
black rot, 351–52
entomosporium leaf spot, 353
fabraea leaf spot, 349–50
fire blight, 345–47
flyspeck, 350–51
frogeye leaf spot, 351–52
"pear decline," 348–49
pear leaf spot, 349–50
powdery mildew, 363–64
rust diseases, 347–48
scab diseases, 341–45
sooty blotch, 350–51
viruses and phytoplasmas, 364
white rot, 351–52
foliar nutrient concentrations, **264**
"June drop," 331, 405, 466
pests and diseases, 119–22
thinning recommendations, **467**
potash of sulfate, 278–79
potassium (K), **98**, 103–5
application of, 105
basics, 253
nutrient status, **256**
ways to add, 103–4
potassium chloride, 104
potassium sulfate, 104, 278–79
powdery mildew, 340, 363–64, 381
precipitation and humidity, 25, 26
predators, noninsect/non-mite, 378–79
preparing for planting, 235–43
amendments, adding, 243
digging the planting holes, 237–39
planting rows/marked rope, 235, 236, *236*
planting rows/tractor, 235–36, 237, *237*
prepping the trees, 239–40
preplanting year, seasonal tasks, 521
prepping trees for planting, 239–40
prices/terms of sales, 510–12
PRI Disease Resistant Apple Breeding Program, 125
production years, seasonal tasks, 524
profitability
costs, assessing, 510–12
depreciation and, 514
propagation of fruit trees
bud sport mutations/clones, 120–21
controlling size, 47–51, *48*, *49*, *50*
prune
harvesting, 491–92
thinning, 481–82
training systems, 59–61, **60**, 62, 463–64
varieties, 194
pruning
crop load and, 467, 468
cuts
locations for, 413–18
branches at trunk, removing, *413*, 413–14, *414*
large limb, removing, 414, *414*
/plant responses, 411–12
dormant, 409, 417, *417*
reasons for, 409
step-by-step, 418
summer, 410–11, 415
sunlight, spurs and, 180
sweet cherry trees, 447–48, 449, *449*
tools, 418–20
chain saws, 419
hand shears, 418–19, *420*
loppers, 419, *420*
pneumatic, hydraulic, and electric, 419–20
pruning saws, 419, *420*
Prunus domestica. *See* European plums (*Prunus domestica*)

Prunus necrotic ringspot, 364
Prunus persica. See peach (*Prunus persica*)
Prunus tomentosa. See Nanking cherry (*Prunus tomentosa*)
psyllids, 381, 385
publicity, 515–17
purchasing trees, 230–34
ball-and-burlap trees, 234
bare root trees, 231–32
in containers, 232–33
root-bag trees, 233–34
purple spot, 476
PYO. *See* U-pick farm
pyrethrum products, 385

Q

quad-V system, 461, *461*, 462
quicklime (burnt lime), 93
quince (*Cydonia oblonga*), 137–40, *138*
basics, 139
customer demand, 140
environment concerns, 139
harvesting, 489
pests and diseases, 139–40
pollination, 139
thinning, 477–78
training systems, 56, **56**, 444–45
varieties, suggested, 169–70
quince rust, 347

R

raised beds, 82–84, *83*
raptors and owls, 373, 374
recertification, 38
recordkeeping
orchard practices, 284
scouting and, 328, *329*
refrigeration, 483
replant disease, 38
research, worldwide, 7–8
Rhizobium radiobacter, 242
roads. *See* access roads
roadside stands, 503–4, 516–17
rock phosphate, 102, 279
Rodale, Bob and Ardath, 9
Rodale, J. I., 8
Rodale Institute, 8, 9
root-bag trees, 233–34
root diseases, 364–67
rooting compounds, 144
root rot, 28, 315
roots
structure of, 89, *89*
tree prepping and, 239, 240
rootstocks
apple, 50, *50*, 128–31, 429, 434
apricot, 181
cherry, 45, 50, *50*, 183–85, 184, 357, 364, 447, 448
dwarfing, 45, 46, 47, 319, 446
interstem and, 239
non-dwarfing, 49
peach, 191
pear, 137
production of, *48*
selection of, 49
size-controlling, 288
soil types/drainage and, 30–31
tree size and, 45, 50, *50*
virus-free, 364
Western sandcherry (*Prunus besseyi*), 199
root suckers, layering, 47, *48*
rot organisms, harvest and, 482, 485
rototilling, 81, 115
Roundup, 116
rubbery wood, 364
rust diseases, 347–48

S

safety
as priority, 66
U-pick farms, 502
"salamanders," 67
sales
quick and professional, 519
terms of, 510–12
by weight/volume, 518
"sandwich" rows, 296, 378
sanitation, 373, 377. *See also* cultural practices
saskatoon, *150*, 150–51
basics, 151
climate and habitat, 150
culinary varieties, 175–76
harvesting, 489
ornamental varieties, 176

saskatoon *(continued)*
pests and diseases, 151
planting, 150–51
pollination, 151
pruning, 446
thinning, 478
training systems, 56, **56**, 57, 445–46
sawflies. *See* flies, sawflies, and midges
scab diseases, pome fruits, 341–45. *See also* apple scab
pear scab, 344–45
scab-resistant apple varieties, 343–44
scaffold branches, 59, 417, *417*, 428–29, 433
scald disorder, 483
scales, 120, 381, 385, 386, 406
scion wood
for fruit variety, 47, 49, *49*
virus-free, 364
scouting, 327–30, *329*, 375–76
seeds, trees grown from, 120
semidwarf fruit trees, 51
set profit approach, 512
shaker devices, 469–70, 479
shells, 279
shoots
fruit formation on, 189
growth of, 255
tissue testing, 260
shothole blight, 362
shothole borer, 399–400
Silent Spring (Carson), 9
site history/neighborhood, 38–39
site preparation, case study on, 84
site selection
birds and, 374
disease management and, 326
size control. *See* interstem
"slender axis" training system, 54, 425
slender spindle design, 431–33, *432*
"slick soils," 276
slope management, 72
sodium nitrate, 279
soil(s). *See also* amendments
apple trees and, 124
characteristics of, 77
ecosystems, 250
orchard floor management and, *324*
organic matter in, 34
particle size distribution (PSD) analysis, 28
pH, 32–33, 90–96
lowering, 94–96, **95**
nutrient availability and, 91, *91*, 259
raising, 92–93
salinity of, 33, 96–97
standards, amendments and, 89–90
structure, erosion control and, 310
soil analysis report, *27*
testing, 27–28, 88, 259–60
types and drainage, 28–32. *See also* drainage
classification of soil types, 28–30, *29*
crops and, 30–31
examining your soil, 31–32
hard pans/clay layers, *30*, 31
soil profile, 30, *30*
soil adaptability. *See* rootstocks
soil aggregates, 290
Soil and Health Foundation, 8
The Soil Association, 8
"soil conditioner," 276, 281
soil organic matter (SOM), 289–91. *See also* alley crops; companion crops; green manure crops
biological activity and, 290–91
CEC and nutrient uptake, 289–90
nitrogen, sulfur and, 290
soil aggregates and, 290
solarization, 115–16
sooty blotch, 350–51
soybean meal, 271
soybean oil, 474
spider mites, 390
spindle and axe training system, 454–57
growth, management of, 455–57
mature trees, management of, 457
using system, 455
spindle support systems, 424
Spinosad, 383–84
split canopy training systems, 423, 456
spreader bars, 415, 416, *416*
sprinkler systems, 36, 66–67
sprouts, 415

spurs
broken, picking fruit and, 485
exposure to sunlight, 180
formation of, 333, 433
pruning and, 442
water sprouts and, 415
square corner, laying out, 87, *87*
staggered planting. *See* offset planting
staking the orchard, 85–88
buffer strips and roads, 86
right angles for planting blocks, 86–88, *87*
standard fruit trees, 51
starch-iodine test, 484–85, *485*
steam weeders, 295
steep ladder training system, 448, 450–51, *451*
Steiner, Rudolf, 7
stems, pulling out, 485
sticky cards, 375–76, *376*, 398
sticky spheres, red, 376, *376*, 396, 397
sticky traps, pheromone-baited, 376, 396, 397
sticky wraps, 394, 407
stone fruit(s). *See also specific stone fruit*
defined, 15
diseases, 355–62
bacterial canker, 357–58
bacterial spot, 358–59
black knot, 359–60
brown blight, 355–57
peach leaf curl, 360
peach scab, 360–61
plum leaf scald, 361–62
plum pocket, 360
powdery mildew, 363–64
shothole blight, 362
viruses and phytoplasmas, 364
foliar nutrient concentrations, **264**
harvesting, 490–92
thinning recommendations, **467**
storing fruit, 482, 483, 485–86
string thinners, 470–72, *471*, 479
stub cuts, 447–48, 449, *449*
sucking pests, 386–90
aphids, 387–88
leafhoppers, 388–99
mealybugs, 388
mites, 389–90
scales, 406
sugar beet lime, 279
sulfate of potash-magnesia, 104
sulfate of potassium magnesia, 277
sulfur, 279–80
adding, **95**, 95–96, 105
as fungicide, 333, 337
overuse of, 14
preplant rates, **98**
soil organic matter (SOM) and, 290
thinning and, 474–75, 478
sul-po-mag, 277
summer pruning, 410–11, 415
sunburned limbs, 415, 417
supported apple trees, 429–36
installing trellises, 429–30
slender spindle design, 431–33
tall spindle design, 433–36, *434*
training trees to trellises, *430*, 431
surfactants, 296
Surround, 263, 340, 385–86
"sustainable agriculture," 9
sweet cherry, 185–86. *See also* training sweet cherry trees
varieties, suggested, 204–7
Swiss sandwich system, 74–75, **317**, 317–18

T

tall spindle design, 54, 433–36, *434*
Tanglefoot sticky wrap, 407
tarnished plant bug, 384
tart cherry, 186–87. *See also* training cherry trees
varieties, suggested, 208–9
Tatura training system, 456. *See also* open Tatura pear system
temperature concerns. *See also* climate considerations; frost
apples, 123–24
Asian pears, 133–34
European pears, 135–36
temperatures, important, 26
tent caterpillars, 394–95
terminal bud, *427*
tetracycline, 340
thermal weeders, 293–95

thinning
 apples, 473–75
 apricots, 478
 chemical, 472, 480
 cherries, 478–79
 by hand, 468
 loquats, 475–76
 mayhaw/medlar/saskatoon, 478
 mechanical, 469–72, *470*, *471*, 479
 natural, 466–67
 peaches/nectarines, 479–81
 pears, 477
 pole, 469
 pome fruit, **467**
 pruning and, 467
 quince, 477–78
 stone fruit, **467**
thinning cuts, 412, *412*, 447, 449, *449*
thrips, 120, 381, 383, 384, 407
tie-downs, 415, *416*
tilth, 77
tissue testing, 260–63
 analysis of test results, 262–63, **264**
 collecting procedures, 261–62
 tips on, 261
 washing leaf samples, 263
tools. *See* pruning tools
topography, 37–38
total soluble solids (TSS), 487
toxic elements/chemicals, 38–39
tractor, advent of, 287–88
training, 51–62. *See also* trees on supports
 apple trees, **52–53**, 52–54
  freestanding, central leader design, 425–26, 427, *427*
  freestanding, open center design, 426, *428*, 428–29
  supported apple trees, 429–36, *430*, *432*, *434*
 apricot trees, 59–61, **60**, 462
 bush cherries, 463
 cherry trees, 57–58, **58**. *See also* training sweet cherry trees
  tart cherry trees, 457–58
 freestanding trees, 421–22
 loquat trees, 56, **56**, 442–43
 mayhaw trees, 56, **56**, *443*, 443–44
 medlars, 56, **56**, 444
 minor pome trees, **56**, 56–57
 peach and nectarine trees, 59–61, **60**, 189–90, 459–62
  open center training, 459–60
  perpendicular-V training system, *460*, 460–61, 462
  quad-V system, *461*, 461, 462
 pear trees, **54–55**, 54–55, 436–42
  freestanding, 437
  high-density systems, 437–42, *439*, *440*, *441*
  to a trellis, 430, *430*, 437
 planting trees and, 243–45
 plum and prune trees, 59–61, **60**, 62, 463–64
 quince trees, 56, **56**, 444–45
 saskatoons, 56, **56**, 57, 445–46
 spindle support systems, 424
 split canopies, 423, 456
 sunlight, spurs and, 180
 sweet cherry trees, 446–57
  heading cuts, 449, *449*
  Spanish bush, *453*, 453–54
  spindel and axe system, 454–57
  steep ladder system, 448, 450–51, *451*
  stub cuts, 447–48, 449, *449*
  thinning cuts, 447, 449, *449*
  Vogel central leader, *452*, 452–53
 vertical axis designs, 423
trap crop, 403
traps
 for insects, 375–76, *376*
 for small mammals, 372
tree density, 288
tree size, 43–51
 controlling, 47–51, *48*, *49*, *50*
 large trees, 44–45, *45*
 right small size and number, 46–47
 small trees, case for, 44–46, *45*
trees on supports, 422–25
 trellis systems, 423–24
tree vigor, controlling, 433
trellises
 installing, for apple trees, 429–30
 in place before planting, 235
 systems, 423–24

training to, apple/pear trees, *430*
water sprouts and, 415

U

United Kingdom, 8
U-pick farm, 498–99, 500–503, 516–17
U.S. Department of Agriculture (USDA), 9, 10–11
Plant Hardiness Zone Map, 20, 22–24, *23*, *24*
U.S. Environmental Protection Agency (EPA), 297–98
U.S. National Organic Plan, 380
U.S. National Organic Program, 299, 339
utilities, 40–41

V

value-added products, 495, 496, 517
variable costs, 513
variety vs. cultivar, 122
vegetation, biological management of, 299
vertical axis designs, 54, 423, 424–25
verticillium, 365–66
V-hedge pear system, 438, 439, *439*
viruses, 364
Vogel central leader system, *452*, 452–53

W

washing fruit, 483
water core, 157, 477, 483, 484
water sprouts
branches with/without, 417
removal of, 415
website, 515–16
weed control, 113–16, 291–99
cultivation, mechanical, 114–15
flame weeders, 294
herbicides, nonorganic, 116
mulches/weed fabrics, 246–47, *247*
organic herbicides, 114
solarization, 115–16
steam weeders, 295
thermal controls, 114
thermal weeders, 293–95
weeder geese, 299
weighting branches, 415, 416, *416*, 416, 428
Western sandcherry (*Prunus besseyi*), 187, 211
whip-and-tongue grafting, *48*
whips, 231, *231*, 425, 426, 427, *427*
white clover, 304–5, 321
whiteflies, 381, 383, 384
white rot, 351–52
wholesaling, 507–9
whorls of branches, 426, 453
windbreaks
buffer strips and, 86
fencing and, 74
pesticide drift and, 63
plant bugs and, 405
wild fruit trees in, 39
winds and, 27, 71
winds, 26–27
Wonder Weeder, 322
wood ashes, 92, 102, 104, 280
wood borers, 120
wood rot, 415
worm castings, 280

Z

zeolites, 113
zinc (Zn), **258**
Zone Map, Plant Hardiness, 20, 22–24, *23*, *24*
zoning laws, 41

Other Storey Titles You Will Enjoy

The Complete Compost Gardening Guide, by Barbara Pleasant & Deborah L. Martin.

Everything a gardener needs to know to produce the best nourishment for abundant, flavorful vegetables.

320 pages. Paper. ISBN 978-1-58017-702-3.
Hardcover. ISBN 978-1-58017-703-0.

The Fruit Gardener's Bible, by Lewis Hill & Leonard Perry.

A complete guide to growing fruits and nuts in the home garden.

320 pages. Paper. ISBN 978-1-60342-567-4.
Hardcover. ISBN 978-1-60342-984-9.

Greenhorns, edited by Zo & Ida Bradbury, Severine von Tscharner Fleming, and Paula Manalo.

Fifty original essays written by a new generation of farmers.

256 pages. Paper. ISBN 978-1-60342-772-2.

Landscaping with Fruit, by Lee Reich.

A complete, accessible guide to luscious landscaping — from alpine strawberry to lingonberry, mulberry to wintergreen.

192 pages. Paper. ISBN 978-1-60342-091-4.
Hardcover with jacket. ISBN 978-1-60342-096-9.

Reclaiming Our Food, by Tanya Denkla Cobb.

Stories of more than 50 groups across America that are finding innovative ways to provide local food to their communities.

320 pages. Paper. ISBN 978-1-60342-799-9.

Starting & Running Your Own Small Farm Business, by Sarah Beth Aubrey.

A business-savvy reference that covers everything from writing a business plan and applying for loans to marketing your farm-fresh goods.

176 pages. Paper. ISBN 978-1-58017-697-2.